# Diagnostic Molecular Microbiology

## PRINCIPLES AND APPLICATIONS

# Diagnostic Molecular Microbiology



## PRINCIPLES AND APPLICATIONS

*Edited by*

**David H. Persing, M.D., Ph.D.**
Section of Clinical Microbiology, Mayo Clinic, Rochester, Minnesota

**Thomas F. Smith, Ph.D.**
Section of Clinical Microbiology, Mayo Clinic, Rochester, Minnesota

**Fred C. Tenover, Ph.D.**
National Center for Infectious Diseases, Centers for Disease Control, Atlanta, Georgia

**Thomas J. White, Ph.D.**
Roche Molecular Systems, Alameda, California

AMERICAN SOCIETY FOR MICROBIOLOGY
WASHINGTON, D.C.

**Library of Congress Cataloging-in-Publication Data**

Diagnostic molecular microbiology : principles and applications / edited by
David H. Persing . . . [et al.].
p. cm.
Includes index.
ISBN 1-55581-056-X
1. DNA probes—Diagnostic use. 2. Diagnostic microbiology—Technique. 3. Nucleic acid hybridization. 4. Communicable diseases—Diagnosis—Laboratory manuals. I. Persing, David H.
QR69.D4D5 1993
616.9′047583—dc20 92-38523
CIP



This book was coedited and chapter 1 was cowritten by Fred C. Tenover in his private capacity. No official support or endorsement by CDC is intended or should be inferred.

Cover illustration by Tomo Narashima

# Contents

## SECTION 2. Viral Pathogens

# Contributors

**Dharam V. Ablashi**
Laboratory of Cellular and Molecular Biology, National Institutes of Health, Bethesda, MD 20892

**Thomas Adam**
Institut Pasteur, F-75724 Paris, France

**Allen J. Aksamit, Jr.**
Department of Neurology and Department of Biochemistry and Molecular Biology, Mayo Clinic, Rochester, MN 55905

**Burt E. Anderson**
Division of Viral and Rickettsial Diseases, National Center for Infectious Diseases, Centers for Disease Control, Atlanta, GA 30333

**Robert D. Arbeit**
Department of Medicine, Boston University School of Medicine, Boston, MA 02130

**Jorge Arévalo**
Department of Physiology and Instituto de Medicina Tropical "Alexander von Humboldt," Universidad Peruana Cayetano Heredia, Lima 100, Peru

**Michel Arthur**
Unité des Agents Antibactériens, Institut Pasteur, 75724 Paris Cedex 15, France

**Jaber Aslanzadeh**
Department of Laboratory Medicine, University of Connecticut Health Center, Farmington, CT 06030

**David Bacchus**
Laboratory of Tumor Cell Biology, National Institutes of Health, Bethesda, MD 20892

**Heidi M. Bauer**
Department of Infectious Diseases, Roche Molecular Systems, Alameda, CA 94501

**Zwi N. Berneman**
Laboratory of Tumor Cell Biology, National Institutes of Health, Bethesda, MD 20892

**Linda D. Bobo**
Department of Medicine, Johns Hopkins University, Baltimore, MD 21205

**Thomas Bodmer**
Institute for Medical Microbiology, University of Berne, 3010 Berne, Switzerland

**Jacques Bonnet**
Institut de Biochimie Cellulaire et Neurochimie du Centre National de la Recherche Scientifique, 33077 Bordeaux, Cedex, France

**Cheryl A. Bopp**
Enteric Diseases Branch, National Center for Infectious Diseases, Centers for Disease Control, Atlanta, GA 30333

**Erik C. Böttger**
Institut für Medizinische Mikrobiologie, Medizinische Hochschule Hannover, 3000 Hannover, Germany

**Madama Bouaré**
DEAP École Nationale de Medicine et de Pharmacie, BP 1805 PEG Bamako, Mali, West Africa

**Charles A. B. Boucher**
Department of Virology, University of Amsterdam, Amsterdam, The Netherlands

**Barbara H. Bowman**
Roche Molecular Systems, Alameda, CA 94501

**Janice T. Brown**
Baxter Prognostics, Issaquah, WA 98027

**Thomas H. Callaway**
Center for Molecular Biology, Roche Biomedical Laboratories, Research Triangle Park, NC 27709

**Lee Ann Campbell**
Department of Pathobiology, University of Washington, Seattle, WA 98195

**M. Donald Cave**
Medical Research Service, J. L. McClellan Memorial Veterans Hospital, Little Rock, AR 72205

**George D. Cimino**
HRI Research Inc., Concord, CA 94520

**C. Graham Clark**
Laboratory of Parasitic Diseases, National Institute of Allergy and Infectious Diseases, Bethesda, MD 20892

**Jonathan P. Clewley**
Virus Reference Division, Central Public Health Laboratory, London NW9 5HT, United Kingdom

**Stewart Cole**
Laboratory of Bacterial Molecular Genetics, Pasteur Institute, 75724 Paris 15, France

**Richard W. Cone**
Department of Laboratory Medicine, University of Washington, Seattle, WA 98105

**Mario Corbellino**
Laboratory of Tumor Cell Biology, National Institutes of Health, Bethesda, MD 20892

**Jack T. Crawford**
Mycobacteriology Laboratory, Centers for Disease Control, Atlanta, GA 30333

**John E. Donelson**
Department of Biochemistry, University of Iowa, and Howard Hughes Medical Institute, Iowa City, IA 52242

**Elizabeth A. Dragon**
Department of Diagnostics Development, Roche Molecular Systems, Branchburg, NJ 08876-1760

**Kathleen D. Eisenach**
Medical Research Service, J. L. McClellan Memorial Veterans Hospital, Little Rock, AR 72205

**Mark J. Espy**
Department of Laboratory Medicine and Pathology, Mayo Clinic, Rochester, MN 55905

**David H. Evans**
Department of Molecular Biology and Genetics, University of Guelph, Guelph, Ontario N1G 2W1, Canada

**Patricia I. Fields**
Enteric Diseases Branch, National Center for Infectious Diseases, Centers for Disease Control, Atlanta, GA 30333

**E. Z. Freedman**
Department of Microbiology, Duke University Medical Center, Durham, NC 27710

**Maurice Garret**
Institut de Biochimie Cellulaire et Neurochimie du Centre National de la Recherche Scientifique, 33077 Bordeaux Cedex, France

**Ulf B. Göbel**
Institut für Medizinische Mikrobiologie und Hygiene, Albert Ludwigs Universität Freiburg, D-7800 Freiburg, Germany

**Yu Gong**
HRI Research Inc., Concord, CA 94520

**Sylvie Goussard**
Laboratoire Sondes Froides bioMérieux, Ecole Normale Supérieure, 69634 Lyon/Cédex 07, France

**Vera Gouvea**
Division of Microbiology, Food and Drug Administration, Washington, DC 20204

**Barbara Graf**
Institut für Medizinische Mikrobiologie und Hygiene, Albert Ludwigs Universität Freiburg, D-7800 Freiburg, Germany

**L. M. Graves**
Division of Bacterial and Mycotic Diseases, National Center for Infectious Diseases, Centers for Disease Control, Atlanta, GA 30333

**Larry Greenfield**
Roche Molecular Systems, Alameda, CA 94501

**Kay S. Greisen**
Department of Infectious Diseases, Roche Molecular Systems, Alameda, CA 94501

**Paul H. Gumerlock**
Department of Internal Medicine, University of California Davis Medical Center, Sacramento, CA 95817

**Cynthia A. Gustaferro**
Department of Infectious Diseases and Internal Medicine, Mayo Clinic, Rochester, MN 55905

**Paul V. Haydock**
MicroProbe Corp., Bothell, WA 98021

**John E. Hearst**
Department of Chemistry, University of California, Berkeley, CA 94720

**John Hierholzer**
Respiratory and Enteric Viruses Section, Division of Viral and Rickettsial Diseases, Centers for Disease Control, Atlanta, GA 30333

**J. Hinnebusch**
Laboratory of Vectors and Pathogens, Rocky Mountain Laboratories, National Institute of Allergy and Infectious Diseases, Hamilton, MT 59840

**Ann C. Hobson**
Department of Laboratory Medicine, University of Washington, Seattle, WA 98105

**Paul Imboden**
Institute for Medical Microbiology, University of Berne, 3010 Berne, Switzerland

**Clark B. Inderlied**
Department of Pathology, University of Southern California Molecular Clinical Microbiology Laboratory, Childrens Hospital Los Angeles, Los Angeles, CA 90027

**Rocío Inga**
Department of Physiology and Universidad Peruana Cayetano Heredia, Instituto de Medicina Tropical "Alexander von Humboldt," Lima 100, Peru

**Stephen T. Isaacs**
HRI Research Inc., Concord, CA 94520

**Ruth F. Jarrett**
Department of Veterinary Pathology, Veterinary School, University of Glasgow, Glasgow G61 1QH, Scotland, United Kingdom

**Pamela J. Johnson**
Nucleic Acid Systems, Chiron Corporation, Emeryville, CA 94608

**W. M. Johnson**
Bureau of Microbiology, Laboratory Centre for Disease Control, Ottawa, Ontario, Canada K1A 0L2

**Olen Kew**
Respiratory and Enteric Viruses Section, Division of Viral and Rickettsial Diseases, Centers for Disease Control, Atlanta, GA 30333

**Julia A. Kiehlbauch**
Enteric Diseases Branch, National Center for Infectious Diseases, Centers for Disease Control, Atlanta, GA 30333

**Louis V. Kirchhoff**
Department of Internal Medicine, University of Iowa, and Department of Veterans Affairs Medical Center, Iowa City, IA 52242

**Philip Kirschner**
Institut für Medizinische Mikrobiologie, Medizinische Hochschule Hannover, 3000 Hannover, Germany

**Mary E. Klotman**
Laboratory of Tumor Cell Biology, National Institutes of Health, Bethesda, MD 20892

**Sharon A. Kochik**
MicroProbe Corp., Bothell, WA 98021

**Shirley Kwok**
Department of Infectious Diseases, Roche Molecular Systems, Alameda, CA 94501

**Brendan A. Larder**
Department of Molecular Sciences, Wellcome Research Laboratories, Kent BR3 3BS, United Kingdom

**Diane U. Leong**
Department of Infectious Diseases, Roche Molecular Systems, Alameda, CA 94501

**Céline Lévesque**
Département de Biochimie, FSG, Université Laval, and Laboratoire et Service d'Infectiologie, Centre de Recherche du CHUL, Ste.-Foy, Quebec G1V 4G2, Canada

**Lily Lin**
HRI Research Inc., Concord, CA 94720

**Shyh-Ching Lo**
Department of Infectious and Parasitic Diseases Pathology, Armed Forces Institute of Pathology, Washington, DC 20306-6000

**Martín López**
Department of Physiology and Instituto de Medicina Tropical "Alexander von Humboldt," Universidad Peruana Cayetano Heredia, Lima 100, Peru

**Jeffery S. Loutit**
Department of Medicine, Stanford University, Stanford, CA 94305

**Michael Lovett**
Department of Biochemistry, McDermott Center, University of Texas Southwestern Medical Center in Dallas, Dallas, TX 75235

**Paolo Lusso**
Laboratory of Tumor Cell Biology, National Institutes of Health, Bethesda, MD 20892

**Claude Mabilat**
Laboratoire Sondes Froides bioMérieux, Ecole Normale Supérieure, 69634 Lyon Cédex 07, France

**Roberta Madej**
PCR Training Department, Roche Molecular Systems, Alameda, CA 94501

**M. Michele Manos**
Roche Molecular Systems, Alameda, CA 94501

**Joel N. Maslow**
Department of Medicine, Boston University School of Medicine, Boston, MA 02130

**Ghassan M. Matar**
Division of Bacterial and Mycotic Diseases, National Center for Infectious Diseases, Centers for Disease Control, Atlanta, GA 30333

**Konstantinos Mathiopoulos**
Laboratory of Malaria Research, National Institutes of Health, Bethesda, MD 20892

**Michael McClelland**
California Institute of Biological Research, La Jolla, CA 92037

**Glenn McConkey**
Laboratory of Malaria Research, National Institutes of Health, Bethesda, MD 20892

**Bruce J. McCreedy**
Center for Molecular Biology, Roche Biomedical Laboratories, Research Triangle Park, NC 27709

**Thomas McCutchan**
Laboratory of Malaria Research, National Institutes of Health, Bethesda, MD 20892

**Michael McDonough**
Respiratory and Enteric Viruses Section, Division of Viral and Rickettsial Diseases, Centers for Disease Control, Atlanta, GA 30333

**Albrecht Meier**
Institut für Medizinische Mikrobiologie, Medizinische Hochschule Hannover, 3000 Hannover, Germany

**Ken Metchette**
HRI Research Inc., Concord, CA 94520

**W. Meyer**
Department of Microbiology, Duke University Medical Center, Durham, NC 27710

**Wakio Minamide**
Shionogi Biomedical Osaka Laboratory, Shionogi and Co., Ltd., Osaka 566, Japan

**T. G. Mitchell**
Department of Microbiology, Duke University Medical Center, Durham, NC 27710

**Catherine Molinas**
Laboratoire de Biologie Moléculaire, Pavillon Leriche, C.H.U. Purpan, 31059 Toulouse Cedex, France

**Kazuhisa Murakami**
Shionogi Research Laboratories, Shionogi and Co., Ltd., Osaka 553, Japan

**Yoshikazu Nakamura**
Department of Tumor Biology, Institute of Medical Science, University of Tokyo, Tokyo 108, Japan

**Kevin A. Nash**
Department of Pathology, University of Southern California Molecular Clinical Microbiology Laboratory, Childrens Hospital Los Angeles, Los Angeles, CA 90027

**Ørjan Olsvik**
Enteric Diseases Branch, National Center for Infectious Diseases, Centers for Disease Control, Atlanta, GA 30333

**Masao Omata**
Second Department of Medicine, Faculty of Medicine, Tokyo University, Tokyo 113, Japan

**Larry R. Pease**
Department of Immunology, Mayo Clinic, Rochester, MN 55905

**David H. Persing**
Department of Laboratory Medicine and Pathology and Division of Infectious Diseases, Department of Medicine, Mayo Clinic, Rochester, MN 55905

**Tanja Popovic**
Enteric Diseases Branch, National Center for Infectious Diseases, Centers for Disease Control, Atlanta, GA 30333

**Naraporn Prayoonwiwat**
Departments of Neurology and Immunology, Mayo Clinic, Rochester, MN 55905

**Elisabeth Puchhammer-Stöckl**
Institute of Virology, University of Vienna, 1095 Vienna, Austria

**Justin D. Radolf**
Departments of Internal Medicine and Microbiology, University of Texas Southwestern Medical Center, Dallas, TX 75235

**David A. Relman**
Departments of Medicine (Infectious Diseases and Geographic Medicine), and Microbiology and Immunology, Stanford University School of Medicine, Stanford, CA 94305-5402

**Moses Rodriguez**
Departments of Neurology and Immunology, Mayo Clinic, Rochester, MN 55905

**José R. Romero**
Department of Pediatrics, University of Colorado Health Sciences Center, Denver, CO 80262

**Harley A. Rotbart**
Departments of Pediatrics and Microbiology/Immunology, University of Colorado Health Sciences Center, Denver, CO 80262

**Paul H. Roy**
Département de Biochimie, Faculté des Sciences et de Génie, Université Laval, and Laboratoire et Service d'Infectiologie, Centre de Recherche du Centre Hospitalier de l'Université Laval, Ste.-Foy, Quebec G1V 4G2, Canada

**Paul N. Rys**
Department of Laboratory Medicine and Pathology, Mayo Clinic, Rochester, MN 55905

**Jeffrey S. Sartin**
Division of Infectious Diseases, Mayo Clinic, Rochester, MN 55905

**Gary K. Schoolnik**
Departments of Medicine and Microbiology and Immunology, Howard Hughes Medical Institute, Stanford University Medical School, Stanford, CA 94305

**Tom G. Schwan**
Laboratory of Vectors and Pathogens, Rocky Mountain Laboratories, National Institute of Allergy and Infectious Diseases, Hamilton, MT 59840

**Joseph Silva, Jr.**
Department of Internal Medicine, University of California Davis Medical Center, Sacramento, CA 95817

**Jean-Luc Simard**
Laboratoire et Service d'Infectiologie, Centre de Recherche du Hospitalier de l'Université Laval, Ste.-Foy, Quebec G1V 4G2, Canada

**Alexander M. Slutsky**
Department of Medicine, Boston University School of Medicine, Boston, MA 02130

**Thomas F. Smith**
Department of Laboratory Medicine and Pathology, Mayo Clinic, Rochester, MN 55905

**John J. Sninsky**
Department of Infectious Diseases, Roche Molecular Systems, Alameda, CA 94501

**Joanne P. Spadoro**
Department of Diagnostics Development, Roche Molecular Systems, Branchburg, NJ 08876-1760

**Nancy A. Strockbine**
Enteric Diseases Branch, National Center for Infectious Diseases, Centers for Disease Control, Atlanta, GA 30333

**Bala Swaminathan**
Division of Bacterial and Mycotic Diseases, National Center for Infectious Diseases, Centers for Disease Control, Atlanta, GA 30333

**Masami Tagawa**
First Department of Medicine, Chiba University School of Medicine, Chiba 260, Japan

**Yajarayma J. Tang**
Department of Internal Medicine, University of California Davis Medical Center, Sacramento, CA 95817

**J. W. Taylor**
Department of Plant Biology, University of California, Berkeley, CA 94720

**Amalio Telenti**
Institute for Medical Microbiology, University of Berne, 3010 Berne, Switzerland

**Fred C. Tenover**
Nosocomial Pathogens Laboratory Branch, Hospital Infections Program, National Center for Infectious Diseases, Centers for Disease Control, Atlanta, GA 30333

**Lucy S. Tompkins**
Department of Medicine and Department of Microbiology and Immunology, Stanford University, Stanford, CA 94305

**S. D. Tyler**
Bureau of Microbiology, Laboratory Centre for Disease Control, Ottawa, Ontario, Canada K1A 0L2

**Elizabeth R. Unger**
Department of Pathology and Laboratory Medicine, Emory University, Atlanta, GA 30322

**Mickey S. Urdea**
Nucleic Acid Systems, Chiron Corporation, Emeryville, CA 94608

**Joan L. Valentine**
Department of Laboratory Medicine, The Johns Hopkins Hospital, Baltimore, MD 21287-7093

**Jos A. M. van de Klundert**
Department of Medical Microbiology, University Hospital, 2300 RC Leiden, The Netherlands

**John S. Vliegenthart**
Rijkskeuringsdienst van Waren, 4461 XN Goes, The Netherlands

**Richard Y.-H. Wang**
Department of Infectious and Parasitic Diseases Pathology, Armed Forces Institute of Pathology, Washington, DC 20306-6000

**Judith B. Weiss**
Department of Infectious Diseases, Roche Molecular Systems, Alameda, CA 94501

**John Welsh**
California Institute of Biological Research, La Jolla, CA 92037

**Thomas J. White**
Roche Molecular Systems, Alameda, CA 94501

**Judith C. Wilber**
Nucleic Acid Systems, Chiron Corporation, Emeryville, CA 94608

**Jolene G. Wong-Lee**
Genelabs Technologies, Inc., Redwood City, CA 94063

**Alan T. Wortman**
Baxter Prognostics, Issaquah, WA 98027

**Osamu Yokosuka**
First Department of Medicine, Chiba University School of Medicine, Chiba 260, Japan

**Wangdong Zhang**
Department of Molecular Biology and Genetics, University of Guelph, Guelph, Ontario N1G 2W1, Canada

# Foreword

The human mind has a tendency to forget the difficulties and problems encountered by people involved in scientific discoveries that occurred 100 years ago or more. Louis Pasteur is rightfully credited for espousing the germ theory of disease based largely on his pioneering work on diseases of wine and silkworms and confirming Koch's earlier work on the germ responsible for anthrax in domestic animals. Pasteur and Koch both met with trials, tribulations, and ridicule, but through persistence they convinced their peers to accept their data. Although many people had made several prior discoveries, Pasteur and Koch are usually considered to be the founders of medical microbiology. Ultimately, Ferdinand Cohn, senior to both Pasteur and Koch, became so impressed with their work that in 1872 he published a three-volume work on bacteria which many believe was responsible for establishing microbiology as a science. Cohn made the first systematic attempt at the classification and taxonomy of bacteria. Using the published work of Pasteur, Koch, and others, Cohn advocated Koch's postulates to be the sine qua non in unequivocally demonstrating a given organism as the etiologic agent of a recognized infectious disease. Among other things, this meant that the responsible agent had to be isolated in pure culture, to be identified, to produce the same disease in an experimental animal, and to be isolated again in pure culture.

Skepticism, doubt, and scorn were expressed by many, but in due course, the germ theory of disease and the clear need for the etiologic agent to be isolated in pure culture from an appropriate clinical specimen became the accepted doctrine. This opened up a "golden era" during which media, staining procedures, chemical tests to detect end products of metabolism, and serologic tests to identify microbial antigens were developed and adopted by medical microbiologists for use in their laboratories for research and diagnostic purposes.

With the pure-culture dogma firmly entrenched, the popular research project was to isolate and identify the agents of as many recognized infectious diseases as possible. This same trend exists today, albeit in a somewhat different fashion. Today, as in the mid- to late 19th century, the distinction between theoretical and applied research is increasingly blurred. Pasteur expressed this view when he stated that, "There are not two sciences. There is only science and the application of science; these two activities are linked as the fruit is to the tree." Even though some still hold to a sharp distinction between basic and applied science, the net result has been, over time, a bountiful yield of well-characterized etiologic agents of bacterial, mycotic, and viral diseases of humans, animals, and plants. Interest in pathogenesis also began to mount, and again a surge of new information, basic and applied, began to appear in increasing quantitites.

One finding of particular significance was the demonstration of in vivo capsular-type transformation between pneumococci by Griffith in 1928 and the in vitro confirmation of pneumococcal-type transformation and the finding by Avery et al. in the 1940s that the "transforming factor" was DNA. This DNA was shown to be responsible for the immunologic specificity of the pneumococcal capsular polysaccharides, which in turn played an important role in the pathogenicity of a given type of pneumococcus. This research, and much of the related research that followed, was a part of the initial investigation into microbial molecular biology.

Tremendous discoveries and advances in many directions followed. In the

20th century we began to witness one breakthrough after another. However, we tenaciously clung to the pure-culture concept, even though it became increasingly evident that it was difficult or impossible to cultivate many microorganisms by existing techniques. Of equal if not greater concern was the time required to isolate and identify the etiologic agent by the best methods available. Irrespective of this concern, the advances in molecular biology and the accompanying biotechnological developments were legion; new journals and books appeared at a rapid pace to provide suitable outlets for dissemination of a tremendous volume of basic and applied findings. Small commercial companies appeared at an uncontrollable rate to produce the reagents and instruments needed for molecular-biological research and the resulting technical applications.

Recombinant DNA technology emerged as a major product of these research efforts, and the applications of this technology are still being vigorously pursued. DNA-DNA hybridization became a reality in the 1970s, and practical applications were rapidly developed. Similarly, nucleic acid sequencing and nucleic acid amplification were two of several very important breakthroughs that received wide acclaim in the 1980s. These achievements of biotechnology resulted in many applications in a number of scientific disciplines. The biomedical applications are definitely among the forerunners in that the diagnosis of many types of diseases, including infectious diseases, has been made possible. This is particularly so for infectious diseases for which the causative agent grows poorly or has not yet been isolated.

These applications are readily being adopted by scientists as useful research and diagnostic tools. We are well advised to refrain from showing the skepticism and doubt of many of our predecessors in the 19th century. Rather, these technologies are commended to the attention of clinical microbiologists as significant and important additions to the currently available methods. This is a time for learning the principles and mastering the techniques of applied molecular biology in clinical microbiology. To assist us, the American Society for Microbiology enlisted the aid of a knowledgeable group of scientists to write a text designed to provide extensive coverage of the principles and applications of molecular biology in the diagnosis of infectious diseases.

This book is a solid success; it vividly captures and describes in clear detail the current and future role of molecular biotechnology in the clinical microbiology laboratory. This is eloquently phrased by David Persing—in chapter 3—who states, "The promise of in vitro nucleic acid amplification techniques is the replacement of the process of biological amplification—growth in culture—with the enzymatic duplication and amplification of specific nucleic acid sequences." This he refers to as the "PCR as petri dish." It seems that we have come full circle to the point where the pure-culture concept can, in part, be laid aside.

Albert Balows

# Preface

The emerging field of diagnostic molecular microbiology applies the principles of nucleic acid hybridization to the detection and characterization of pathogenic microorganisms associated with infectious disease. Although the widespread use of DNA probes for the diagnosis of infectious diseases has been discussed for many years, the last decade witnessed nearly explosive growth in the number and variety of specific applications. Recent techniques such as the polymerase chain reaction may eventually replace biologic amplification (i.e., growth on artificial medium) with enzymatic duplication of nucleic acids in vitro. Organisms that are difficult or impossible to cultivate now are new targets for the diagnostic arsenal of the clinical microbiologist. New organisms, many of which have never been cultivated on artificial medium, are being identified, and their associations with human disease are being elucidated. In only a few years, diagnostic molecular microbiology has evolved from a mere technologic curiosity into a field of vast breadth and complexity, with the potential to revolutionize the science and practice of medicine.

It is fitting that the tools of molecular biology should now focus on the detection of microorganisms, because molecular biology owes its existence to microorganisms. The identification of DNA as the transforming factor grew out of the seminal work by Griffith, Avery, and MacLeod in which avirulent *Pneumococcus* mutants were transformed into virulent clones by a bacterial extract (1, 5). The fundamental principles of molecular genetics were elucidated by the study of bacterial genetics, bacterial phages, and bacterial enzymes (3, 4, 6, 7, 9). The birth of modern biotechnology in the late 1970s depended directly on the DNA replication machinery and synthetic capacity of *Escherichia coli* (2). Even the polymerase chain reaction, the method hailed as the greatest development in molecular biology in the last decade, derives much of its power from a DNA polymerase isolated from the thermophilic bacillus *Thermus aquaticus* (8).

This book is divided into two major parts in accord with its twofold mission: to provide a theoretical and a practical framework for understanding the new technology. In the first part we provide an up-to-date summary of technologic developments in this rapidly evolving field, with a view toward practical implementation of molecular techniques in the clinical microbiology laboratory. In chapter 1, the use of conventional DNA probes for the detection and characterization of pathogenic microorganisms is discussed and specific applications are presented. In chapter 2, molecular methods for strain typing are reviewed in detail, because these techniques are likely to be used in the future by many clinical and public health laboratories for epidemiologic investigations. In chapter 3, we review the power and the pitfalls of in vitro nucleic acid amplification techniques, and in chapter 4, we provide practical guidelines for their implementation. Methods for sample preparation (chapter 6) and analysis of amplification products (chapter 7) are described, as are procedures developed for the inactivation of the products of the amplification reactions to help avoid false-positive results (chapter 5). Chapter 8 (Laboratory Design and Work Flow) was prepared by individuals working in laboratories in which these techniques are performed on a routine basis. Last but not least, the important issue of quality control for nucleic acid amplification techniques is addressed in chapter 9.

The second part of this book is a compilation of specific practical applications of molecular diagnostic techniques for the diagnosis of infectious diseases. Essentially, it is a compendium of molecular recipes from around the world. In this section we have attempted to provide a representative but by no means exhaustive collection of techniques for the specific detection and characterization of bacterial, viral, fungal, and parasitic pathogens. It is our hope that by compiling these applications in a single reference, we will illustrate the broad scope of molecular diagnostic techniques and provide a practical starting point for the implementation of these protocols in the clinical microbiology laboratory.

We are most indebted to the many contributors who made this book possible. Their willingness to participate in this endeavor was fueled by unprecedented enthusiasm. The best evidence of this enthusiasm was their overwhelming response to our invitation to contribute. We are also indebted to our spouses, families, and colleagues, who have contended with late meals and canceled appointments in order that the significant undertaking of this book be accomplished.

DAVID H. PERSING

## References


1. **Avery, O. T., C. M. MacLeod, and M. McCarty.** 1944. Studies on the chemical nature of the substance inducing transformation of pneumococcal types: induction of transformation by a desoxyribonucleic acid fraction isolated from pneumococcus type III. *J. Exp. Med.* **79:**137–158.
2. **Cohen, S. N., A. C. Y. Chang, H. W. Boyer, and R. B. Helling.** 1973. Construction of biologically functional bacterial plasmids *in vitro*. *Proc. Natl. Acad. Sci. USA* **70:**3240–3244.
3. **d'Herelle, M. F.** 1917. Sur un microbe invisible antagoniste des bacilles dysentériques. *C.R. Acad. Sci.* **165:**373–375.
4. **Ellis, E. L., and M. Delbrück.** 1939. The growth of bacteriophage. *J. Gen. Physiol.* **22:**365–384.
5. **Griffith, F.** 1928. The significance of pneumococcal types. *J. Hyg.* **27:**113–159.
6. **Meselson, M. S., and C. M. Radding.** 1975. A general model for genetic recombination. *Proc. Natl. Acad. Sci. USA* **72:**358–361.
7. **Meselson, M., and F. W. Stahl.** 1958. The replication of DNA in *Escherichia coli*. *Proc. Natl. Acad. Sci. USA* **44:**671–682.
8. **Saiki, R. K., D. H. Gelfand, S. Stoffel, S. J. Scharf, R. Higuchi, G. T. Horn, K. B. Mullis, and H. A. Erlich.** 1988. Primer-directed enzymatic amplification of DNA with a thermostable DNA polymerase. *Science* **239:**487–491.
9. **Twort, F. W.** 1915. An investigation on the nature of ultra-microscopic viruses. *Lancet* **ii:**1241–1243.

PART I

# Principles of Diagnostic Molecular Microbiology

# Nucleic Acid Probes for Detection and Identification of Infectious Agents

# 1

FRED C. TENOVER AND ELIZABETH R. UNGER

The first practical application of recombinant DNA technology in the field of infectious diseases was reported by Moseley et al. in 1980 (63). In this study, enterotoxigenic *Escherichia coli* was detected by DNA-DNA hybridization in stool samples that had been spotted onto filters and incubated overnight on MacConkey agar plates to allow growth (63). The DNA probe test was sensitive and specific compared with traditional cell culture assays. In the decade that followed, DNA probes capable of identifying various bacterial pathogens became available commercially. In addition, nucleic acid probes directed against viral, fungal, parasitic, and helminthic pathogens were described (91). Probes to viral pathogens were used with increasing frequency in histopathology laboratories to identify organisms within the cells and tissues they infect (98).

Today, probes occupy an important niche in many diagnostic and surgical pathology laboratories. The increase in the number of molecular biology sections in departments of pathology and laboratory medicine across the United States and in other countries attests to the importance of this technology. Although commercially produced kits were the only access to this technology for many laboratories in the mid-1980s, the development of oligonucleotide synthesizers changed this by allowing laboratories to make synthetic oligonucleotide probes of virtually any sequence. Thus, laboratories now have more latitude to evaluate probes described in the literature and to design their own probes on the basis of published sequence data. This chapter reviews the basic technology underlying probe use and published data regarding commercially available probe kits, and it explores the advantages and disadvantages of using this technology in the clinical laboratory.

## Definitions

Nucleic acid probes are segments of DNA or RNA that have been labeled with enzymes, antigenic substrates, chemiluminescent moieties, or radioisotopes and can bind with high specificity to complementary sequences of nucleic acid (Fig. 1). Probes can be directed to either DNA or RNA targets and can be 20 to thousands of bases long. Oligonucleotide probes (often defined as probes with fewer than 50 bp) can be chemically synthesized and purified with relative ease by any of several commercially available instruments. Oligonucleotide probes have the advantage of hybridizing more rapidly to target molecules (10), and they can, under stringent conditions, detect the change of a single nucleotide within a given nucleic acid sequence (101).

**Figure 1.** Denaturation of duplex DNA into single strands by heating, and combination of one of the single-stranded DNA segments with the labeled probe (hybridization).

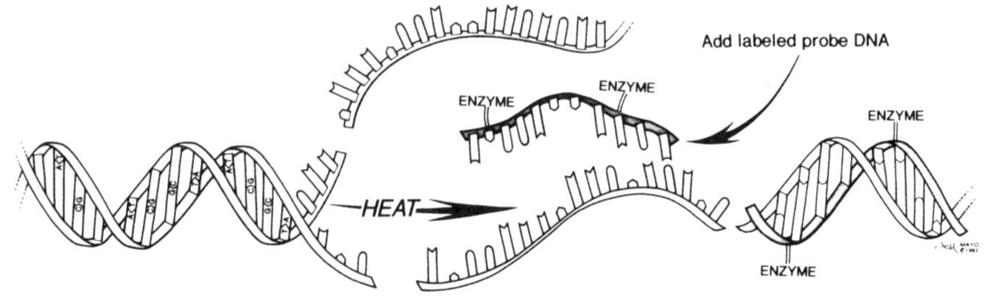


## Stringency

Stringency relates to the number of mismatched base pairs that can be tolerated when two nucleic acid molecules come together to form a double-stranded molecule. Stringency is affected by several variables, including the temperature, salt concentration, and pH of the hybridization reaction (10, 84). The higher the stringency of the reaction (such as is achieved by using buffers of low salt concentration or by conducting the hybridization reactions at high temperature), the less likely it is for mismatched base pairs to stay together as a stable double-stranded molecule (10). For example, under conditions of low stringency (such as with the use of higher salt concentrations), a probe directed to 16S rRNA sequences of *Staphylococcus aureus* might bind to similar sequences in other staphylococcal species if these organisms are in the hybridization reaction. This results in a false-positive reaction. However, under conditions of high stringency, stable binding occurs only with the rRNA of *S. aureus* and no false-positive results are observed. Thus, to avoid false-positive reactions, hybridization conditions—particularly ionic strength, pH, and temperature—must be carefully controlled.

## Reporter Molecules

Most early DNA probes were labeled with $^{32}P$, and investigators used either autoradiography or liquid scintillation counting to detect positive reactions (91). However, in most commercial probe kits, radioactive isotopes for detection of target sequences have been replaced by enzymes, affinity labels, and chemiluminescent molecules (5, 15, 41, 42).

The direct labeling of oligonucleotide probes with enzymes, such as alkaline phosphatase or horseradish peroxidase, was introduced in the mid-1980s (41). Alkaline phosphatase labels are available for some commercially prepared probes, such as those for herpes simplex virus, human papillomavirus, cytomegalovirus, and Epstein-Barr virus.

Affinity labels, such as biotin (15) and digoxigenin (42), may be incorporated into probes by enzymatic (e.g., nick translation or random-priming techniques) or nonenzymatic (e.g., photobiotin) methods (84). Many of these methods for labeling DNA or RNA are available in the form of commercially prepared kits.

After the hybridization reaction is completed, an enzyme linked to streptavidin or a similarly labeled antibody is used to detect the moiety attached to the probe. The detection is completed through the use of colorimetric or chemiluminescent substrates. A considerable array of nonradioactive detection systems are available for hybridization reactions to accommodate the needs of most clinical laboratories.

Chemiluminescent markers are chemical groups that release light when exposed to certain substrates after the hybridization reaction is complete. The light emitted can be detected with X-ray film or a luminometer. Such assays have sensitivity ranges equal to or greater than those of $^{32}$P-labeled probes, have long shelf lives, and do not necessitate the additional safety and control measures required by use of $^{32}$P (5). Chemiluminescent labels, such as acridinium esters, have been incorporated into many commercially available probe kits and are used in a hybridization protection assay format that is rapid and highly efficient (5).

## Hybridization Formats

### Solution Phase Hybridizations

In solution (or liquid) phase hybridizations, both the target nucleic acid and the probe are free to interact in the reaction mixture, which speeds the rate at which duplex molecules are formed. The format requires relatively little sample nucleic acid, although optimal results are achieved when the target nucleic acid is extracted and purified prior to use in the hybridization assay (54). For the probe to be effective, it must be single stranded and incapable of self-annealing. Hybridized probes can be detected in several ways. In one approach, hybrids are digested with S1 nuclease, which is specific for single-stranded DNA and leaves only duplex DNA, which is then recovered from solution quantitatively by precipitation with trichloroacetic acid or by binding to hydroxyapatite columns (which are specific for double-stranded DNA) (10). However, nuclease protection methods are too cumbersome for most clinical laboratories, which must routinely handle large numbers of samples.

A novel variant of the solution hybridization reaction is the hybridization protection assay (5). In this assay, a chemiluminescent acridinium ester attached to the probe is subjected to alkaline hydrolysis after the hybridization reaction unless the probe forms a stable hybrid with the target molecule that results in formation of a double-stranded nucleic acid molecule protecting the chemiluminescent ester from hydrolysis. Unbound esterified probes are subjected to degradation and do not produce light when exposed to the inducer at the termination of the assay. Hybridization is assessed and quantitated with a luminometer. This assay format allows multiple samples to be processed with relative ease.

### Solid Support Hybridization Assays

Nucleic acids bound to a solid surface are still available to participate in hybridization reactions. However, the sensitivity of solid phase hybridization reactions tends to be lower than that of liquid hybridization because not all sequences are available for hybridization (10, 84, 91). However, this technique greatly facilitates the handling of multiple samples. The dot blot and sandwich hybridization assays (Fig. 2) permit qualitative detection of the presence of an infectious agent in a clinical sample. Difficulties can arise, however, in the

**Figure 2.** The three formats for hybridization assays used in clinical laboratories.

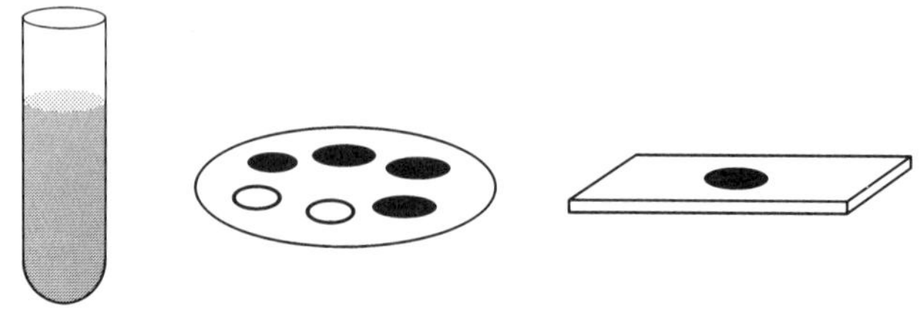


interpretation of low positive signals and with the unpredictable backgrounds, which can be high with some types of samples.

The highly specialized Southern hybridization assay, which allows the investigator to determine the size of the restriction fragment to which a probe hybridizes, requires purified sample DNA of high molecular weight (88). The sample is digested with a restriction endonuclease, the fragments are separated by size by gel electrophoresis, and the nucleic acid is transferred to a membrane for hybridization. Although this technique is used extensively for the detection and typing of human papillomavirus infections, its complexity, the requirement for a relatively large quantity of sample, and the time required for completion of the assay have limited its application to only a few clinical assays.

In situ hybridization assays, in which whole cells or tissue sections affixed to microscope slides are taken through the hybridization process, are a variant of solid phase hybridization assays. This technology, which has become an important tool in surgical pathology, is reviewed later in the chapter.

All hybridization assays require careful probe selection, optimization of hybridization conditions (including reaction and washing temperatures, salt concentration, and pH), and strict attention to quality control to achieve the sensitivity and specificity required for clinical diagnostic work. Empirical verification of reaction conditions is required to exclude the possibility of undesired cross-reactivity between closely related organisms. The recent introduction of oligonucleotide probes has shortened the time of hybridization assays and increased the sensitivity of the assay because large quantities of short sequences can drive hybridization reactions to completion more quickly than probes of 500 to 1,000 bp.

## Indications for Use of Probe-Based Tests in the Laboratory

Nucleic acid hybridization tests often decrease the time necessary to identify fastidious microorganisms. They also allow laboratories to increase the number of types of pathogens that can be easily detected and identified (91). Infectious agents such as parvovirus B19 (109) or enterotoxigenic and enterohemorrhagic strains of *E. coli* (26, 46, 63), for which traditional culture techniques are either not available or not practical because of the biochemical similarity of pathogenic and nonpathogenic strains, respectively, are examples of organisms that have been successfully detected by using probes.

Although probe tests may be more expensive than traditional biochemical

**Table 1.** Examples of commercially available probes

| Type of assay | Organism | Commercial source |
|---|---|---|
| Direct detection in clinical samples | Bacteria | |
| | *Chlamydia trachomatis* | Gen-Probe |
| | *Gardnerella vaginalis*[a] | MicroProbe |
| | Group A streptococci | Gen-Probe |
| | *Legionella pneumophila* | Gen-Probe |
| | *Neisseria gonorrhoeae* | Gen-Probe |
| | *Trichomonas vaginalis*[a] | MicroProbe |
| | Viruses | |
| | Human papillomavirus[b] | Digene |
| Culture confirmation assays | Bacteria | |
| | Campylobacters[c] | Gen-Probe |
| | Enterococci | Gen-Probe |
| | Group B streptococci | Gen-Probe |
| | *Haemophilus influenzae* | Gen-Probe |
| | *Listeria monocytogenes* | Gen-Probe |
| | *Mycobacterium tuberculosis* complex | Gen-Probe |
| | *Mycobacterium avium* | Gen-Probe |
| | *Mycobacterium avium* complex | Gen-Probe |
| | *Mycobacterium intracellulare* | Gen-Probe |
| | *Mycobacterium gordonae* | Gen-Probe |
| | *Neisseria gonorrhoeae* | Gen-Probe |
| | Fungi | |
| | *Histoplasma capsulatum* | Gen-Probe |
| | *Blastomyces dermatitidis* | Gen-Probe |
| | *Coccidioides immitis* | Gen-Probe |
| | *Cryptococcus neoformans* | Gen-Probe |
| | Viruses | |
| | Human papillomavirus[d] | Digene |

[a] Both organisms are detected by a single paddle.
[b] Direct detection of human papillomavirus in samples.
[c] Thermophilic campylobacters, including *C. jejuni*, *C. coli*, and *C. lari*.
[d] Typing of isolates detected by the ViraPap test.

tests, their rapid turnaround times can translate into an overall savings both to the hospital and to the patient by providing a timely diagnosis. Rapid answers also can prevent invasive diagnostic procedures, such as open-lung biopsies in transplant patients to search for causes of pneumonia; unnecessary therapeutic regimens, such as ganciclovir for cytomegalovirus infections; or stays in expensive isolation rooms, such as for patients with suspected tuberculosis while awaiting the identification of acid-fast bacilli. Thus, probes can have many positive effects on decreasing the time to detection and identification of pathogens routinely encountered in the microbiology laboratory. Examples of commercially available DNA probes are presented in Table 1. In addition, in situ hybridization can be used to assess the host response to a particular infectious agent and can often be used to determine the extent of the infection by surveying samples of tissue from various organ systems.

## Probes to Enteric Pathogens

Microbiologists have been awaiting the introduction of commercial probe kits for enteric pathogens since the initial report by Moseley et al. (63) on the successful use of probes for enterotoxigenic *E. coli*. Unfortunately, they are still waiting. Although probes for sexually transmitted diseases and respiratory pathogens have become mainstays in many laboratories, technologists who perform stool tests must still endure long hours of reviewing plates of selective agars, subculturing enrichment broths, and setting up numerous biochemical and latex agglutination tests. These tests are performed in addition to enzyme-linked immunosorbent assays (ELISAs) for rotaviruses and cell culture assays for detection of *Clostridium difficile* toxin. Although DNA probes were described for virtually all common bacterial, viral, and protozoan enteric pathogens (26, 41, 46, 59, 63, 85, 92, 93, 106), the overall lack of sensitivity of these assays precluded their widespread use directly on stool samples. One commercial probe that is available is a culture confirmation assay for the thermophilic campylobacters—*Campylobacter jejuni*, *Campylobacter coli*, and *Campylobacter lari*—although the assay does not differentiate among these organisms (93).

The problem posed by the relatively small number of organisms that often are present in the stools of patients with enteric pathogens can be circumvented by isolating nucleic acid directly from the stool specimen prior to probe analysis (16). Although several nucleic acid extraction systems are commercially available, they are expensive, and the technique has not gained in popularity. For developed countries, nucleic acid amplification techniques may well be the answer (31, 67, 83).

For developing countries, practical filter hybridization techniques have been described for identifying several of the diarrhea-causing *E. coli* strains. One economical yet highly effective biotin-based probe assay was described by Gicquelais et al. (34). Although the assay does not take advantage of oligonucleotide probes (all probes except one are at least 1 kb long), the assay, which uses lysozyme and modest amounts of proteinase K to decrease background signals, is straightforward and has worked well for almost all of the organisms investigated. The probe results for enteroinvasive *E. coli* were occasionally difficult to interpret because of the high levels of background hybridization even with the added cleaning procedure. The key to making the assay economical was the incorporation of Whatman 541 filters in place of nitrocellulose (26, 52).

One other novel application of probe technology is for biotyping strains of *Vibrio cholerae* (1). As cholera becomes more widespread throughout the world, this type of procedure can prove particularly valuable during outbreaks, especially when other probe techniques are already in use.

## Food-Borne Diseases

Another use of probes is to identify the agents of food-borne intoxications and the specific toxins involved. Van Damme-Jongsten et al. (100) described a series of probes for typing toxigenic isolates of *Clostridium perfringens*, and Neill et al. (66) described a probe system for identifying the various enterotoxin genes and toxic shock syndrome toxin 1 from *S. aureus*. Both systems presume that the organisms have been isolated in pure culture prior to the use of the probes. Such assays could prove of particular value to state or public health laboratories that are commonly involved in investigations of food-borne disease.

## Respiratory Pathogens

### *Mycobacterium Probes*

Probe technology has had a major impact on the identification of respiratory pathogens, particularly mycobacteria. Probes are now considered state of the art for the culture confirmation of *Mycobacterium tuberculosis*, *M. avium*, *M. intracellulare*, and *M. gordonae* (17, 36, 48, 74). Although high-pressure liquid chromatography is also an excellent reference method, it requires considerably more equipment and expertise than the use of probes does (11). Seventy-eight public health laboratories in the United States now use probes for identification of *M. tuberculosis*.

Currently, there is a single type of commercially available probe assay for the identification of mycobacteria. Each of the kits that is available for the detection of various organisms or *Mycobacterium* complexes is a liquid hybridization assay that uses chemiluminescent technology for detection of hybrids. The assays show high sensitivity and specificity, although each kit has a few drawbacks. For example, some isolates of *Mycobacterium avium* fail to react with the current Gen-Probe Accuprobe method (48, 74). Recent reports, however, suggest that the new Gen-Probe kit for *M. avium* may demonstrate improved sensitivity.

The major advantage of probes is the ability to use them in conjunction with the BACTEC broth culture system for rapid identification of mycobacteria, especially from respiratory samples (17, 18, 74). This combination of systems can decrease the time to identify mycobacteria by several weeks compared with traditional culture and biochemical identification methods. Probes are also helpful in detecting mixed mycobacterial infections, particularly when organisms such as *M. tuberculosis* and an organism from the *M. avium* complex, or two organisms from the *M. avium* complex, are both present in the same sample (17).

The *M. tuberculosis* probe will react with all members of the *M. tuberculosis* complex, including *M. bovis*, *M. simiae*, and *M. africanum*. This prompted some laboratory directors to consider the probe identification as presumptive until confirmed by biochemical methods, even when the starting material was from a colony morphologically consistent with *M. tuberculosis*. Because *M. bovis* and *M. africanum* are infrequent causes of human infections in the United States and because *M. simiae* occurs only in nonhuman primates, many investigators think that the probe identification, when starting from a colony, can be considered definitive as long as the report issued states that the organism is a member of the *M. tuberculosis* complex.

### *Legionella and Mycoplasma Probes*

A probe for the direct detection and identification of *Legionella* species in respiratory samples has been available for several years (24, 27, 72, 102). This probe is capable of detecting all *Legionella* species rather than just *Legionella pneumophila*, the most common pathogen of the genus. However, the sensitivity of the probe assay (~70%) is only slightly better than that of direct fluorescent-antibody tests and thus is not high enough to replace the culture technique (24, 72). Nonetheless, several hospitals that have endemic problems with legionellosis found the probes valuable for screening specimens during outbreaks (72). The *Legionella* probe is labeled with $^{125}I$.

Until recently there was a commercial probe available for the detection and identification of *Mycoplasma pneumoniae* (25, 43); however, this probe is no longer available. Although various probes for other respiratory pathogens have been described, including a probe for direct detection of *Haemophilus influenzae* in clinical samples (53), these have not been used to any significant degree in clinical laboratories.

### Probes to Periodontal Pathogens

A relatively new application of probe technology is for the detection and identification of organisms that cause periodontal disease. There have been several recent investigations into the utility of probes in this area, including those by Dix et al. (23), Loesche et al. (50), and Moncla et al. (60). In the study by Loesche et al. (50), probes were developed for *Porphyromonas gingivalis*, *Treponema denticola*, and *Bacteroides forsythus* and compared with culture, immunologic reagents, and detection of benzoyl-DL-arginine-naphthylamide as indicators of periodontal disease. The probes proved superior to culture in identifying patients with disease. Oligonucleotide probes were also described for several of these organisms. Thus, probe technology may be a novel way of approaching an important area of infectious disease that has long been neglected. These probes, however, are not yet commercially available.

### Sexually Transmitted Pathogens

There has been considerable interest for many years in the rapid detection of the agents of sexually transmitted diseases. Totten et al. (96) were the first to develop probes for the direct detection of *Neisseria gonorrhoeae* in clinical samples. This assay, based on the detection of the cryptic plasmid carried by most gonococcal strains, worked well and had a sensitivity and specificity of approximately 90%. Rossau et al. (81) developed probes to *N. gonorrhoeae* that were based on 16S rRNA sequences. Because there are up to 10,000 copies of rRNA in bacteria, this approach, which was first developed by Kohne et al. (44a), makes use of the natural amplification of target sequences present in the organism. The 16S rRNA approach has been replicated for many organisms and shows considerable promise because of the increased number of target molecules present per organism.

The greatest market for commercial probe kits has been for the detection and identification of *N. gonorrhoeae* and *Chlamydia trachomatis*. Probes for *N. gonorrhoeae* come in two formats: one for culture confirmation and the other for direct detection of the organism from endocervical or urethral swabs. The *C. trachomatis* probe kit is only for direct detection of the organism in clinical samples. Both *N. gonorrhoeae* and *C. trachomatis* probe tests can be performed on material obtained from a single swab.

Lewis et al. (47) evaluated the *N. gonorrhoeae* probe in the culture confirmation format and found it to have high sensitivity and specificity, producing no false-negative or false-positive results. However, the culture had to be less than 96 h old to yield an accurate result. Colonies from cultures older than 120 h in this study often produced less than the 50,000 relative light units necessary for a positive test result. A study by Limberger et al. (49) on gonococcal cultures mailed to a state health department showed even more impressive results; many nonviable cultures still demonstrated positive probe results even after being submitted to extremes of temperature during mailing. Clinical data and probe competition studies confirmed the accuracy of the Gen-Probe assay, even under these adverse conditions.

The probe for the direct detection of *N. gonorrhoeae* in clinical samples was reviewed by Panke et al. (70). This study showed that the Gen-Probe kit produced results within acceptable limits for performance if the interpretive criteria were amended to include an indeterminate zone. In the study, tests yielding values between 5 and 10 were repeated until two clearly negative or positive values were obtained. This category contained 27% of the values. The authors did not mention whether there was a learning curve associated with the use of the test and whether the number of indeterminate values decreased after several weeks of performing the test.

The *C. trachomatis* direct probe test was evaluated by several groups. The test did not receive a favorable review from Woods et al. (105). These investigators thought that the test did not have adequate sensitivity in asymptomatic or low-prevalence populations to warrant its overall use. This is in contrast to the results reported by Yang et al. (108) and Lees et al. (45). Both of these groups found that the Gen-Probe PACE 2 *C. trachomatis* assay had acceptable sensitivity and specificity. However, both groups expressed reservations. For example, Yang et al. (108) described an indeterminate ratio range from 0.7 to 1.5 and recommended that tests producing results in this range be repeated until consistent positive or negative results were seen. Part of the variation in the sensitivity of the test was attributed to sample collection methods. Lees et al. (45) found that the PACE 2 assay had 86.2% sensitivity and 99.9% specificity in the study population, which manifested a *C. trachomatis* prevalence of 3.2%. The differences in the results of the studies may be due to variations in the culture methods used by the different laboratories or to other technical differences in the procedures used for collecting and transporting specimens.

A more recent study by Ossewaarde et al. (68) that compared the results of culture, the Gen-Probe assay, and the Chlamydiazyme ELISA with the results of the polymerase chain reaction (PCR) by using two sets of primers showed the superiority of PCR for detecting *C. trachomatis* and demonstrated that the probe assay had a sensitivity of approximately 64%, equal to that of Chlamydiazyme and slightly lower than that of the culture technique. The specificity of the probe assay, however, was 99%.

Although data on the sensitivity and specificity of the *Chlamydia* probe assay for evaluating specimens from males are difficult to find in the literature, most laboratory directors who use the probe assay suggest that its sensitivity for urethral cultures from males is generally lower than that observed in females but that the specificity remains high.

The fact that both the gonococcal and chlamydial assays can be performed on material collected from a single swab is a distinct advantage for laboratories that process large numbers of specimens. However, no allowance is made for assessing the quality of the sample before performing the assay, such as can be done when a direct fluorescent-antibody assay is used.

The other sexually transmitted agents for which probes have become important are the human papillomaviruses (7, 12, 51, 55, 103, 110). Because cell culture methods are not available for these viruses, probe tests have become an important diagnostic tool. Several different kits both to detect and to type the viruses are now commercially available (77). Whether the use of the probe assays will become commonplace has yet to be determined (see the section on in situ hybridization).

Probes to other sexually transmitted agents, such as *Haemophilus ducreyi* (71) and molluscum contagiosum virus (30), have been described but have not been used to any great degree in clinical laboratories.

### Other Genitourinary Tract Pathogens and Diseases

Urinary tract infections are common in humans and are a frequent reason for seeking medical attention. Many novel devices and kits have been developed to increase the speed and accuracy of diagnoses of urinary tract infections. Among the non-culture-based methods currently available is a commercially prepared probe kit for the detection of bacteria and yeasts in urine. The system (Flash-Track) was recently reviewed by Koenig et al. (44), who found it to be expensive and labor intensive. Unfortunately, only urine specimens with colony counts of at least $10^4$ were evaluated. An adaptation of this system may have value for screening blood products (see below).

Probes were also described for the direct detection of cytomegalovirus in urine (65). However, because of the need for high-speed centrifugation, these assays are likely to be too difficult for most clinical laboratories to undertake.

***Bacterial Vaginosis***

Bacterial vaginosis is a common complaint of young women (28). Currently, the diagnosis is made on the basis of clinical criteria and a few simple laboratory tests (28, 80, 89). *Gardnerella vaginalis* was associated with the disease early on, and some studies (89) suggested that DNA probes were useful in identifying the presence of *G. vaginalis* in vaginal samples. Later studies (28), however, showed that approximately 60% of women in a control group also had this organism in their vaginal flora.

A recent study by Sheiness et al. (86) suggested that DNA probes can be used to quantitate the number of *G. vaginalis* organisms present in vaginal washes and that there is a strong correlation between bacterial vaginosis and the presence of at least $10^7$ *G. vaginalis* organisms per ml of fluid. This novel approach may be of value for increasing the accuracy and specificity of the diagnosis of bacterial vaginosis. The study also shows that probe assays can be used to make quantitative assessments of specific organisms present in a sample. A nonradioactive commercial assay for the simultaneous detection of *G. vaginalis* and *Trichomonas vaginalis* in vaginal fluid is now available (39).

## Diagnosis of Bacterial Sepsis

Identification of the organisms that cause bacterial sepsis is one of the key functions of the clinical microbiology laboratory. Rarely do the results of other microbiologic tests have the same influence on patient management, particularly in terms of guiding antimicrobial therapy. However, probes have had relatively little impact on this aspect of microbiology. As with enteric pathogens, the fact that many blood cultures with positive results may contain only 1 to 10 organisms per ml at the time of collection (107) deterred the development of probes for direct analysis of blood. However, a series of probe tests have been marketed as culture confirmation assays (20, 22) and they have been applied to the identification of organisms after amplification in BACTEC blood culture vials.

Davis and Fuller (21) were the first to apply commercial probes to the identification of group B streptococci, *S. aureus*, and enterococci from BACTEC blood culture vials. The method they used was relatively simple. After a low-speed centrifugation step to remove erythrocytes, centrifugation at a higher speed was performed to pellet the organisms. The pellet was washed several times, and a Gram stain was performed to indicate which probes would be most appropriate. For example, the enterococcus, pneumococcus, and group B streptococcus probes were used when gram-positive organisms appeared in pairs or short to moderate-length chains, whereas the *S. aureus* probe was used when gram-positive cocci appeared in grapelike clusters. The *H. influenzae* probe was reserved for Gram stains showing gram-negative pleomorphic rods or coccobacillary forms. The cell pellet was used as the source of nucleic acid in the probe tests. The test, which can be completed in 30 to 40 min, was shown to be 100% specific. This specific indication is not described in the package insert of the product because this was not the original intent of this test. The use of probes to screen positive blood culture broths, however, is a novel and effective use of probe technology. Probes were used in a similar fashion in conjunction with the Isolator blood culture system to detect bloodstream infections caused by *Salmonella typhi* (82).

### Probes to Other Blood-Borne Pathogens

Several other pathogens are also found in the bloodstream, including viral, fungal, and protozoan agents. Although probes have been used to detect viral agents in blood, including human immunodeficiency virus (79) and hepatitis B virus (9), these indications will probably give way to nucleic acid amplification technology (83), which is more sensitive than probe methods. However, there are some novel applications for probes, including the detection of organisms such as parvoviruses (109), that may continue to be of value.

Probes continue to be used in surveys of developing countries for assessment of malaria (6, 56) and *Babesia* infections (58). For these purposes, probes are of particular value because they permit the efficient screening of hundreds of samples, which would be impossible to perform by microscopy.

### Screening Blood and Blood Products before Transfusion

The occurrence of a series of cases of transfusion-related bacteremia with *Yersinia enterocolitica* and various other organisms (38, 94) prompted the suggestion of using a commercially available universal probe system, originally designed for screening urine cultures, for screening blood and blood products for microorganisms before transfusion. The probe test incorporates a series of probe sequences common to all bacteria and yeasts. Preliminary studies (75) suggested that this technique might have utility, although the number of organisms, especially *Y. enterocolitica* and *Enterobacter agglomerans*, present in contaminated packed erythrocytes has been relatively small (3).

### Probes to Fungi

Commercial probes for the identification of fungi grown in culture were introduced in the late 1980s. These probes focus on *Cryptococcus neoformans* and the dimorphic fungi *Blastomyces dermatitidis*, *Histoplasma capsulatum*, and *Coccidioides immitis*. As technologists with specific training and expertise in the identification of pathogenic molds become less common in clinical laboratories, these probes may prove of value in the identification of these pathogenic fungi. Although other probes that hybridize to fungal DNA have been reported (14, 19), they have been used primarily for epidemiologic studies and typing and not for primary identification. A recent report (76) of a DNA probe for *Cryptococcus neoformans* stressed that even though it cannot be used for primary detection of fungi in respiratory samples, it has excellent utility as a typing tool and could be developed as a specific probe to be used in conjunction with a universal fungal primer pair in the PCR assay.

## Probes to Protozoa and Helminths

Probes that recognize protozoa and helminths other than the diarrheal and blood-borne pathogens include those to organisms such as *Leishmania* species (104) and the filarial worm *Brugia malayi* (87); all of these probes have been described, but none is commercially available.

## Probes to Antimicrobial Resistance Genes

More than two dozen probes to antimicrobial resistance genes have been described in the literature, covering beta-lactam, aminoglycoside, tetracycline, glycopeptide, chloramphenicol, macrolide, sulfa, trimethoprim, and heavy-metal resistance genes (90). However, the probe that appears to have the most value to the clinical laboratory at present is the one directed against the *mec* resistance determinant, which encodes the novel penicillin-binding protein responsible for most methicillin resistance in staphylococci (2, 13, 29). Such a probe could be used with a rapid detection system for screening colonies of staphylococci on plates for methicillin resistance or in a fashion similar to that described by Davis and Fuller (21) for screening pellets from blood culture vials when a Gram stain of the material indicates the likely presence of staphylococci. Although there are some isolates of *S. aureus* that are resistant to methicillin as a result of altered penicillin-binding proteins (95) and not the presence of the *mec* determinant, such isolates are rarely seen in clinical laboratories and produce few false-negative results.

Gram-negative organisms contain literally dozens of resistance genes that encode resistance to the beta-lactam family of antimicrobial agents, which includes the penicillins and cephalosporins. Although probes to each resistance determinant have not been described, most of these resistance genes are derived from one of five ancestral genes encoding the TEM, SHV, CARB, PSE, or OXA β-lactamases. Probes that hybridize to sequences common to each of these families of genes, such as the TEM- or SHV-type β-lactamases, have been described and could be used to indicate strains potentially resistant to ampicillin (4, 40, 69). Some ampicillin resistance genes, however, encode altered porins that decrease the access of the drug to the site of action. Because the production of altered porins is often the result of random mutation in the genes encoding the porin molecules, it would not be feasible to make probes to detect these often subtle genetic changes. However, this resistance phenotype is relatively infrequent and is usually recognized only in isolates from patients who have been on long-term antimicrobial therapy. Thus, the number of false-negative results due to this problem should be quite small.

There is a problem, however, with trying to develop probes to detect the chromosomally mediated ampicillin resistance determinants that are inducible or become activated when patients receive therapy. Such chromosomal determinants are present in most strains of the family *Enterobacteriaceae* (8), thus posing the problem of distinguishing between clinically susceptible and resistant organisms. Developing a probe to detect and quantitate mRNA produced by a resistance gene would indicate gene expression and thus presumably resistance. Such a system, however, has yet to be described.

Given the limitations noted above, the use of probes as an alternate method of susceptibility testing will probably not be feasible for all combinations of organism and antimicrobial agent. For a limited number of combinations, however, when organisms are present in normally sterile body sites and have relatively few modes of resistance, the use of probes may be feasible. For example, the detection of penicillinase-producing (i.e., ampicillin-resistant) *N. gonorrhoeae* in urethral samples from males using a combination of a diagnostic probe and a resistance determinant probe (in this case a probe to *N. gonorrhoeae* and a probe to the TEM β-lactamase gene) has been reported by Perine et al. (73) to be an effective probe pair. Other such combinations are likely.

Even though the development of probes to monitor resistance to antibacterial agents was relatively successful, the application of this technology to assessment of antifungal, antiviral, or antiparasitic resistance received less attention. One particularly interesting success was the use of probe technology for assessment of chloroquine resistance in *Plasmodium falciparum*. McLaughlin et al. (57) used a probe to *P. falciparum* to monitor levels of parasitemia in children receiving chloroquine therapy. Although the probe was used to detect the organism and not the resistance determinant, it provided information about the ability of the organism to proliferate in the presence of drug; this information would not otherwise have been available if a skilled microscopist was not available.

## In Situ Hybridization

In situ hybridization assays follow the same general principles as those for liquid or filter hybridization. In situ hybridization, however, occurs within the morphologic context of specific tissues or cells. The introduction of morphology to the context of the hybridization assay places this technology at the junction between anatomic and clinical pathology. The information provided by the results of the hybridization assay, when viewed in conjunction with the analysis of tissue morphology and host response, provides confirmation that the infectious agent identified is involved in either an active or latent infection. The morphologic component, however, takes this form of hybridization out of the familiar realm of the clinical microbiologist. In situ hybridization represents a unique approach to the diagnosis of infection.

The following section contains a brief review of the principles of hybridization that are unique to the in situ format, a discussion of the advantages and disadvantages of different detection systems, the potential role of automation, and an overview of current diagnostic applications.

### Preparation of Tissues for Hybridization: A Delicate Balance

In situ hybridization requires that target nucleic acid be liberated from its cellular surroundings in such a way as to be available for hybridization while preserving the cellular morphology for subsequent interpretation and analysis. Although harsh chemicals and proteolytic enzymes can be used to extract DNA from organisms submitted to the clinical microbiology laboratory, such methods would destroy the cellular morphology, which must be preserved if the disease state is to be interpreted accurately.

Nucleic acids exist in a complex environment of proteins, carbohydrates, and lipids. The complexity of the arrangement results in the histologic detail used by anatomic pathologists in their diagnostic evaluation. The same complexity of arrangement results in relative entrapment of potential nucleic acid targets (54). Fixatives, especially cross-linking agents such as formalin, paraformaldehyde, and glutaraldehyde, increase the morphologic preservation of tissues and cells but decrease the availability of target nucleic acids. Treating tissue sections or cells with proteases increases the availability of the target but can also result in a decreased morphologic framework and must therefore be used with caution. The practical consequence of this compromise between availability of target and morphology is that the probes must be relatively small to favor tissue penetrance. Genetically complex probes may be used (such as whole genomic probes or large plasmid probes), and in many cases they contribute to increased sensitivity; however, some method of labeling that decreases the average fragment length to 300 bases or fewer must be used.

The sensitivity of in situ hybridization is limited by the accessibility of target within individual cells. However, the ability to distinguish among individually infected cells is also the strength of the technique. Thus, in the setting of one highly infected cell in a sea of uninvolved cells, in situ hybridization has a sensitivity advantage over extraction-based methods, which average the nucleic acid content of all cells in the sample.

In clinical settings, routinely processed formalin-fixed paraffin-embedded tissue sections often form the starting material for in situ hybridization assays. Unfortunately, the limits of routine fixation and processing are quite wide. Marked variations can be noted within a laboratory between biopsy specimens, routine surgical specimens, and autopsy tissue. Interlaboratory variation can be even more significant. This means that the optimal conditions of tissue pretreatment must be empirically determined by each laboratory for each tissue assayed.

The use of endogenous positive controls greatly facilitates the optimization process. An endogenous positive control probe is one that should be detected in each tissue if the sample has been adequately fixed and prepared. For human tissues, human placental DNA is a useful control probe. Each nucleus should yield strongly positive results if the conditions of the assay are optimized. Use of endogenous positive control probes allows the laboratory to monitor the preservation of nucleic acids throughout the sample. Inadequately preserved or overfixed tissues with unavailable nucleic acid targets may thus be eliminated from reporting; this decreases false-negative results. Controls must be fixed and processed in the same manner as the test material. For example, the results achieved with alcohol-fixed whole cells will be quite different from those obtained with formalin-fixed paraffin-embedded 5-μm tissue sections.

## Detection Systems

Original applications of in situ hybridization, developed in a research setting, involved radiolabeled probes and used autoradiography for detection. These methods are still favored by many investigators. Results achieved with autoradiography are generally accepted as the standard for maximal sensitivity. Isotopic labels that have been applied to in situ hybridization include $^{35}S$, $^{3}H$, $^{125}I$, and $^{32}P$, each introduced into the probe through enzymatic incorporation of radioactive nucleotides. High-energy isotopes, such as $^{125}I$ and $^{32}P$, permit short exposure times, but the high energy also results in significant scatter of signal, precluding accurate localization of positive signals. The probes are also quite unstable because of the degradation caused by the high energy of the label and because the isotope has a short half-life. The highest degree of detail for signal localization is achieved with $^{3}H$, which also has a long half-life. Although tritiated probes are stable, they frequently necessitate exposure times of 30 days to achieve sufficient signal. A practical compromise is achieved with $^{35}S$ labels. Because autoradiographic detection relies on exposure of a silver emulsion applied over the tissue, the reaction product is not in the same plane of section as the tissue. This results in the loss of fine morphologic features. One advantage of autoradiography, however, is that the signal can be evaluated quantitatively. The number of grains of exposed silver may be linearly related to the number of copies of target.

*Nonradioactive Detection Methods*

The advantages of nonradioactive detection systems for clinical application are paramount. Chief among the advantages is the stability of the probes. The probe is the central reagent in these assays, analogous in importance to the primary antibody in an immunochemical assay. A well-characterized, stable probe is absolutely essential for a clinical assay. Nonradioactive detection systems are generally faster than autoradiography, result in superior signal localization, are less hazardous to laboratory personnel and the environment, and eliminate the need for expensive disposal and monitoring systems. These advantages have resulted in the increased application and refinement of nonradioactive methods. When fully optimized, these methods can achieve sensitivity comparable to that of autoradiographic methods. It is, however, difficult to achieve quantitative results with these systems.

With nonradioactive probes, the detection phase of the reaction is as essential to a successful assay as is the hybridization reaction itself. Color reagents must be carefully optimized to minimize background and to maximize signal. Various nonradioactive labels are available. Biotin is the prototype label, and similar results are achieved with other haptens such as digoxigenin and sulfone. These groups are detected with antibody or avidin reagents coupled to a signal-generating compound. Signal can be generated by an enzyme coupled with a colorimetric substrate or from a fluorescent reporter molecule. Probes can also be labeled directly with fluorescent tags or enzymes, but direct labeling systems have, in general, been less sensitive than indirect systems. Affinity probes use indirect signal generation. This means that the specific activity of the probe (i.e., the number of affinity groups incorporated per probe) is not as important as the availability of these affinity groups for subsequent detection after hybridization.

Methods that optimize incorporation of label do not necessarily result in maximal generation of signal. Because the affinity labels are relatively bulky compared with isotopic labels, steric hindrance to hybridization can result from too much incorporation of affinity label. Such problems must be recognized and avoided.

One advantage of affinity labels is that various signal-generating systems may be used for the same label. For example, with a biotinylated probe, avidin may be tagged with a fluorescent reporter group or coupled to an enzyme, such as alkaline phosphatase or horseradish peroxidase. Each of these enzymes in turn has various substrates that vary in the color, sensitivity, and stability of their products. Maximal sensitivity has been reported with systems utilizing alkaline phosphatase with BCIP-NBT (5-bromo-4-chloro-3-indoylphosphate–Nitro Blue Tetrazolium) substrate (McGadey reagent). Wide variations in sensitivity of colorimetric methods are, to a large extent, attributable to the choice and optimization of the detection protocol.

## Interpretation

After completion of the assay, counterstaining, and sealing with coverslips, the results of in situ hybridization are evaluated most frequently by light microscopy, usually in a qualitative fashion. To take advantage of the morphologic context, the evaluator should be an experienced histopathologist. The

**Table 2.** Examples of infectious agents detected by in situ hybridization

| | |
|---|---|
| Viral agents | Bacterial agents |
| Adenovirus | *Chlamydia trachomatis* |
| Cytomegalovirus | *Legionella* species |
| Epstein-Barr virus | *Haemophilus influenzae* |
| Herpes simplex virus | Fungal agent |
| Hepatitis B virus | *Pneumocystis carinii* |
| Human papillomavirus | |
| Human immunodeficiency virus | |
| JC virus | |
| B19 virus | |

results of in situ hybridization must be interpreted with caution; the appearance of negative control slides is fully as important as that of the positive control slide. Attention to the stringency of the hybridization and wash conditions has the same importance in the in situ assay as in other formats.

## Automation

The technical burden of processing multiple slides with multiple probes can be significant and results in the introduction of some variation between assays. The introduction of automation contributed greatly to the application of this technology. The first and most practical automated system (Code-on Stainer; Instrument Laboratories, Lexington, Mass.) uses capillary gap technology for application of reagents and computer-driven tabletop robotics for transportation of slides from station to station (97). This allows handling of 30 slide pairs (60 slides), controlling time and temperature of each step of the assay, confining any hazardous reagents to a workstation, and minimizing use of technicians' time. Automation also permits the inclusion of more controls in each assay. Automation of in situ hybridization was first described in 1988 (97) and subsequently was used in studies of human papillomavirus (103), human immunodeficiency virus (99), and Epstein-Barr virus (33, 37) in clinical material.

## Applications

As with other probe assays, the in situ hybridization format is best suited for infectious agents for which culture and serologic methods are difficult, expensive, or not readily available. The requirement of a tissue sample limits the application of this technology to some degree. However, histologic and cytologic evaluation of tissue and cellular samples can indicate the presence of infectious agents that were not expected on clinical grounds. Thus, in situ hybridization can confirm a diagnosis suggested on morphologic grounds without the need for additional clinical samples. Table 2 lists several infectious agents for which in situ assays have been reported.

Despite the potential clinical utility of these tests, clinical application of this technology is not widespread. This may be attributed to the cost of commercially available probes and the complexity of the technology. For herpes simplex virus and cytomegalovirus, many investigators report that immunocytochemical methods of detection have equivalent or superior sensitivity (32, 35). The decreased cost of an immunoassay compared with a hybridization assay also must be taken into account.

At present, the detection and typing of human papillomavirus by colorimetric in situ hybridization is the assay most likely to find diagnostic application (7, 12). Human papillomavirus is unique in that culture methods do not exist and typing is dependent on the genetic component of the virus (12). Immunohistochemical methods do not permit distinction among human papillomavirus types. Although more than 60 types have been reported, the types with clinical importance are those found in the urogenital tract (55). Epidemiologic studies have linked human papillomavirus to cervical carcinomas (51, 64) and precursor lesions (103, 110). In situ studies were useful in detecting the virus within neoplastic cells. Because certain viral types were found most consistently in carcinomas (55), laboratories are exploring the possibility of incorporating human papillomavirus typing with morphologic assessment as an additional predictor of clinical behavior. In situ analysis for human papillomavirus also allows a more objective assessment of viral change in the setting of inflammation (78). The long-term clinical implication of human papillomavirus of any type is not clear, and the clinical utility of human papillomavirus testing and typing is still being determined.

In situ hybridization also was applied extensively to the recognition and study of Epstein-Barr virus in various tissues (61, 62).

## Future Directions

The current success of probe-based assays suggests that their use in the clinical laboratory will continue to increase. Procedures for automating hybridization assays are in various stages of development, although at present such machines are directed toward in situ hybridization assays rather than processing of routine microbiologic samples.

Although the number of commercial probe kits continues to increase, so does the recognition of the power of nucleic acid amplification-based assay systems for the detection of small numbers of organisms in clinical samples. Indeed, sensitivity has often been the drawback to the more widespread use of probe technology, especially for the detection of enteric pathogens in stools or of *M. tuberculosis* in respiratory secretions. However, amplification assays have their own inherent problems, discussed in detail in other chapters of this volume. Thus, probes have achieved a certain niche in the clinical laboratory for the detection or identification of infectious-disease agents in culture confirmation assays, for direct identification of organisms in clinical samples, and for detection of organisms in tissue samples by in situ hybridization.

## References


1. **Alm, R. A., and P. A. Manning.** 1990. Biotype-specific probe for *Vibrio cholerae* serogroup O1. *J. Clin. Microbiol.* **28:**823–824.
2. **Archer, G. L., and E. Pennell.** 1990. Detection of methicillin resistance in staphylococci by using a DNA probe. *Antimicrob. Agents Chemother.* **34:**1720–1724.
3. **Arduino, M. J., L. A. Bland, M. A. Tipple, S. M. Aguero, M. S. Favero, and W. R. Jarvis.** 1989. Growth and endotoxin production of *Yersinia enterocolitica* and *Enterobacter agglomerans* in packed erythrocytes. *J. Clin. Microbiol.* **27:**1483–1485.
4. **Arlet, G., and A. Philippon.** 1991. Construction by polymerase chain-reaction and intragenic DNA probes for three main types of transferable β-lactamases (TEM, SHV, CARB). *FEMS Microbiol. Lett.* **82:**19–25.
5. **Arnold, L. J., Jr., P. W. Hammond, W. A. Wiese, and N. C. Nelson.** 1989. Assay formats involving acridinium-ester-labeled DNA probes. *Clin. Chem.* **35:**1588–1594.
6. **Barker, R. H., L. Suebsaeng, W. Rooney, G. C. Alecrim, H. V. Dourado, and D. F.**

**Wirth.** 1986. Specific DNA probe for the diagnosis of *Plasmodium falciparum* malaria. *Science* **231:**1434–1436.

7. **Beckmann, A. M., D. Myerson, J. R. Daling, N. B. Kiviat, C. M. Fenoglio, and J. K. McDougall.** 1985. Detection and localization of human papillomavirus DNA in human genital condylomas by in situ hybridization with biotinylated probes. *J. Med. Virol.* **16:**265–273.
8. **Bergström, S., O. Olsson, and S. Normark.** 1982. Common evolutionary origin of chromosomal beta-lactamase genes in enterobacteria. *J. Bacteriol.* **150:**528–534.
9. **Blum, H. E., A. T. Haase, and G. N. Vyas.** 1984. Molecular pathogenesis of hepatitis B virus infection: simultaneous detection of viral DNA and antigens in paraffin-embedded liver sections. *Lancet* **ii:**771–775.
10. **Britten, R. J., and E. H. Davidson.** 1985. Hybridisation strategy, p. 3–14. *In* B. D. Hames and S. J. Higgins (ed.), *Nucleic Acid Hybridisation: A Practical Approach.* IRL Press, Oxford.
11. **Butler, W. R., D. G. Ahearn, and J. O. Kilburn.** 1986. High-performance liquid chromatography of mycolic acids as a tool in the identification of *Corynebacterium*, *Nocardia*, *Rhodococcus*, and *Mycobacterium* species. *J. Clin. Microbiol.* **23:**182–185.
12. **Caussy, D., W. Orr, A. D. Daya, P. Roth, W. Reeves, and W. Rawls.** 1988. Evaluation of methods for detecting human papillomavirus deoxyribonucleotide sequences in clinical specimens. *J. Clin. Microbiol.* **26:**236–243.
13. **Chambers, H. F.** 1988. Methicillin-resistant staphylococci. *Clin. Microbiol. Rev.* **1:**173–186.
14. **Cheung, L. L., and J. B. Hudson.** 1988. Development of DNA probes for *Candida albicans. Diagn. Microbiol. Infect. Dis.* **10:**171–179.
15. **Chollet, A., and E. H. Kawashima.** 1985. Biotin-labeled synthetic oligodeoxyribonucleotides: chemical synthesis and uses as hybridization probes. *Nucleic Acids Res.* **13:**1529–1541.
16. **Coll, P., K. Phillips, and F. C. Tenover.** 1989. Evaluation of a rapid method of extracting DNA from stool samples for use in hybridization assays. *J. Clin. Microbiol.* **27:**2245–2248.
17. **Conville, P. S., J. F. Keiser, and F. G. Witebsky.** 1989. Mycobacteremia caused by simultaneous infection with *Mycobacterium avium* and *Mycobacterium intracellulare* detected by analysis of a BACTEC 13A bottle with the Gen-Probe kit. *Diagn. Microbiol. Infect. Dis.* **12:**217–219.
18. **Conville, P. S., J. F. Keiser, and F. G. Witebsky.** 1989. Comparison of three techniques for concentrating positive BACTEC 13A bottles for mycobacterial DNA probe analysis. *Diagn. Microbiol. Infect. Dis.* **12:**309–313.
19. **Crowley P., and S. G. Oliver.** 1987. A microculture hybridization technique for the detection of specific DNA sequences in filamentous fungi. *Exp. Mycol.* **11:**70–73.
20. **Daly, J. A., N. L. Clifton, K. C. Seskin, and W. M. Gooch III.** 1991. Use of rapid, nonradioactive DNA probes in culture confirmation tests to detect *Streptococcus agalactiae*, *Haemophilus influenzae*, and *Enterococcus* spp. from pediatric patients with significant infections. *J. Clin. Microbiol.* **29:**80–82.
21. **Davis, T. E., and D. D. Fuller.** 1991. Direct identification of bacterial isolates in blood cultures by using a DNA probe. *J. Clin. Microbiol.* **29:**2193–2196.
22. **Denys, G. A., and R. B. Carey.** 1992. Identification of *Streptococcus pneumoniae* with a DNA probe. *J. Clin. Microbiol.* **30:**2725–2727.
23. **Dix, K., S. M. Watanabe, S. McArdle, D. I. Lee, C. Randolf, B. Moncla, and D. E. Schwartz.** 1990. Species-specific oligodeoxynucleotide probes for the identification of periodontal bacteria. *J. Clin. Microbiol.* **28:**319–323.
24. **Doebbeling, B. N., M. J. Bale, F. P. Koontz, C. M. Helms, R. P. Wenzel, and M. A. Pfaller.** 1988. Prospective evaluation of the Gen-Probe assay for detection of legionellae in respiratory specimens. *Eur. J. Clin. Microbiol. Infect. Dis.* **7:**748–752.
25. **Dular, R., R. Kajioka, and S. Kasatiya.** 1988. Comparison of Gen-Probe commercial

kit and culture technique for the diagnosis of *Mycoplasma pneumoniae* infection. *J. Clin. Microbiol.* **26:**1068–1069.

26. **Echeverria, P., J. Seriwatana, O. Sethabutr, and D. N. Taylor.** 1985. DNA hybridization in the diagnosis of bacterial diarrhea. *Clin. Lab. Med.* **5:**447–462.
27. **Edelstein, P. H., R. N. Bryan, R. K. Enns, D. E. Kohne, and D. L. Kacian.** 1987. Retrospective study on Gen-Probe Rapid Diagnostic System for detection of legionellae in frozen clinical respiratory tract samples. *J. Clin. Microbiol.* **25:**1022–1026.
28. **Eschenbach, D. A., S. Hillier, C. Critchlow, C. Stevens, T. DeRouen, and K. K. Holmes.** 1988. Diagnosis and clinical manifestations of bacterial vaginosis. *Am. J. Obstet. Gynecol.* **158:**819–828.
29. **Fluit, A. C., A. T. A. Box, and J. Verhoef.** 1990. A probe for the detection of methicillin-resistant *Staphylococcus aureus. Eur. J. Clin. Microbiol. Infect. Dis.* **9:**605–608.
30. **Forghani, B., L. S. Oshiro, C. S. Chan, J. W. Hurst, J. Dennis, G. Darai, A. L. Warford, and R. M. Cohen.** 1992. Direct detection of *Molluscum contagiosum* virus in clinical specimens by *in situ* hybridization using biotinylated probe. *Mol. Cell. Probes* **6:**67–77.
31. **Frankel, G., L. Riley, J. A. Giron, J. Valmassoi, A. Friedmann, N. Strockbine, S. Falkow, and G. K. Schoolnik.** 1990. Detection of *Shigella* in feces using DNA amplification. *J. Infect. Dis.* **161:**1252–1256.
32. **Fung, J. C., J. Shanley, and R. C. Tilton.** 1985. Comparison of the detection of herpes simplex virus in direct clinical specimens with herpes simplex-specific DNA probes and monoclonal antibodies. *J. Clin. Microbiol.* **22:**748–753.
33. **Gal, A. A., E. R. Unger, M. N. Koss, and T. S. B. Yen.** 1991. Detection of Epstein-Barr virus in lymphoepithelioma-like carcinoma of the lung. *Mod. Pathol.* **4:**264–268.
34. **Gicquelais, K. G., M. M. Baldini, J. Martinez, L. Maggi, W. C. Martin, V. Prado, J. B. Kaper, and M. M. Levine.** 1990. Practical and economical method for using biotinylated DNA probes with bacterial colony blots to identify diarrhea-causing *Escherichia coli. J. Clin. Microbiol.* **28:**2485–2490.
35. **Gleaves, C. A., D. A. Hursh, D. H. Rice, and J. D. Meyers.** 1989. Detection of cytomegalovirus from clinical specimens in centrifugation culture by in situ DNA hybridization and monoclonal antibody staining. *J. Clin. Microbiol.* **27:**21–23.
36. **Gonzalez, R., and B. A. Hanna.** 1987. Evaluation of Gen-Probe DNA hybridization systems for the identification of *Mycobacterium tuberculosis* and *Mycobacterium avium-intracellulare. Diagn. Microbiol. Infect. Dis.* **8:**69–77.
37. **Guarner, J., C. del Rio, D. Carr, L. E. Hendrix, J. W. Eley, and E. R. Unger.** 1991. Non-Hodgkin's lymphomas in patients with human immunodeficiency virus infection: presence of Epstein-Barr virus by in situ hybridization, clinical presentation, and follow-up. *Cancer* **68:**2460–2465.
38. **Halpin, T. J., S. Kilker, J. Epstein, and M. Tourault.** 1992. Bacterial contamination of platelet pools—Ohio, 1991. *Morbid. Mortal. Weekly Rep.* **41**(3)**:**36–37.
39. **Hillier, S. L., and A. M. Briselden.** 1992. Evaluation of a rapid oligonucleotide test for direct detection of *Gardnerella vaginalis* and *Trichomonas vaginalis* from vaginal specimens, abstr. C-215, p. 456. *Abstr. 92nd Gen. Meet. Am. Soc. Microbiol. 1992.* American Society for Microbiology, Washington, D.C.
40. **Huovinen, S., P. Huovinen, and G. A. Jacoby.** 1988. Detection of plasmid-mediated β-lactamases with DNA probes. *Antimicrob. Agents Chemother.* **32:**175–179.
41. **Jablonski, E., E. W. Moomaw, R. H. Tullis, and J. L. Ruth.** 1986. Preparation of oligonucleotide-alkaline phosphatase conjugates and their use as hybridization probes. *Nucleic Acids Res.* **14:**6115–6128.
42. **Kessler, C.** 1991. The digoxigenin:anti-digoxigenin (DIG) technology—a survey on the concept and realization of a novel bioanalytical indicator system. *Mol. Cell. Probes* **5:**161–205.
43. **Kleemola, S. R. M., J. E. Karjalainen, and R. K. H. Räty.** 1990. Rapid diagnosis of

*Mycoplasma pneumoniae* infection: clinical evaluation of a commercial probe test. *J. Infect. Dis.* **162:**70–75.

44. **Koenig, C., L. J. Tick, and B. A. Hanna.** 1992. Analyses of the FlashTrack DNA probe and UTIscreen bioluminescence tests for bacteriuria. *J. Clin. Microbiol.* **30:**342–345.

44a. **Kohne, D. E., J. Hogan, V. Jonas, E. Dean, and T. H. Adams.** 1986. Novel approaches for the rapid and sensitive detection of microorganisms: DNA probes to rRNA, p. 110–112. *In* L. Leive (ed.), *Microbiology—1986.* American Society for Microbiology, Washington, D.C.

45. **Lees, M. I., D. M. Newnan, and S. M. Garland.** 1991. Comparison of a DNA probe assay with culture for the detection of *Chlamydia trachomatis. J. Med. Microbiol.* **35:**159–161.

46. **Levine, M. M., J.-G. Xu, J. B. Kaper, H. Lior, V. Prado, B. Tall, J. Nataro, H. Karch, and K. Wachsmuth.** 1987. A DNA probe to identify enterohemorrhagic *Escherichia coli* of O157:H7 and other serotypes that cause hemorrhagic colitis and hemolytic uremic syndrome. *J. Infect. Dis.* **156:**175–182.

47. **Lewis, J. S., D. Kranig-Brown, and D. A. Trainor.** 1990. DNA probe confirmatory test for *Neisseria gonorrhoeae. J. Clin. Microbiol.* **28:**2349–2350.

48. **Lim, S. D., J. Todd, J. Lopez, E. Ford, and J. M. Janda.** 1991. Genotypic identification of pathogenic *Mycobacterium* species by using a nonradioactive oligonucleotide probe. *J. Clin. Microbiol.* **29:**1276–1278.

49. **Limberger, R. J., R. Biega, A. Evancoe, L. McCarthy, L. Slivienski, and M. Kirkwood.** 1992. Evaluation of culture and the Gen-Probe PACE 2 assay for detection of *Neisseria gonorrhoeae* and *Chlamydia trachomatis* in endocervical specimens transported to a state health laboratory. *J. Clin. Microbiol.* **30:**1162–1166.

50. **Loesche, W. J., D. E. Lopatin, J. Giordano, G. Alcoforado, and P. Hujoel.** 1992. Comparison of the benzoyl-DL-arginine-naphthylamide (BANA) test, DNA probes, and immunological reagents for ability to detect anaerobic periodontal infections due to *Porphyromonas gingivalis*, *Treponema denticola*, and *Bacteroides forsythus. J. Clin. Microbiol.* **30:**427–433.

51. **Lorincz, A. T., G. F. Temple, R. J. Kurman, A. B. Jenson, and W. D. Lancaster.** 1987. Oncogenic association of specific human papillomavirus types with cervical neoplasia. *JNCI* **79:**671–677.

52. **Maas, R.** 1983. An improved colony hybridization method with significantly increased sensitivity for detection of single genes. *Plasmid* **10:**296–298.

53. **Malouin, F., L. E. Bryan, P. Shewciw, J. Douglas, D. Li, H. Van Den Elzen, and J.-R. Lapointe.** 1988. DNA probe technology for rapid detection of *Haemophilus influenzae* in clinical specimens. *J. Clin. Microbiol.* **26:**2132–2138.

54. **Matthews, J. A., and L. J. Kricka.** 1988. Analytical strategies for the use of DNA probes. *Anal. Biochem.* **169:**1–25.

55. **McCance, D. J., M. J. Campion, P. K. Clarkson, P. M. Chesters, D. Jenkins, and A. Singer.** 1985. Prevalence of human papillomavirus type 16 DNA sequences in cervical intraepithelial neoplasia and invasive carcinoma of the cervix. *Br. J. Obstet. Gynaecol.* **92:**1101–1105.

56. **McLaughlin, G. L., C. Decrind, R. Dayal-Drager, M. Hassan-King, S. Subramanian, and B. M. Greenwood.** 1991. Optimization of a rapid nonisotopic DNA probe assay for *Plasmodium falciparum* in The Gambia. *J. Clin. Microbiol.* **29:**1517–1519.

57. **McLaughlin, G. L., P. Deloron, A. Y. Huong, C. Sezibera, and G. H. Campbell.** 1988. DNA hybridization for assessment of response of *Plasmodium falciparum* to chloroquine therapy. *J. Clin. Microbiol.* **26:**1704–1707.

58. **McLaughlin, G. L., T. D. Edlind, and G. M. Ihler.** 1986. Detection of *Babesia bovis* using DNA hybridization. *J. Protozool.* **33:**125–128.

59. **Meloni, B. P., A. J. Lymbery, and R. C. A. Thompson.** 1989. Characterization of *Giardia* isolates using a non-radiolabeled DNA probe, and correlation with the results of isoenzyme analysis. *Am. J. Trop. Med. Hyg.* **40:**629–637.

60. **Moncla, B. J., P. Braham, K. Dix, S. Watanabe, and D. Schwartz.** 1990. Use of synthetic oligonucleotide DNA probes for the identification of *Bacteroides gingivalis*. *J. Clin. Microbiol.* **28:**324–327.
61. **Montone, K. T., L. R. Budgeon, and D. J. Brigati.** 1990. Detection of Epstein-Barr virus genomes by *in situ* DNA hybridization with a terminally biotin-labeled synthetic oligonucleotide probe from the EBV *Not I* and *Pst I* tandem repeat regions. *Mod. Pathol.* **3:**89–96.
62. **Montone, K. T., H. Friedman, R. L. Hodinka, D. G. Hicks, J. A. Kant, and J. E. Tomaszewski.** 1992. *In situ* hybridization for Epstein-Barr virus *Not I* repeats in posttransplant lymphoproliferative disorder. *Mod. Pathol.* **5:**292–302.
63. **Moseley, S. L., I. Huq, A. R. M. A. Alim, M. So, M. Samadpour-Motalebi, and S. Falkow.** 1980. Detection of enterotoxigenic *Escherichia coli* by DNA colony hybridization. *J. Infect. Dis.* **142:**892–898.
64. **Muñoz, N., X. Bosch, J. M. Kaldor.** 1988. Does human papillomavirus cause cervical cancer? The state of the epidemiological evidence. *Br. J. Cancer* **57:**1–5.
65. **Musiani, M., M. Zerbini, G. Gentilomi, G. Gallinella, S. Venturoli, D. Gibellini, and M. La Placa.** 1990. Rapid detection of cytomegalovirus DNA in urine samples with a dot blot hybridization immunoenzymatic assay. *J. Clin. Microbiol.* **28:**2101–2103.
66. **Neill, R. J., G. R. Fanning, F. Delahoz, R. Wolff, and P. Gemski.** 1990. Oligonucleotide probes for detection and differentiation of *Staphylococcus aureus* strains containing genes for enterotoxins A, B, and C and toxic shock syndrome toxin 1. *J. Clin. Microbiol.* **28:**1514–1518.
67. **Olive, D. M.** 1989. Detection of enterotoxigenic *Escherichia coli* after polymerase chain reaction amplification with a thermostable DNA polymerase. *J. Clin. Microbiol.* **27:**261–265.
68. **Ossewaarde, J. M., M. Rieffe, M. Rosenberg-Arska, P. M. Ossenkoppele, R. P. Nawrocki, and A. M. van Loon.** 1992. Development and clinical evaluation of a polymerase chain reaction test for detection of *Chlamydia trachomatis*. *J. Clin. Microbiol.* **30:**2122–2128.
69. **Ouellette, M., G. C. Paul, A. M. Philippon, and P. H. Roy.** 1988. Oligonucleotide probes (TEM-1, OXA-1) versus isoelectric focusing in β-lactamase characterization of 114 resistant strains. *Antimicrob. Agents Chemother.* **32:**397–399.
70. **Panke, E. S., L. I. Yang, P. A. Leist, P. Magevney, R. J. Fry, and R. F. Lee.** 1991. Comparison of Gen-Probe DNA probe test and culture for the detection of *Neisseria gonorrhoeae* in endocervical specimens. *J. Clin. Microbiol.* **29:**883–888.
71. **Parsons, L. M., M. Shayegani, A. L. Waring, and L. H. Bopp.** 1989. DNA probes for the identification of *Haemophilus ducreyi*. *J. Clin. Microbiol.* **27:**1441–1445.
72. **Pasculle, A. W., G. E. Veto, S. Krystofiak, K. McKelvey, and K. Vrsalovic.** 1989. Laboratory and clinical evaluation of a commercial DNA probe for the detection of *Legionella* spp. *J. Clin. Microbiol.* **27:**2350–2358.
73. **Perine, P. L., P. A. Totten, K. K. Holmes, E. H. Sng, A. V. Ratnam, R. Widy-Wersky, H. Nsanze, E. Habte-Gabr, and W. G. Westbrook.** 1985. Evaluation of a DNA hybridization method for the detection of African and Asian strains of *Neisseria gonorrhoeae* in men with urethritis. *J. Infect. Dis.* **152:**59–63.
74. **Peterson, E. M., R. Lu, C. Floyd, A. Nakasone, G. Friedly, and L. M. de la Maza.** 1989. Direct identification of *Mycobacterium tuberculosis*, *Mycobacterium avium*, and *Mycobacterium intracellulare* from amplified primary cultures in BACTEC media using DNA probes. *J. Clin. Microbiol.* **27:**1543–1547.
75. **Piper, K., M. E. Brecher, L. Bland, and D. H. Persing.** 1992. Use of a chemiluminescent universal bacterial probe for pretransfusion screening of blood products, abstr. D-125, p. 116. *Abstr. 92nd Gen. Meet. Am. Soc. Microbiol. 1992.* American Society for Microbiology, Washington, D.C.
76. **Polacheck, I., G. Lebens, and J. B. Hicks.** 1992. Development of DNA probes for early diagnosis and epidemiological study of cryptococcosis in AIDS patients. *J. Clin. Microbiol.* **30:**925–930.

77. **Ranki, M., A. W. Leinonen, T. Jalava, P. Nieminen, V. R. X. Soares, J. Paavonen, and A. Kallio.** 1990. Use of the AffiProbe HPV test kit for detection of human papillomavirus DNA in genital scrapes. *J. Clin. Microbiol.* **28:**2076–2081.
78. **Richart, R. M., and G. J. Nuovo.** 1990. Human papillomavirus DNA in situ hybridization may be used for quality control of genital tract biopsies. *Obstet. Gynecol.* **75:**223–226.
79. **Richman, D. D., J. A. McCutchan, and S. A. Spector.** 1987. Detecting human immunodeficiency virus RNA in peripheral blood mononuclear cells by nucleic acid hybridization. *J. Infect. Dis.* **156:**823–827.
80. **Roberts, M. C., S. L. Hillier, F. D. Schoenknecht, and K. K. Holmes.** 1985. Comparison of Gram stain, DNA probe, and culture for the identification of species of *Mobiluncus* in female genital specimens. *J. Infect. Dis.* **152:**74–77.
81. **Rossau, R., M. Duhamel, E. Van Dyck, P. Piot, and H. Van Heuverswyn.** 1990. Evaluation of an rRNA-derived oligonucleotide probe for culture confirmation of *Neisseria gonorrhoeae. J. Clin. Microbiol.* **28:**944–948.
82. **Rubin, F. A., P. D. McWhirter, N. H. Punjabi, E. Lane, P. Sudarmono, S. P. Pulungsih, M. Lesmana, S. Kumala, D. J. Kopecko, and S. L. Hoffman.** 1989. Use of a DNA probe to detect *Salmonella typhi* in the blood of patients with typhoid fever. *J. Clin. Microbiol.* **27:**1112–1114.
83. **Saiki, R. K., D. H. Gelfand, S. Stoffel, S. J. Scharf, R. Higuchi, G. T. Horn, K. B. Mullis, and H. A. Erlich.** 1988. Primer-directed enzymatic amplification of DNA with a thermostable DNA polymerase. *Science* **239:**487–491.
84. **Sambrook, J., E. F. Fritsch, and T. Maniatis.** 1989. *Molecular Cloning: A Laboratory Manual*, 2nd ed. Cold Spring Harbor Laboratory, Cold Spring Harbor, N.Y.
85. **Samuelson, J., R. Acuno-Soto, S. Reed, F. Biagi, and D. Wirth.** 1989. DNA hybridization probe for clinical diagnosis of *Entamoeba histolytica. J. Clin. Microbiol.* **27:**671–676.
86. **Sheiness, D., K. Dix, S. Watanabe, and S. L. Hillier.** 1992. High levels of *Gardnerella vaginalis* detected with an oligonucleotide probe combined with elevated pH as a diagnostic indicator of bacterial vaginosis. *J. Clin. Microbiol.* **30:**642–648.
87. **Sim, B. K. L., J. W. Mak, W. H. Cheong, I. Sutanto, L. Kurniawan, H. A. Marwoto, E. Franke, J. R. Campell, D. F. Wirth, and W. F. Piessens.** 1986. Identification of *Brugia malayi* in vectors with a species-specific DNA probe. *Am. J. Trop. Med. Hyg.* **35:**559–564.
88. **Southern, E. M.** 1975. Detection of specific sequences among DNA fragments separated by gel electrophoresis. *J. Mol. Biol.* **98:**503–517.
89. **Spiegel, C. A., R. Amsel, and K. K. Holmes.** 1983. Diagnosis of bacterial vaginosis by direct Gram stain of vaginal fluid. *J. Clin. Microbiol.* **18:**170–177.
90. **Tenover, F. C.** 1986. Studies of antimicrobial resistance genes using DNA probes. *Antimicrob. Agents Chemother.* **29:**721–725.
91. **Tenover, F. C.** 1988. Diagnostic deoxyribonucleic acid probes for infectious diseases. *Clin. Microbiol. Rev.* **1:**82–101.
92. **Tenover, F. C.** 1989. *DNA Probes for Infectious Diseases*, p. 53–64. CRC Press, Inc., Boca Raton, Fla.
93. **Tenover, F. C., L. Carlson, S. Barbagallo, and I. Nachamkin.** 1990. DNA probe culture confirmation assay for identification of thermophilic *Campylobacter* species. *J. Clin. Microbiol.* **28:**1284–1287.
94. **Tipple, M. A., L. A. Bland, J. J. Murphy, M. J. Arduino, A. L. Panlilio, J. J. Farmer III, M. A. Tourault, C. R. Macpherson, J. E. Menitove, A. J. Grindon, P. S. Johnson, R. G. Strauss, J. A. Bufill, P. S. Ritch, J. R. Archer, O. C. Tablan, and W. R. Jarvis.** 1990. Sepsis associated with transfusion of red cells contaminated with *Yersinia enterocolitica. Transfusion* **30:**207–213.
95. **Tomasz, A., H. B. Drugeon, H. M. de Lencastre, D. Jabes, L. McDougall, and J. Bille.** 1989. New mechanism for methicillin resistance in *Staphylococcus aureus*: clinical

isolates that lack the PBP 2a gene and contain normal penicillin-binding proteins with modified penicillin-binding capacity. *Antimicrob. Agents Chemother.* **33:**1869–1874.

96. **Totten, P. A., K. K. Holmes, H. H. Handsfield, J. S. Knapp, P. L. Perine, and S. Falkow.** 1983. DNA hybridization technique for the detection of *Neisseria gonorrhoeae* in men with urethritis. *J. Infect. Dis.* **148:**462–471.
97. **Unger, E. R., D. J. Brigati, M. L. Chenggis, L. R. Budgeon, D. Koebler, C. Cuomo, and T. Kennedy.** 1988. Automation of in situ hybridization: application of the capillary action robotic workstation. *J. Histotechnol.* **11:**253–258.
98. **Unger, E. R., L. R. Budgeon, D. Myerson, and D. J. Brigati.** 1986. Viral diagnosis by in situ hybridization: description of a rapid simplified colorimetric method. *Am. J. Surg. Pathol.* **10:**1–8.
99. **Unger, E. R., F. W. Chandler, M. L. Chenggis, C.-Y. Ou, D. T. Warfield, and P. M. Feorino.** 1989. Demonstration of human immunodeficiency virus by colorimetric in situ hybridization: a rapid technique for formalin-fixed paraffin-embedded material. *Mod. Pathol.* **2:**200–204.
100. **Van Damme-Jongsten, M., J. Rodhouse, R. J. Gilbert, and S. Notermans.** 1990. Synthetic DNA probes for detection of enterotoxigenic *Clostridium perfringens* strains isolated from outbreaks of food poisoning. *J. Clin. Microbiol.* **28:**131–133.
101. **Wallace, R. B., J. Shaffer, R. F. Murphy, J. Bonner, T. Hirose, and K. Itakura.** 1979. Hybridization of synthetic oligodeoxyribonucleotides to ΦX174 DNA: the effect of single base pair mismatch. *Nucleic Acids Res.* **6:**3543–3557.
102. **Wilkinson, H. W., J. S. Sampson, and B. B. Plikaytis.** 1986. Evaluation of a commercial gene probe for identification of *Legionella* cultures. *J. Clin. Microbiol.* **23:**217–220.
103. **Wilson, R. W., M. L. Chenggis, and E. R. Unger.** 1990. Longitudinal study of human papillomavirus infection of the female urogenital tract by in situ hybridization. *Arch. Pathol. Lab. Med.* **114:**155–159.
104. **Wirth, D. F., and D. M. Pratt.** 1982. Rapid identification of *Leishmania* species by specific hybridization of kinetoplast DNA in cutaneous lesions. *Proc. Natl. Acad. Sci. USA* **79:**6999–7003.
105. **Woods, G. L., A. Young, J. C. Scott, Jr., T. M. H. Blair, and A. M. Johnson.** 1990. Evaluation of a nonisotopic probe for detection of *Chlamydia trachomatis* in endocervical specimens. *J. Clin. Microbiol.* **28:**370–372.
106. **Wren, B. W., C. L. Clayton, N. B. Castledine, and S. Tabaqchali.** 1990. Identification of toxigenic *Clostridium difficile* strains by using a toxin A gene-specific probe. *J. Clin. Microbiol.* **28:**1808–1812.
107. **Yagupsky, P., and F. S. Nolte.** 1990. Quantitative aspects of septicemia. *Clin. Microbiol. Rev.* **3:**269–279.
108. **Yang, L. I., E. S. Panke, P. A. Leist, R. J. Fry, and R. F. Lee.** 1991. Detection of *Chlamydia trachomatis* endocervical infection in asymptomatic and symptomatic women: comparison of deoxyribonucleic acid probe test with tissue culture. *Am. J. Obstet. Gynecol.* **165:**1444–1453.
109. **Zerbini, M., M. Musiani, S. Venturoli, G. Gallinella, D. Gibellini, G. Gentilomi, and M. La Placa.** 1990. Rapid screening for B19 parvovirus DNA in clinical specimens with a digoxigenin-labeled DNA hybridization probe. *J. Clin. Microbiol.* **28:**2496–2499.
110. **zur Hausen, H.** 1989. Papillomaviruses as carcinomaviruses. *Adv. Virol. Oncol.* **8:**1–26.

# Molecular Typing Methods

2

Bala Swaminathan and Ghassan M. Matar

## Molecular Typing Methods: Definition, Applications, and Advantages

The concept of a clonal connection between isolates of microorganisms from a common-source outbreak is important in the epidemiology of infectious diseases. Many outbreaks of infections result from exposure to a common source of the etiologic agent. In general, these infectious microorganisms are clonal; that is, they are the progeny of a single cell and thus are genetically identical or nearly so. At the bacterial species level, there is sufficient genetic diversity to allow identification of different clones or clonal groups (strains that have a high degree of genetic relatedness) among isolates of the same species collected from different sources and sites and at different times. This is called subtyping; it is done by examining a set of bacterial isolates for several characteristics that allow discrimination below the species level.

Subtyping has broad applications in public health bacteriology. It is essential for determining whether previously recognized virulent clones are present in a set of strains, for recognizing outbreaks, for identifying the source of infection in outbreak and sporadic disease settings, and for ensuring that immunization programs lead to eradication of the target pathogen instead of replacement with a variant subtype.

In the past, subtyping was accomplished by testing for one or several phenotypic markers. Examples of these conventional subtyping procedures include biotyping, serotyping, bacteriophage typing, and bacteriocin typing. Although these methods are valuable for strain characterization and are still in widespread use, they have several shortcomings. These tests are organism specific; each organism requires the use of a unique set of reagents and procedures. Serotyping and phage typing require specialized reagents (e.g., absorbed sera, phages, and bacterial strains for propagating phages) that may be available to only a few specialized reference laboratories. Bacterial isolates from nonclinical sources such as food, environment, or indwelling catheters frequently cannot be typed by serotyping and phage typing.

Hybridoma technology enabled us to subtype bacterial strains by using monoclonal antibodies prepared against cell surface antigens. Monoclonal antibodies introduced a high degree of standardization into serology-based subtyping. This method has been used successfully for subtyping some *Legionella* spp. and *Neisseria meningitidis* (1, 36). Once the reagents are prepared and standardized, the subtyping procedure is extremely simple and can be done easily in a minimally equipped laboratory. Nevertheless, this approach requires the preparation of several monoclonal antibodies directed against variable epitopes of each organism to be subtyped.

Antimicrobial resistance patterns, although based on phenotypic traits, have been highly standardized and are useful for subtyping many bacterial isolates, particularly those involved in nosocomial transmissions. Because factors responsible for antibiotic resistance are frequently encoded by plasmid DNA, an association between resistance patterns and plasmid profiles is often apparent. *Pseudomonas cepacia* and some species of coagulase-negative staphylococci, however, are exceptions because they have unstable resistance patterns (69).

Molecular typing methods are based on the physical characterization of molecules (fatty acids, proteins, carbohydrates, nucleic acids, and other chemicals) produced by bacteria and are therefore universally applicable. Several factors influenced the development of the molecular typing methods: (i) the tremendous advances made in the separation of micromolecules and macromolecules, (ii) the wide availability of sophisticated and reasonably priced instruments that allow separation of complex chemical mixtures and identification of the separated components, (iii) the availability of personal computers and the ability to interface these computers with analytical instruments to allow data acquisition and data analysis, and (iv) the extraordinary developments in nucleic acid technology that resulted in the widespread availability of these sophisticated methods to modestly equipped laboratories.

Molecular typing methods revolutionized epidemiology. Some notable examples of the application of molecular subtyping methods to epidemiology are the tracing of a large salmonellosis epidemic to the distribution of contaminated marijuana (68) and the tracking of a virulent clone of *N. meningitidis* group A as it spread essentially unchanged from China to Africa through the Indian subcontinent and the Middle East countries in 20 years (13). Although these methods have high discriminating power and high reproducibility, they must be standardized and carefully validated. As with any conventional typing method, this is done by using a carefully selected set of strains, some of which are homogeneous (by epidemiologic and other microbiologic criteria) and some of which are heterogeneous.

## Molecular Typing Methods

Molecular typing methods may be categorized into three broad groups on the basis of the type of macromolecules targeted for subtyping: lipopolysaccharide (LPS) and fatty acid-based methods, protein-based methods, and nucleic acid-based methods.

### LPS- and Fatty Acid-Based Methods

Analysis of LPS by sodium dodecyl sulfate-polyacrylamide gel electrophoresis (SDS-PAGE) is a relatively simple method for evaluating heterogeneity among strains within species of gram-negative bacteria. LPS analysis has been useful for subtyping nonserotypeable *Haemophilus influenzae* strains that cause otitis media (51) and for subtyping *Eikenella corrodens* (12).

Cellular fatty acid analysis by gas-liquid chromatography is a rapid and reliable method for the identification of bacteria at the species level (77). It has also been applied to the subtyping of a few bacterial species (16, 77). Although it is attractive because of the relatively simple sample preparation method and the high degree of automation that has been engineered into the commercially available systems, its discriminating power is generally inadequate for subtyping. Further, its utility is diminished in that fatty acid profiles are significantly affected by the growth history of the bacterial culture (e.g., growth medium and incubation temperature).

### Protein-Based Methods

Whole-cell and outer membrane protein profiles determined by SDS-PAGE are also useful for subtyping bacteria. These methods were applied to the characterization of several gram-negative bacteria, notably *H. influenzae* type b (4, 23).

Multilocus enzyme electrophoresis (MEE) is a special type of protein-based subtyping method whose results can be directly correlated with the genotype (71). This method involves the determination of the mobilities in a starch gel matrix of a selected set of constitutive metabolic enzymes of bacterial strains. Variations in the mobility of a constitutive enzyme for different strains of the same species under these conditions can be attributed to isozymes (functionally similar forms of an enzyme, including all polymers of subunits produced by different gene loci or by different alleles at the same locus) or to allozymes (variants of polypeptides representing different allelic alternatives at the same gene locus). The basic assumption in MEE is that changes in the mobility of enzymes are a result of charge differences due to amino acid substitutions in the polypeptide sequence and that these charge differences reflect changes in the DNA encoding the polypeptide (62).

MEE is performed as follows. Freshly grown bacterial cells (packed-cell volume, approximately 1 ml) are suspended in a breaking buffer, and cells are ruptured to release the cytoplasmic contents in a French pressure cell, in a vortex mixer with glass beads, or by sonication. The soluble cell extract is subjected to electrophoresis on 1-cm-thick starch gels. After electrophoresis for 4 to 8 h, the starch gels are sliced into 2-mm-thick slices, and each slice is treated with an appropriate color-forming substrate for a particular enzyme. After staining, the position of the enzyme is visible on the gel slice as a discrete band of a certain color. By repeating this procedure on three or four gels, information for approximately 20 enzymes is obtained. The relative mobility of an enzyme is determined by comparison with a set of standards (extracts from control strains), and an allele number is assigned for each enzyme of a strain arbitrarily or by comparison with previously determined alleles. When this procedure is completed, a bacterial strain can be described by a string of numbers corresponding to the allele designations for each of the enzymes evaluated. MEE is a useful procedure and was used extensively for the study of bacterial population and evolutionary genetics and for epidemiology of infectious diseases (6, 79, 80).

### Nucleic Acid-Based Methods

Nucleic acid-based subtyping was first applied to study the epidemiology of a nosocomial outbreak of *Klebsiella pneumoniae* infection (58). Initial applications involved primarily the detection and characterization of plasmids in bacteria. Developments in nucleic acid isolation, separation, and amplification since 1975 have led to the application of nucleic acid-based subtyping methods to various specific epidemiologic problems. These include restriction endonuclease

(RE) analysis of plasmid or genomic DNA, probe-based fingerprinting, and polymerase chain reaction (PCR)-based fingerprinting. The basic principles of each of these techniques, their strengths and weaknesses, and their general applicability are discussed in this chapter. Detailed descriptions of methodology and the applications of these techniques to specific microorganisms can be found in later chapters.

### *DNA Sequencing*

Direct comparison of genomic DNA sequences of bacterial strains is the best means of quantitatively determining whether two strains are similar or different. It is possible to analyze short segments of DNA for sequence similarities and differences. The combination of PCR to amplify genomic DNA fragments (50) and an automated DNA-sequencing procedure involving fluorescent dye-labeled terminators (Applied Biosystems, Foster City, Calif.) to directly sequence the PCR-amplified DNA fragment (70) allows a 300- to 500-bp DNA fragment to be sequenced in 24 h. Dean et al. (17) determined the nucleotide sequences of three variable segments for the major outer membrane protein gene of *Chlamydia trachomatis* and suggested that this method may be an important means of studying the molecular epidemiology of *Chlamydia* spp. Although direct comparison of nucleotide sequence differences may someday lead to a definitive typing method, DNA sequencing is still too expensive and too complex to be applied in practical situations to large numbers of bacterial isolates. However, improvements in automation of target amplification and direct sequence analysis may someday lead to practical implementation of this method in the clinical laboratory (chapter 3).

### *Chromosomal and Extrachromosomal DNA Restriction Profiles*

**Chromosomal DNA restriction analysis.** Restriction site analysis provides a cost-effective means of determining DNA sequence variations. This method involves comparison of the number and the size of fragments produced by digestion of DNA with a RE, which is an enzyme that cuts DNA at a constant position within a specific recognition site usually composed of 4 to 6 bp. Because of the high specificity of REs, complete digestion of a given DNA with a specific RE provides a reproducible array of fragments. These fragments, usually ranging in size from 20,000 to 1,000 bp, can be separated by agarose gel electrophoresis and visualized by staining with ethidium bromide.

Variations in the array of fragments generated by a specific RE are called restriction fragment length polymorphisms (RFLPs). RFLPs can result from sequence rearrangements, insertion or deletion of DNA, or base substitution within the RE cleavage sites. The first two types of changes can be observed with all REs, whereas the third type can be observed only with REs for which the recognition site encompasses the substitution site. Common REs and their recognition sites are shown in Table 1.

The selection of a RE for use in RFLP analysis is based on two important criteria. First, the restriction fragments must be suitable for analysis in terms of size and frequency. Best results are obtained with restriction fragments of 1,000 to 15,000 bp. Second, the fragments in this size range should not be too numerous, to avoid overlapping bands that may obscure differences. Most investigators select the best RE(s) by empirically screening several enzymes with a set of epidemiologically characterized strains of the bacterial species of interest.

Forbes et al. (21) devised a method to select the appropriate enzyme in a

**Table 1.** Common REs and their recognition sites[a]

| RE | Recognition sequence[b] | Temp (°C) | Buffer[c] | Comments |
|---|---|---|---|---|
| *Eco*RI | 5′ G′AATT C 3′<br>3′ C TTAA′G 5′ | 37 | 3 | Star activity[d] |
| *Cla*I | 5′ AT′CG AT 3′<br>3′ TA GC′TA 5′ | 37 | 4 | Sensitive to methylation |
| *Sma*I | 5′ CCC′GGG 3′<br>3′ GGG′CCC 5′ | 25 | 4 | KCl required<br>Half-life is 15 min at 37°C |
| *Bam*HI | 5′ G′GATC C 3′<br>3′ C CTTG′G 5′ | 37 | *Bam*HI | Star activity in buffers of low ionic strength or with glycerol concn of >5% |
| *Hin*dIII | 5′ A′AGCT T 3′<br>3′ T TCGA′A 5′ | 37 | 2 | Star activity<br>Requires a duplex DNA of >10 bases to cut |
| *Hin*fI | 5′ G′ANT C 3′<br>3′ C TNA′G 5′ | 37 | 2 | Cleaves single-stranded DNA slowly<br>Star activity |
| *Taq*I | 5′ TCGA 3′<br>3′ AGCT 5′ | 37 | *Taq*I | Cleaves single-stranded DNA slowly<br>Sensitive to methylation |
| *Hae*III | 5′ Pu GCGC′Py 3′<br>3′ Py′CGCG Pu 5′ | 37 | 2 | |
| *Hha*I | 5′ G CG′C 3′<br>3′ C′GC G 5′ | 37 | 2 | Cleaves single-stranded DNA at 5% the rate of double-stranded DNA |
| *Sfi*I | 5′ GGCCN NNN′NGGCC 3′<br>3′ CCGGN′NNN NCCGG 5′ | 37 | 2 | |
| *Nhe*I | 5′ G′CTAG C 3′<br>3′ C GATC′G 5′ | 37 | 4 | |
| *Apa*I | 5′ G GGCC′C 3′<br>3′ C′CCGG G 5′ | 37 | 4 | Inhibited by high salt concentration |

[a] Source of information: New England Biolabs catalog, 1992.
[b] Symbols: N, any nucleotide; Pu, purine; Py, pyrimidine; ′, cleavage site.
[c] Buffer 2: 50 mM NaCl, 10 mM Tris HCl, 10 mM $MgCl_2$, 1 mM dithiothreitol (pH 7.9 at 25°C). Buffer 3: 150 mM NaCl, 6 mM Tris HCl (pH 7.9), 6 mM $MgCl_2$, 6 mM 2-mercaptoethanol. Buffer 4: 50 mM potassium acetate, 20 mM Tris acetate, 10 mM magnesium acetate, 1 mM dithiothreitol (pH 7.9 at 25°C). Buffer *Bam*HI: 150 mM NaCl, 50 mM Tris HCl, 10 mM $MgCl_2$, 1 mM dithiothreitol (pH 7.9 at 25°C). Buffer *Taq*I: 10 mM NaCl, 10 mM Tris HCl, 10 mM $MgCl_2$, 10 mM 2-mercaptoethanol (pH 8.4 at 25°C).
[d] An enzyme exhibits star activity whenever it cleaves DNA sequences that are similar but not identical to their defined recognition sequence under extreme nonstandard conditions. These conditions include high glycerol concentration (>5%, vol/vol), high units-to-microgram of DNA ratio (>100 U/μg), low ionic strength (<25 mM), high pH (>pH 8), presence of organic solvents, and substitution of $Mg^{2+}$ with other divalent cations.

more objective manner. They use published nucleotide sequences for the bacterial species under study (or a closely related species) to determine the mono-, di-, and trinucleotide frequencies. The moles percent G+C content is calculated from the mononucleotide frequencies and compared with the published moles percent G+C for the species to determine whether the sequences studied are representative of the entire genome. The mean fragment length for a particular RE is determined from the dinucleotide and trinucleotide frequencies. The mean fragment length in nucleotides is the reciprocal of the frequency of a particular recognition sequence for a given RE and genome. By using the frequency ($p$) of dinucleotides and trinucleotides and second-order Markov chains, the mean fragment length for a restriction enzyme with a 6-base recognition sequence ($N_1N_2N_3N_4N_5N_6$) is given by:

$$\frac{p\ (N_2N_3) \cdot p\ (N_3N_4) \cdot p\ (N_4N_5)}{p\ (N_1N_2N_3) \cdot p\ (N_2N_3N_4) \cdot p(N_3N_4N_5)\ \cdot\ p\ (N_4N_5N_6)}$$

The basic steps in RFLP analysis are (i) growth of bacteria and isolation and purification of DNA, (ii) treatment of the DNA with one or more REs separately or together under optimized conditions to obtain complete digestion, (iii) separation of the restriction fragments by electrophoresis on agarose gels, and (iv) visualization of the separated fragments by staining with ethidium bromide.

Isolation and purification of DNA are critical steps in RFLP analysis. The DNA must be sufficiently intact and should be free of impurities that may inhibit REs and lead to partial or no cutting. DNA isolation procedures vary depending on the bacterial species. Genomic DNA is easily isolated from most gram-negative bacteria by lysozyme pretreatment followed by lysis with a nonionic detergent (e.g., Triton X-100) or an ionic detergent (e.g., SDS). The lysate is extracted successively with phenol and chloroform to remove proteins and lipid-soluble materials. We use Phase Lock Gel (5 Prime → 3 Prime, Boulder, Colo.) for the phenol-chloroform extractions. Phase Lock Gel migrates under centrifugal force to form a seal between the organic and aqueous phases. The organic phase and interface material are trapped below the Phase Lock Gel, and the barrier is sufficiently stable that the aqueous phase can easily be decanted or pipetted into a fresh tube. DNA and RNA in the aqueous phase are precipitated with ethanol or isopropanol, pelleted by centrifugation, washed with 70% ethanol, dried, and resuspended in a buffer of low ionic strength (usually 10 mM Tris–1 mM EDTA [pH 8.0]).

Occasionally, problems are encountered with this procedure for the isolation of genomic DNA from encapsulated gram-negative bacteria, such as *H. influenzae* and *N. meningitidis*, or from bacteria that synthesize large quantities of polysaccharides (*Pseudomonas* spp.). The capsular polysaccharide is carried with the DNA in the aqueous phase during the phenol-chloroform extractions and coprecipitated with the DNA. This problem may be overcome by removing the polysaccharides in the cell lysate with hexadecyltrimethylammonium bromide in the presence of 0.7 M NaCl (82).

It is difficult to lyse gram-positive bacteria by the lysozyme and detergent method. Some gram-positive bacteria can be lysed efficiently by pretreatment of the cells with specific lytic enzymes (lysostaphin for *Staphylococcus aureus*) or general-purpose lytic enzymes (mutanolysin, achromopeptidase). A procedure developed in our laboratory (38), in which pancreatic lipase is used to pretreat the cells before lysis with 2% SDS, allowed us to isolate DNA efficiently from several gram-positive bacteria.

The efficiency of digestion of genomic DNA by REs depends on the purity of the DNA. Contaminants (proteins, polysaccharides, phenol, chloroform, ethanol, EDTA, SDS, and inappropriate salt concentration) inhibit the activity of REs and lead to incomplete digestion of the DNA. The inhibitory effects may be overcome by increasing the number of enzyme units in the reaction, increasing the reaction volume to dilute the contaminants, increasing the time of incubation, or making successive additions of enzyme after incubation for 1 to 3 h. Occasionally, addition of the polycation spermidine (1 to 2.5 mM) overcomes the inhibitory effect of negatively charged contaminants on the RE. However, because spermidine precipitates DNA at 4°C, it must be added after the reaction mix has been incubated for a few minutes at the appropriate temperature for the restriction reaction. Finally, some REs are inhibited by methylation of nucleotides within the recognition site. There is no solution to this problem; REs that are not affected by methylation of nucleotides must be evaluated. RFLP of PCR-amplified DNA is not susceptible to this problem because PCR-amplified DNA is not methylated.

The advantages of chromosomal DNA restriction analysis are that it is universally applicable, sensitive because the entire genome is evaluated for RFLP, and relatively easy to perform. Its disadvantage is that the genomic restriction fragments are usually too numerous and too closely spaced for scanning by densitometers or image acquisition devices. Consequently, although this method is useful for comparing a few strains, it is not the method of choice for large-scale comparisons. Forbes et al. (21) overcame some of the problems of this method by dividing the fingerprints into sections on the basis of the size range of fragments and then comparing the fragments in a few of the sections to establish similarities or differences. The methods described below involve the use of several alternative strategies to decrease the complexity of the chromosomal DNA restriction analysis.

**Plasmid analysis.** A simple yet powerful method for identifying outbreak strains of bacteria is plasmid analysis. Plasmids are autonomous, self-replicating extrachromosomal DNA elements that are carried by many bacteria but are not essential (in most species) for normal bacterial growth. Plasmids may code for virulence factors or carry antimicrobial resistance genes. The development of rapid and inexpensive techniques for extracting plasmid DNA and separating plasmids on the basis of their size by agarose gel electrophoresis led to the widespread use of plasmids in epidemiologic investigations. Because plasmids of identical size may vary in their nucleotide sequence, treatment of the isolated plasmid DNA with a RE (usually an enzyme that recognizes a specific 4-base sequence) before electrophoretic separation allows one to determine whether plasmids of identical size from different strains are in fact the same plasmid.

Plasmid profiles have been useful for subtyping several bacterial species, including *Escherichia coli*, *Salmonella* serotypes (including *Salmonella* serotype typhi), *Shigella*, *Campylobacter*, *Vibrio cholerae*, *H. influenzae*, *N. gonorrhoeae*, *N. meningitidis*, *S. aureus*, and *Legionella* species (32, 45). Among the notable applications of plasmid profiling in molecular epidemiology is the identification of a clone of *Salmonella* serotype muenchen, which was spread in several midwestern states of the United States by contaminated marijuana (68). The epidemic isolates contained unique plasmids of 4.6 and 11 kb that were not found in control strains of the serotype isolated in previous years or from sources unrelated to marijuana. Plasmid profiling was also helpful in the investigation of the largest food-borne salmonellosis outbreak, which occurred in Illinois in 1985 and affected an estimated 150,000 to 200,000 persons (32, 45). Because plasmid profiling is a relatively simple method, more than 100 outbreak-related isolates could be examined in 1 week. For detailed discussions of plasmid analysis and its epidemiologic applications, we recommend the review by Mayer (45) and the chapter on plasmid profiles in *Clinical Microbiology Procedures Handbook* (35).

Although plasmid profiles are valuable in epidemiologic investigations and may be the most discriminating method of subtyping for some bacteria, they are not universally applicable. Many bacterial species harbor plasmids infrequently; even when plasmids are present in some bacteria, there is often little heterogeneity in their nucleotide sequences, rendering them useless for subtyping. Also, plasmids are gained or lost during the evolutionary process and thus may not be satisfactory for long-term follow-up studies. Plasmid profiling may be most useful when applied in conjunction with another subtyping method (45) and has been particularly effective in the investigation of nosocomial outbreaks (29, 42, 85).

**Hybridization techniques.** Attempts have been made to simplify genomic DNA RFLP analyses by decreasing the number of fragments to be compared. One approach is to probe the chromosomal DNA restriction fragments with one or more labeled DNA fragments. In this procedure, the restriction fragments separated by agarose gel electrophoresis for RFLP determination are transferred to nitrocellulose or nylon membranes by Southern blotting and hybridized with the labeled probe(s). Only the genomic DNA restriction fragments that hybridize with the probe are visible, thus greatly simplifying the fingerprint. Previously, the probes were almost always labeled with high-energy radioisotopes (e.g., $^{32}P$), but the use of nonradioactive reporter molecules such as biotin or digoxigenin has become more common. Probes used for subtyping include randomly cloned genomic DNA fragments, probes derived from genes coding for putative virulence factors, insertion sequences, bacteriophage DNA, and 16S+23S rRNA.

*Randomly cloned probes.* Several probes made by shotgun cloning of genomic DNA fragments are evaluated in initial experiments to select those that recognize relatively stable chromosomal regions and are of discriminatory value. Randomly cloned probes are more discriminatory than rRNA for the subtyping of *Brucella* species and *Legionella pneumophila* (26, 71). Their disadvantages are that a new set of probes is required for each bacterial species and useful probes are selected only after a large number of clones from a genomic DNA library have been screened.

*Probes based on specific virulence factors.* DNA probes derived from nucleotide sequences coding for putative virulence factors or from sequences upstream or downstream of virulence-associated structural genes have been used for molecular subtyping. This method is not universally applicable; probes have to be developed for each bacterial species. Further, the selection of suitable subtyping probes must be done empirically by screening several nucleotide sequences. For example, the genes coding for alginate and elastase in *P. aeruginosa* are highly conserved and thus are not useful for subtyping, whereas a probe derived from the 5′-end region of the gene coding for exotoxin A is useful for subtyping *P. aeruginosa* isolates from patients with cystic fibrosis (41).

*Probes derived from insertion sequences.* Insertion sequences (ISs) are transposable repetitive DNA elements found in many prokaryotic and eukaryotic organisms. The presence of repeated DNA elements facilitates chromosomal rearrangements and may aid in accelerating evolution. Bacterial IS elements usually have a sequence of 10 to 40 bp at one end, which is repeated in an inverse orientation at the other end. IS*986* of *Mycobacterium tuberculosis* is 1,358 bp long and is flanked by inverted repeats of 30 bases. A high degree of RFLP is observed when chromosomal digests of strains belonging to the *M. tuberculosis* complex are probed with a 500-bp *Eco*RI fragment or a 386-bp *Bam*HI-*Xho*I fragment of a plasmid containing IS*986* (30). The utility of the latter probe for subtyping strains of the *M. tuberculosis* complex was demonstrated by van Soolingen et al. (72). The IS-based subtyping of the *M. tuberculosis* complex is particularly useful because ribosomal DNA RFLP analysis (see ribotyping below) does not discriminate well among strains of the organisms. For IS*986*, the frequency of transposition is low and the RFLPs are stable after several passages in vitro or in vivo (30, 72). The disadvantage of IS-based subtyping is that it is not universally applicable.

*Bacteriophage M13-based probe.* The gene III coat protein of bacteriophage M13, which contains two clusters of a 15-bp repeat, was used successfully as a probe for the determination of genomic RFLPs in *E. coli* (33) and methicillin-resistant *S. aureus* (76). Whether this probe will be useful for subtyping other bacterial species has not been determined.

*Ribosomal DNA RFLP analysis (ribotyping).* Ribotyping involves the fingerprinting of genomic DNA restriction fragments that contain all or part of the genes coding for the 16S and 23S rRNA. Conceptually, ribotyping is similar to probing restriction fragments of chromosomal DNA with cloned probes (randomly cloned probes or probes derived from a specific coding sequence such as virulence factor). However, ribotyping has two major advantages: first, because the genes coding for rRNA are highly conserved, a single probe can be used to subtype all eubacteria; second, because most bacteria contain multiple ribosomal operons, a reasonable number of fragments is obtained after probing to allow interspecies and intraspecies discrimination (25, 66). For example, *E. coli* contains seven ribosomal operons (*rrn* loci) distributed throughout the chromosome. The RE *Sal*I cuts once within the 16S rRNA gene and once within the 23S rRNA gene to give a 2.7-kb internal fragment containing portions of genes coding for 16S and 23S rRNA. One obtains 15 fragments by use of Southern blot on a *Sal*I digest of *E. coli* chromosomal DNA with 16S+23S rRNA of *E. coli*: 14 fragments representing the left and right halves of seven operons and 1 conserved internal fragment from all seven operons (31).

DNA isolation, restriction, electrophoresis, and transfer to nitrocellulose or nylon matrix are accomplished by using standard methods (59). DNA is then probed after transfer. Several labeling options are available. *E. coli* 16S+23S rRNA may be purchased commercially (Boehringer Mannheim Biochemicals, Indianapolis, Ind.; Sigma Chemical Co., St. Louis, Mo.). The rRNA can be labeled with $^{32}P$ by using T4 polynucleotide kinase (2). An acetylaminofluorene-labeled 16S+23S rRNA kit is commercially available from Eurogentec S.A., Campus du Sart-Tilman, Liège, Belgium. The kit includes acetylaminofluorene-labeled rRNA, monoclonal antibody to acetylaminofluorene, and a secondary antibody labeled with alkaline phosphatase.

Recently, a modification of the enhanced chemiluminescence labeling system was used to produce sensitive rRNA probes (protocol 7.3). cDNA can also be synthesized from the rRNA template by using reverse transcriptase, which can be labeled with biotin or digoxigenin (3, 37). Plasmid pkk3535, constructed by cloning the *rrnB* operon of *E. coli* into pBR322 (8), may be used for ribotyping. The entire plasmid or the ribosomal operon insert can be labeled with biotin or digoxigenin by using commercially available kits (24). However, caution must be exercised when whole-plasmid probes are used because the vector sequences may hybridize to homologous elements in bacterial genomic DNA.

The matrix of hybridization (nitrocellulose or nylon) is chosen on the basis of the characteristics of the probe and the type of label used. Nitrocellulose filters work well with radioactive probes but are easily damaged. Nylon filters are sturdy and do not break or tear. We obtained excellent results with Magnagraph (Micron Separations, Westboro, Mass.) nylon membranes and digoxigenin-labeled probes.

Standard procedures are followed for hybridization, which is usually done at 37°C in 41% formamide. After hybridization, the filters are rinsed twice in 2× SSC (1× SSC is 0.15 M NaCl plus 0.015 M sodium citrate) and twice in 0.1% SDS and washed twice with 0.1× SSC–0.1% SDS, 0.5× SSC–0.1% SDS, or 1× SSC–0.1% SDS, depending on the moles percent G+C content of the bacteria being studied (Fig. 1).

Ribotyping has been useful for subtyping diverse bacteria including species of

**Figure 1.** Schematic illustration of RFLP of genomic DNA and ribotyping.

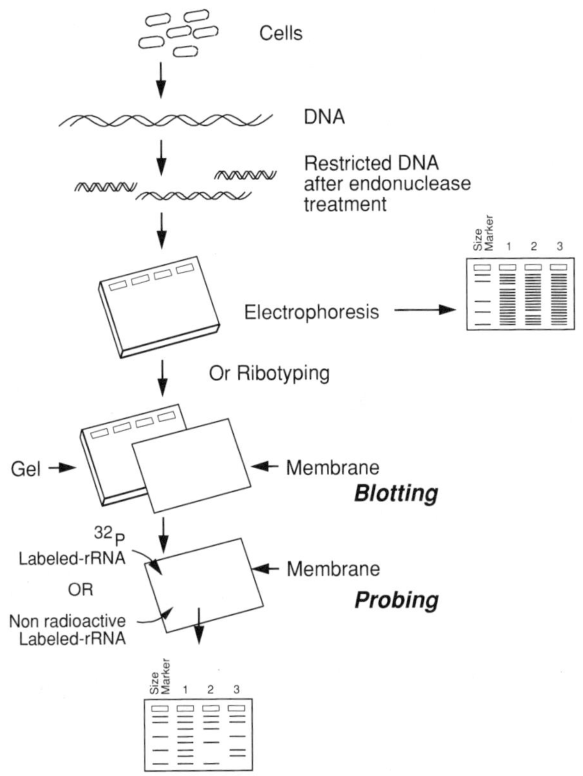


the family *Enterobacteriaceae*, *Haemophilus* spp., *N. meningitidis*, legionellae, *Listeria monocytogenes*, *Campylobacter* spp., staphylococci, and leptospiras (24, 25, 31, 37, 40, 66). For some bacteria (e.g., *Salmonella* serotype typhi), ribotyping may be the only subtyping method that offers sufficient discrimination for use in epidemiologic investigations (2, 52). The ribotype of a strain appears to be stable after in vitro and in vivo passage (40); therefore, this method may be used for long-term epidemiologic studies. Figure 2 shows the application of ribotyping to subtype isolates of *N. meningitidis* serogroup C (82).

A disadvantage of ribotyping is that multiple steps are required (DNA isolation, restriction, electrophoresis, Southern blotting); this is time-consuming and labor intensive. Also, the appropriate RE(s) must be determined for each bacterial species. Use of multiple enzymes (two or three enzymes) for ribotyping increases the discriminating power of the method (74, 82, 84) for many bacteria but also multiplies the work load. Some bacteria, such as *Mycobacterium* spp. and *Mycoplasma* spp., contain only one or two *rrn* loci. Ribotyping may not be useful for these organisms.

**Pulsed-field gel electrophoresis.** Pulsed-field gel electrophoresis (PFGE) allows the generation of simplified chromosomal restriction fragment patterns without having to resort to probe hybridization methods. In this method, REs that cut DNA infrequently are used to generate large fragments of chromosomal DNA, which are then separated by special electrophoretic procedures.

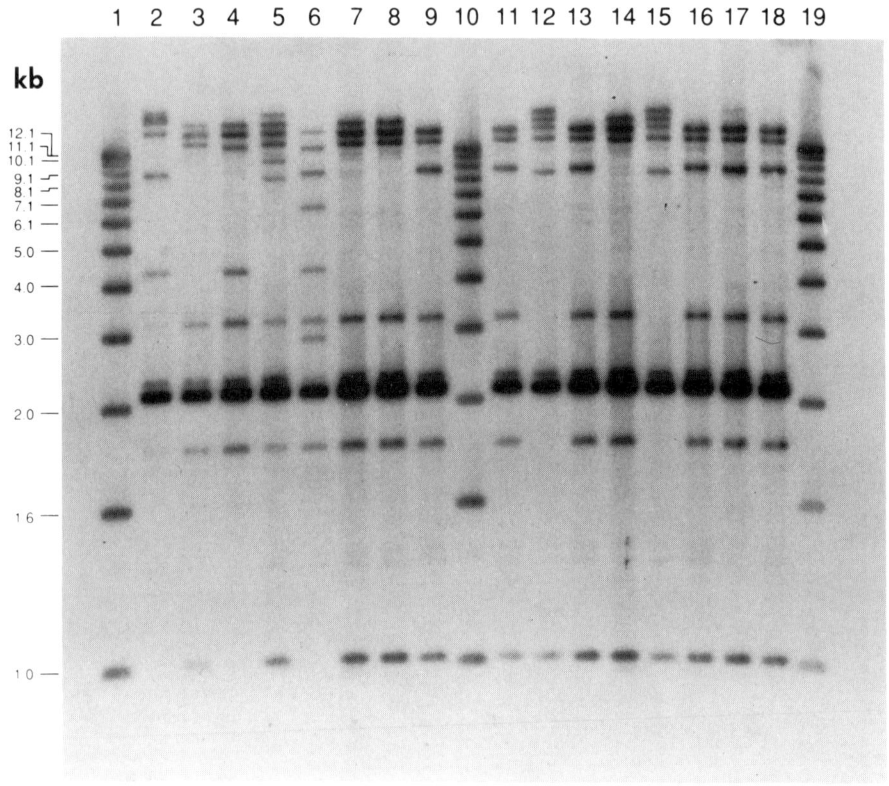


**Figure 2.** Ribotyping of 16 isolates of *N. meningitidis* group C. Lanes: 1, 10, and 19, molecular size marker (1-kb ladder); 2 to 9 and 11 to 18, *Eco*RI restriction fragments of *N. meningitidis* group C DNA probed with $^{32}$P-labeled 16S+23S *E. coli* rRNA.

DNA fragments larger than 40 kb are not efficiently resolved by conventional agarose gel electrophoresis. These DNA fragments are too large to be sieved through even the largest pores of the gel matrix; consequently, they migrate in the electrical field at approximately the same rate, called the limiting mobility (20). However, these fragments can be resolved by cyclically altering the orientation of the electric field during electrophoresis (61).

Because intact DNA is required for PFGE analysis, conventional DNA isolation methods that cause shearing of DNA are not suitable. DNA isolation procedures for PFGE involve the entrapment of bacteria in agarose plugs. The cells are lysed by treatment of the agarose plugs with enzyme and detergent. While the large chromosomal DNA remains entrapped in the agarose gel matrix, other contaminants are removed by extensive washing or dialysis (20). Restriction of the DNA is performed with enzymes chosen to yield a relatively small number of large restriction fragments. These enzymes recognize specific 8-base sequences (e.g., *Not*I, *Sfi*I) or selected 6-base sequences (e.g., *Sma*I, *Apa*I), depending on the moles percent G+C content of the bacterial species under investigation. Agarose plugs containing the restricted DNA are subjected to PFGE.

Various pulsed-field techniques are used to separate restricted DNA fragments. The simplest of the techniques is field inversion gel electrophoresis, a system that uses periodic inversion of an electric field, uniform in both forward and reverse directions, with the forward pulse set slightly longer than the reverse

**Figure 3.** Schematic illustration of PFGE.

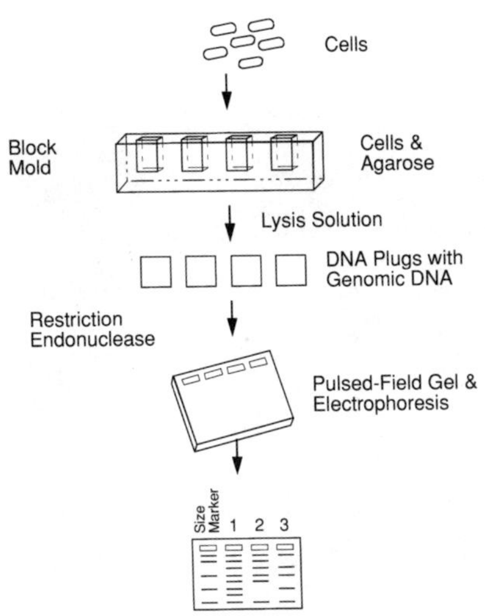


pulse (11). Orthogonal-field-alternation gel electrophoresis was devised for separating large DNA fragments by subjecting them alternately to two orthogonal electric fields (10). These two techniques are well suited for separating fragments with fewer than $10^6$ bp, especially in the 50- to 750-kbp range. DNA fragments with a wider base range require field-angle alternation electrophoresis for optimal separation.

The transverse alternating-field electrophoresis system (22) and contour-clamped homogeneous electric field gel electrophoresis (15, 75) are examples of field-angle alternation-based electrophoretic methods. In transverse alternating-field electrophoresis, the electric field is oriented transversely to the gel, inducing the DNA fragments to move in a straight line with no horizontal distortion. The contour-clamped homogeneous electric field gel electrophoresis system uses an array of hexagonal electrodes to generate uniform electric fields at 120° to each other, inducing DNA fragments to move along a straight line with no distortion. These techniques allow the separation of fragments ranging from 200 kbp to 12 Mbp with sharp bands and straight lanes (3, 63). By using a combination of low field strengths, low concentrations of agarose, long switching intervals, and extended periods of electrophoresis, DNA molecules up to 12 Mbp in length can be resolved. Shorter switching intervals may be used to resolve fragments with fewer than 200 kbp (Fig. 3 and 4).

Several conditions must be optimized in performing electrophoretic separation of very high molecular weight DNA fragments. The separation of large DNA molecules in agarose gels is affected by several factors such as agarose concentration, buffer concentration, temperature, pulse times, voltage, and total electrophoresis run time. The first-generation PFGE systems required extensive experimentation to achieve the ideal conditions for obtaining acceptable separation of large DNA molecules by PFGE. Recently introduced second-generation PFGE systems (Q-Life, Kingston, Ontario, Canada; CHEF Mapper, Bio-Rad Laboratories, Richmond, Calif.) offer built-in algorithms that automatically select the appropriate conditions for a particular separation.

**Figure 4.** PFGE of nine isolates of *Listeria monocytogenes*. Lanes: 1, lambda phage *c*I857*S7* DNA ladder; 2 to 10; *Listeria monocytogenes* DNA restricted with *Apa*I and subjected electrophoresis in a CHEF DR II PFGE unit (initial pulse time, 1 s; final pulse time, 20 s; running time, 22 h).

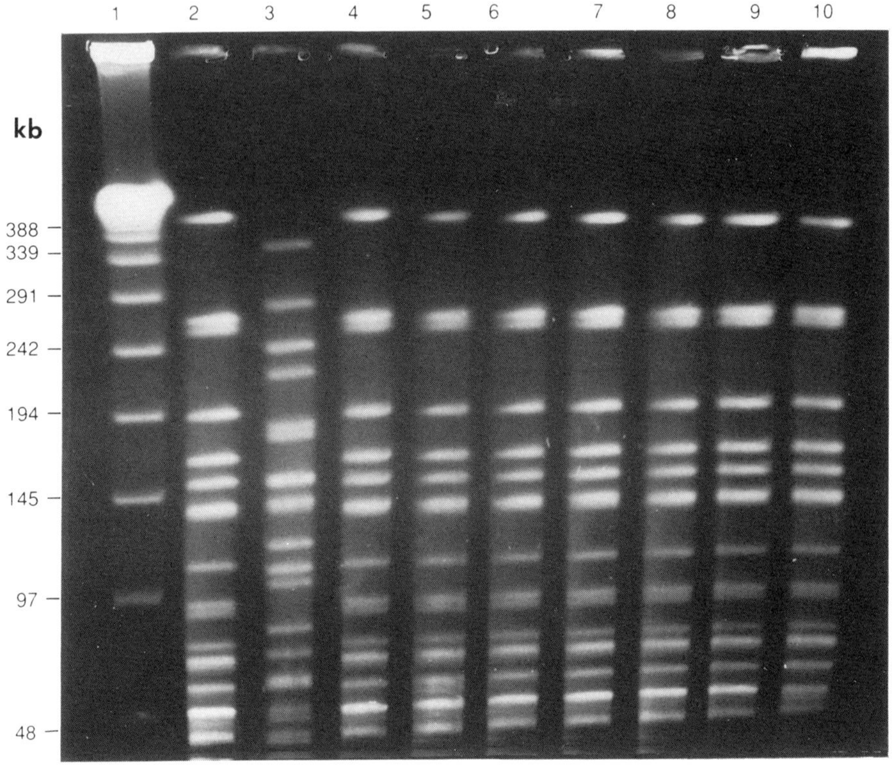


PFGE has been applied to subtyping of several gram-positive and gram-negative bacteria, including *Listeria monocytogenes* (7) (Fig. 4), *N. meningitidis* (9), pseudomonads (27), staphylococci (55), *Streptococcus faecium* (49), shigellae (65), *K. pneumoniae* (43), *Serratia marcescens* (18), *Campylobacter jejuni* (83), and mycobacteria (14). For some bacteria (e.g., *Listeria monocytogenes*), PFGE has greater discriminating capacity than ribotyping and other probe-based RFLP methods.

The disadvantages of PFGE are the long and tedious procedures required for isolating and restricting the genomic DNA by the agarose plug method, the requirement of large quantities of expensive enzymes and reagents for DNA isolation and restriction, and the complexity of experimentally optimizing the electrophoretic conditions. These have prevented the extensive use of PFGE. However, new developments in instrumentation, as seen in the second-generation PFGE systems, and simplification of protocols will render PFGE useful for subtyping microorganisms in the near future.

***PCR-Based Subtyping***

PCR offers several advantages over other nucleic acid-based subtyping methods: speed, the need for only a few cells of the microorganism, and a protocol involving fewer steps. Several PCR-based subtyping methods have been reported in the past 2 years. However, extensive comparisons of these methods with other established molecular typing methods have not yet been completed. Therefore, the suitability of the PCR-based methods for epidemiologic investigations has not been demonstrated adequately, although initial evaluations indicate their potential for this application.

PCR-based methods may be categorized into two types. The first type (PCR-RFLP) involves amplifying a known sequence, cutting with REs, and comparing restriction fragments of the amplified DNA from different strains. The second type involves randomly amplifying segments of the target DNA by using a single primer that does not have any known homology to the target sequence. The latter method does not require restriction of the amplicon, because several fragments of various sizes are generated during the PCR.

The PCR-RFLP method offers significant advantages over conventional DNA restriction analyses. Genetic differences are determined by PCR amplification of a target locus (usually 1 to 2 kb), followed by cutting of the amplicon with a RE with a 4-base recognition sequence. No Southern blotting or probing is necessary. Another significant advantage of PCR-RFLP over genomic DNA restriction is that problems of poor restriction of genomic DNA as a result of DNA base modifications (methylation) are not encountered (Fig. 5).

Virulence genes and ribosomal operons have been targeted for PCR-RFLP subtyping. PCR-RFLP of four virulence-associated genes of *Listeria monocytogenes* allowed the identification of different serotypes (74). PCR-RFLP of the gene coding for the major outer membrane protein of *Chlamydia trachomatis* distinguished all 18 serotypes (60).

Of ribosomal genes, those coding for 16S rRNA are useful for identification to the species level (28) but may not be useful for subtyping. The intergenic spacer region between the genes coding for 16S and 23S rRNA is useful for identifying eubacteria (5) and for subtyping *N. meningitidis* (47) and *Rochalimaea henselae* (Fig. 6) (44), the etiologic agent of epithelioid angiomatosis and the suspected etiologic agent of cat scratch disease. Kostman et al. (39) observed RFLPs in the amplified DNA from *P. cepacia* isolates by using PCR primers that amplified the intergenic 16S to 23S spacer region. This allowed them to discriminate between strains without restricting the amplified DNA. They suggested that the multiple bands after amplification represented spacer regions of different lengths in different ribosomal operons of *P. cepacia.*

Versalovic et al. (73) used consensus primers in the PCR to amplify DNA sequences located between successive repetitive elements (repetitive extragenic palindromic elements and enterobacterial repetitive intergenic consensus sequences) in gram-negative bacteria. The repetitive extragenic palindromic primers differentiated among laboratory strains of *E. coli*, suggesting that this method may have the potential for subtyping gram-negative enteric bacteria.

Simple, reproducible fingerprints of genomic DNA can be generated in a PCR by using single primers that have no known homology to the target sequence (78, 81). This method is called random amplification of polymorphic DNA (RAPD) or arbitrarily primed PCR (protocol 7.5). The amplification is performed at low stringency (annealing at 36 to 45°C and ≥2 mM $MgCl_2$), allowing the primer to anneal to several locations on the two strands of target DNA. Primers 10 bases long and with at least 40 mol% G+C have been useful for RAPD subtyping (81).

RAPD subtyping has been applied successfully to *Clostridium difficile*, *Listeria monocytogenes*, and *Campylobacter jejuni* (46, 48, 79). The advantages of this method are that no prior sequence information about the target is needed,

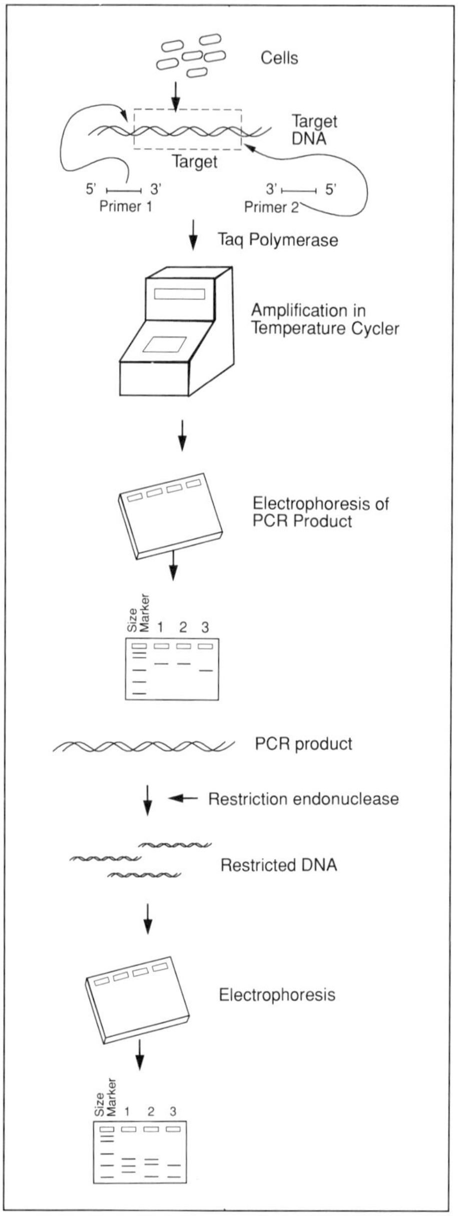


**Figure 5.** Schematic illustration of PCR-RFLP.

a primer from a universal set of 10-mer oligonucleotides can be used, and restriction of the amplicon is not needed. RAPD is probably the simplest DNA-based subtyping method to date, if a temperature-cycling instrument is available.

## Data Acquisition and Analysis

Except for nucleic acid sequencing data, all other nucleic acid-based subtyping data yield DNA fragments of different sizes. Each lane in a gel, membrane, or autoradiogram represents the fingerprint of one strain and consists of a few

**Figure 6.** PCR-RFLPs of the 16S-23S spacer region of the *Rochalimaea* rRNA operon. Lanes 1 and 15, molecular size marker (1-kb ladder); 2 to 14, PCR-amplified *Rochalimaea* 16S-23S spacer sequence of rDNA restricted with *Hae*III.

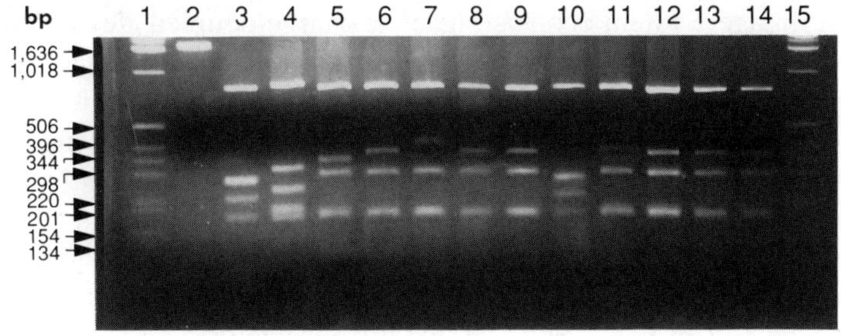


(plasmid profiles, probe-based typing, PFGE, PCR-RFLP, and RAPD) to several hundred fragments of different sizes (genomic DNA fingerprints). Methods of analysis and interpretation of these data depend on the objectives to be achieved by subtyping.

If the objective is to compare a limited number of strains to determine whether one strain is different from one or several other strains, simple visual examination of the data may suffice. If the objective is to compare profiles of a large number of strains to aid in retrospective or prospective epidemiologic studies or for determining genetic relationships, considerable thought must be given to the design of the experiments; acquisition of data from the gels, filters, or autoradiograms; and statistical analyses of the data. This is because such comparisons entail matching of profiles obtained on different gels, filters, or autoradiograms. Run-to-run variations in the mobilities of DNA fragments must be normalized against a standard(s) to allow comparisons between gels.

Data may be acquired in several ways. Agarose gels stained with ethidium bromide are usually photographed, and the photographic negative or print generally is used for analysis. Autoradiograms and filters on which DNA fragments have been visualized by colorimetric methods can be used directly for data acquisition.

For visual analysis, the first step is to determine the relative size of each fragment in a lane by comparing it with a molecular size standard run on the same gel. The molecular size standard should be chosen carefully to bracket all the fragments in the sample lanes. It is important to locate molecular size standards at or near each end to correct for unequal migrations of DNA at each end. To correct for more rapid migration of DNA in the end lanes compared with the interior of the gel as a result of temperature differences during electrophoresis (smiling effect), molecular size standards also should be run in the middle of the gel. One option is to place the standards at the two ends and after every four samples in a 20-well gel. The distance migrated from the origin by each fragment of the standard is measured with a ruler and recorded. The distance migrated by each fragment in the sample lane is also recorded. Because all distances are measured from the origin, it is important to mark the origin properly on filters and autoradiograms. The estimated molecular sizes of the restriction fragments for each sample are calculated. The relationship between the size and electrophoretic mobility of a DNA fragment is not linear. Several curve-fitting algorithms have been developed for determining DNA fragment sizes; of these, the local reciprocal method (19) and the cubic spline interpolation (56) have the highest accuracy (57).

Data may also be acquired by using a digitizer or a flatbed scanner connected to a computer and running an appropriate software program. For large-scale data acquisition, an image acquisition system is recommended. In our laboratory, we use the BioImage System (consisting of a charge-coupled device camera connected to a Sun Microsystems computer running on a UNIX operating system; BioImage, Millipore Corp., Bedford, Mass.) equipped with Whole Band Analysis software for data acquisition from gels, filters, and autoradiograms. The Whole Band Analysis package of the BioImage System includes a sophisticated fragment size normalization algorithm and a lane-matching program that can match a pattern (fingerprint) to any archived pattern in the data base and print as many as eight patterns at a time. The molecular sizes determined with the BioImage System can be put into an ASCII file for further analysis on personal computers or mainframe computers.

For large-scale strain comparisons, it is useful to create dendrograms showing relatedness among strains in a set. A matrix of similarity coefficients is created by using an appropriate window size, which is usually maintained at 5 to 10% of the fragment size (54). A dendrogram can be generated from these by the unweighted pair group method with arithmetic averages (64) and SAS programs (SAS Institute, Cary, N.C.). An example of a dendrogram is shown in Fig. 7. Strains that have several restriction fragments in common will cluster close to each other in the dendrogram, with their relatedness index approaching 0.0. If two strains have several unique restriction fragments (very few restriction fragments in common), the relatedness index will approach 1.0.

A different approach has been used by some investigators to generate dendrograms. They compile a numeric profile of a strain by scoring for the presence or absence of all fragments found in the study group. If a strain contains a particular fragment, it receives a score of 1 for that fragment; the absence of a particular fragment is denoted by a score of 0. The numeric string generated in this manner is used to determine relatedness values among strains of the unweighted pair group method with arithmetic averages and SAS programs (82).

Another approach to analysis of ribotyping data is the use of correspondence analysis (53). Correspondence analysis is an eigenvector method of ordination similar to principal-component analysis. It differs from the latter in the use of a covariance matrix based on $\chi^2$ distances rather than a covariance matrix based on euclidean distances.

Some investigators consider it inappropriate to use statistical methods to analyze RFLP data. These investigators publish either photographs showing examples of different patterns or normalized patterns generated by a computer system, or both. They contend that RFLP data violate the assumption of independence among characters. If a new restriction site appears between two existing sites, this will be observed in the daughter strain by the disappearance of a large fragment and the appearance of two new smaller fragments. Although the parent and daughter strains share two of three restriction sites, they have no common fragments. Therefore, the use of RFLP data to determine genetic distances may introduce major errors in phylogenetic analyses (67). Our work has shown that dendrograms generated from ribotyping data match those generated from MEE data in general trends and are useful for epidemiologic studies; however, they should not be used to assess the genetic relatedness among strains of a species.

**Figure 7.** Dendrogram of ribotype analysis of *N. meningitidis* serogroup C. ET, enzyme type. Reprinted from the *Journal of Clinical Microbiology* (82) with permission of the publisher.

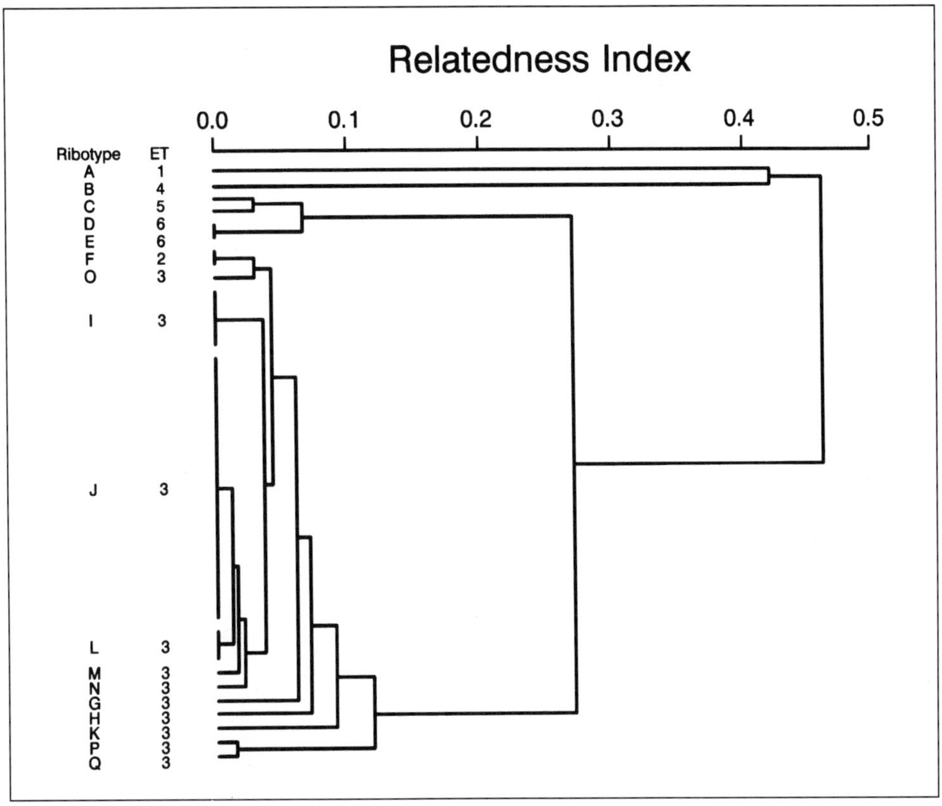


## Evaluation of Molecular Subtyping Methods

New molecular subtyping methods must be evaluated against one another and against established methods by using a set of carefully selected, epidemiologically well-characterized strains. The set must include strains that are expected to be identical (e.g., mother-infant pairs, strains from a common-source outbreak) and strains that are expected to be different (e.g., strains with no epidemiologic association, those that are different by another typing method). The objective of subtyping is to provide laboratory evidence for supporting or rejecting hypotheses developed from epidemiologic observations.

Two criteria are important in comparative evaluation of subtyping methods: reproducibility and discriminating ability. Typeability (the proportion of strains that can be subtyped by a method) is an important factor in determining the usefulness of subtyping methods based on the phenotype (such as serotyping and phage typing). For molecular subtyping methods, typeability is not a useful criterion because essentially all strains are typeable by these methods. Reproducibility is expressed as the percentage of strains classified as the same subtype on repeated testing. Use of appropriate controls to ensure reproducibility of the subtyping results cannot be overemphasized. Incomplete digestion of plasmid or genomic DNA by a RE may cause artifactual bands to appear, falsely showing differences between strains. With universal probes (as in ribotyping), it is often difficult to evaluate the faint bands that appear in a pattern after Southern

**Table 2.** Comparison of approximate cost per sample for four nucleic acid-based subtyping methods[a]

| Component | Cost/sample ($) for: | | | |
|---|---|---|---|---|
| | Restriction of genomic DNA | Ribotyping | PFGE | PCR-RFLP |
| Material | 1.75 | 2.75 | 12.00 | 4.31 |
| Labor[b] | 14.05 | 24.00 | 10.00 | 8.88 |
| Total | 15.80 | 26.75 | 22.00 | 13.19 |

[a] Material and labor costs were computed on the basis of 10 samples per batch for genomic DNA restriction and ribotyping and 18 samples per batch for PFGE and PCR-RFLP.
[b] Labor is calculated on the basis of $10.00/h.

hybridization. Some investigators introduce a second level of discrimination (subsubtype) to describe strains that differ in their ribotype pattern in only one or two faint bands.

For determining the discriminating ability of a new method, Hunter and Gaston (34) recommended the use of Simpson's index of diversity, which is given by the following equation:

$$D = 1 - \frac{1}{N(N-1)} \sum_{j=0}^{j=s} n_j (n_j - 1)$$

where $D$ is the numerical index of discrimination, $N$ is the number of strains in a sample population, $s$ is the number of subtypes described, and $n_j$ is the number of strains belonging to the $j$th subtype.

## Relative Costs of Molecular Subtyping Methods

The relative costs of the nucleic acid-based typing methods are shown in Table 2. The PCR-based methods are the least expensive and can be completed in 1 to 2 days. The chromosomal DNA restriction analysis can be completed in 3 days after the bacterial culture is available in sufficient quantity for extraction of DNA; Southern hybridization with probes requires an additional 2 to 3 days. PFGE protocols require several steps involving long incubations (several hours to overnight); consequently, they require 7 days or longer for completion.

PFGE- and PCR-based subtyping methods require the use of specialized instruments. The electrophoresis system for PFGE costs $5,000 to $17,000. A temperature-cycling instrument for PCR costs $2,000 to $6,000. In addition, methods listed in Table 2 involve the use of microcentrifuges ($2,500), conventional electrophoresis systems ($500 to $1,000), and micropipetting devices ($200 each). A UV illuminator and a black-and-white instant photography system ($1,300 to $6,900) are needed to view and photograph ethidium bromide-stained agarose gels.

## Conclusions

Nucleic acid-based subtyping methods have proved to be extremely useful for epidemiologic investigations. These methods are attractive because they have high discriminating capacity, are reproducible, do not require the use of organism-specific reagents, are applicable for typing all strains, allow objective evaluation of results, allow statistical analyses of the data, and are amenable to automation.

As sensitive molecular subtyping methods are developed, we are faced frequently with the problem of how to interpret the data obtained by these methods. The extreme sensitivity of DNA sequencing and some of the RFLP methods may lead to the observation of minor differences even among strains from a well-defined common source outbreak. Our dilemma is similar to the problem faced by toxicologists, whose ability to interpret the toxicologic significance of chemicals has not kept pace with their ability to detect extremely minute quantities of toxic chemicals by physical and chemical methods. To maintain the utility of the new molecular subtyping methods in epidemiology, their high sensitivity may require us to accept less than a 100% match in comparing the fingerprints of different strains. This is already being done with many phenotypic subtyping methods.

After a decade of active development in molecular subtyping methodology, it is time to comparatively evaluate molecular typing methods for use with bacteria of public health significance in order to identify those that are most useful for each bacterial species of interest. Once promising methods are identified, they should be standardized with respect to laboratory protocols (DNA isolation methods, REs to be used, probe selection for Southern hybridizations, PCR primers, and amplification conditions) and methods used to acquire and analyze data. This standardization could be performed as international collaborative studies with participation by groups active in subtyping the target bacterial species and with the use of a carefully selected, epidemiologically and microbiologically characterized set of strains. Without this kind of collaborative standardization, we will be faced with the problem of data overload and the inability to make comparisons between methods and laboratories to reach meaningful and epidemiologically relevant conclusions.

## References


1. **Abdillahi, H., and J. T. Poolman.** 1987. Whole-cell ELISA for typing *Neisseria meningitidis* with monoclonal antibodies. *FEMS Microbiol. Lett.* **48:**367–371.
2. **Altwegg, M., F. W. Hickman-Brenner, and J. J. Farmer III.** 1989. Ribosomal RNA gene restriction patterns provide increased sensitivity for typing *Salmonella typhi* strains. *J. Infect. Dis.* **160:**145–149.
3. **Baloga, A. O., and S. K. Harlander.** 1991. Comparison of methods for discrimination between strains of *Listeria monocytogenes* from epidemiological surveys. *Appl. Environ. Microbiol.* **57:**2324–2331.
4. **Barenkamp, S. J., R. S. Munson, Jr., and D. M. Granoff.** 1981. Subtyping isolates of *Haemophilus influenzae* type b by outer-membrane protein profiles. *J. Infect. Dis.* **143:**668–676.
5. **Barry, T., G. Colleran, M. Glennon, L. K. Dunican, and F. Gannon.** 1991. The 16S/23S ribosomal spacer region as a target for DNA probes to identify eubacteria. *PCR Methods Applic.* **1:**51–56.
6. **Bibb, W. F., B. Schwartz, B. G. Gellin, B. D. Plikaytis, and R. E. Weaver.** 1989. Analysis of *Listeria monocytogenes* by multilocus enzyme electrophoresis and application of the method to epidemiologic investigations. *Int. J. Food Microbiol.* **8:**233–239.
7. **Brosch, R., C. Buchrieser, and J. Rocourt.** 1991. Subtyping of *Listeria monocytogenes*

serovar 4b by use of low-frequency-cleavage restriction endonucleases and pulsed-field gel electrophoresis. *Res. Microbiol.* **142:**667–675.

8. **Brosius, J., A. Ullrich, M. A. Raker, A. Gray, T. J. Dull, R. R. Gutell, and H. F. Noller.** 1981. Construction and fine mapping of recombinant plasmids containing the rrnB ribosomal RNA operon of *E. coli. Plasmid* **6:**112–118.
9. **Bygraves, J. A., and M. C. J. Maiden.** 1992. Analysis of the clonal relationships between strains of *Neisseria meningitidis* by pulsed field gel electrophoresis. *J. Gen. Microbiol.* **138:**523–531.
10. **Carle, G. F., M. Frank, and M. V. Olson.** 1986. Electrophoretic separations of large DNA molecules by periodic inversion of the electric field. *Science* **232:**65–68.
11. **Carle, G. F., and M. V. Olson.** 1984. Separation of chromosomal DNA from yeast by orthogonal-field-alternation gel electrophoresis. *Nucleic Acids Res.* **12:**5647–5664.
12. **Casey Chen, C.-K., and M. E. Wilson.** 1990. Outer membrane protein and lipopolysaccharide heterogeneity among *Eikenella corrodens* isolates. *J. Infect. Dis.* **162:**664–671.
13. **Caugant, D. A., L. O. Frøholm, K. Bøvre, E. Holten, C. E. Frasch, L. F. Mocca, W. D. Zollinger, and R. K. Selander.** 1986. Intercontinental spread of a genetically distinctive complex of clones of *Neisseria meningitidis* causing epidemic disease. *Proc. Natl. Acad. Sci. USA* **83:**4927–4931.
14. **Charvin, M., N. Rastogi, and V. V. Lévy-Frébault.** 1991. An easy and rapid method for isolation of entire mycobacterial genome for application in pulsed-field gel electrophoresis. *Curr. Microbiol.* **22:**327–331.
15. **Chu, G., D. Vollrath, and R. W. Davis.** 1986. Separation of large DNA molecules by contour-clamped homogeneous electric fields. *Science* **234:**1582–1585.
16. **Coloe, P. J., J. F. Slattery, P. Cavanaugh, and J. Vaughan.** 1986. The cellular fatty acid composition of *Campylobacter* species isolated from cases of enteritis in man and animals. *J. Hyg.* **96:**225–229.
17. **Dean, D., M. Patton, and R. S. Stephens.** 1991. Direct sequence evaluation of the major outer membrane protein gene variant regions of *Chlamydia trachomatis* subtypes D′, I′, and L2′. *Infect. Immun.* **59:**1579–1582.
18. **Denton, C. L., and A. T. McManus.** 1991. Identification of endemic *Serratia marcescens* in a burn ICU by pulsed field electrophoresis, p. 427. *Abstr. 91st Gen. Meet. Am. Soc. Microbiol. 1991.* American Society for Microbiology, Washington, D.C.
19. **Elder, J. K., and E. M. Southern.** 1983. Measurement of DNA length by gel electrophoresis. II. Comparison of methods for relating mobility to fragment length. *Anal. Biochem.* **128:**227–231.
20. **Finney, M.** 1988. Pulsed-field gel electrophoresis, p. 2.5.9–2.5.15. *In* F. M. Ausubel, R. Brent, R. E. Kingston, D. D. Moore, J. A. Smith, J. G. Seidman, and K. Struhl (ed.), *Current Protocols in Molecular Biology*. Greene Publishing and Wiley-Interscience, New York.
21. **Forbes, K. J., K. D. Bruce, J. Z. Jordens, A. Ball, and T. H. Pennington.** 1991. Rapid methods in bacterial DNA fingerprinting. *J. Gen. Microbiol.* **137:**2051–2058.
22. **Gardiner, K., and D. Patterson.** 1988. Transverse alternating electrophoresis. *Nature* (London) **331:**371–372.
23. **Granoff, D. M., S. J. Barenkamp, and R. S. Munson, Jr.** 1982. Subtyping of *Haemophilus influenzae* type b by outer membrane protein profiles. *Semin. Infect. Dis.* **4:**339–345.
24. **Graves, L. M., B. Swaminathan, M. W. Reeves, and J. Wenger.** 1991. Ribosomal DNA fingerprinting of *Listeria monocytogenes* using a digoxigenin-labeled probe. *Eur. J. Epidemiol.* **7:**77–82.
25. **Grimont, F., and P. A. D. Grimont.** 1991. DNA fingerprinting, p. 249–280. *In* E. Stackebrandt and M. Goodfellow (ed.), *Nucleic Acid Techniques in Bacterial Systematics: Modern Microbiological Methods*. John Wiley & Sons, Chichester, England.
26. **Grimont, F., J. M. Verger, P. Cornelis, J. Limet, M. Lefevre, M. Grayon, B. Regnault,**

**J. Van Broeck, and P. A. D. Grimont.** 1992. Molecular typing of *Brucella* with cloned DNA probes. *Res. Microbiol.* **143:**55–65.

27. **Grothues, D., and B. Tümmler.** 1991. New approaches in genome analysis by pulsed-field gel electrophoresis: application to the analysis of *Pseudomonas* species. *Mol. Microbiol.* **5:**2763–2776.
28. **Gurtler, V., V. A. Wilson, and B. C. Mayall.** 1991. Classification of medically important clostridia using restriction endonuclease site differences of PCR-amplified 16S rDNA. *J. Gen. Microbiol.* **137:**2673–2679.
29. **Hanifah, Y. A., K. Hiramatsu, and T. Yokota.** 1992. Characterization of methicillin-resistant *Staphylococcus aureus* associated with nosocomial infections in the University Hospital, Kuala Lumpur. *J. Hosp. Infect.* **21:**15–28.
30. **Hermans, P. W. M., D. van Soolingen, J. W. Dale, A. R. J. Schuitema, R. A. McAdam, D. Catty, and J. D. A. van Embden.** 1990. Insertion element IS*986* from *Mycobacterium tuberculosis*: a useful tool for diagnosis and epidemiology of tuberculosis. *J. Clin. Microbiol.* **28:**2051–2058.
31. **Hinojosa-Ahumada, M., B. Swaminathan, S. B. Hunter, D. N. Cameron, J. A. Kiehlbauch, I. K. Wachsmuth, and N. A. Strockbine.** 1991. Restriction fragment length polymorphisms in rRNA operons for subtyping *Shigella sonnei. J. Clin. Microbiol.* **29:**2380–2384.
32. **Holmberg, S. D., and K. Wachsmuth.** 1989. Plasmid and chromosomal DNA analyses in the epidemiology of bacterial diseases, p. 105–129. *In* B. Swaminathan and G. Prakash (ed.), *Nucleic Acid and Monoclonal Antibody Probes: Applications in Diagnostic Microbiology*. Marcel Dekker, Inc., New York.
33. **Huey, B., and J. Hall.** 1989. Hypervariable DNA fingerprinting in *Escherichia coli*: minisatellite probe from bacteriophage M13. *J. Bacteriol.* **171:**2528–2532.
34. **Hunter, P. R., and M. A. Gaston.** 1988. Numerical index of the discriminatory ability of typing systems: an application of Simpson's index of diversity. *J. Clin. Microbiol.* **26:**2465–2466.
35. **Isenberg, H. D.** 1992. *Clinical Microbiology Procedures Handbook*. American Society for Microbiology, Washington, D.C.
36. **Joly, J. R., Y.-Y. Chen, and D. Ramsay.** 1983. Serogrouping and subtyping of *Legionella pneumophila* with monoclonal antibodies. *J. Clin. Microbiol.* **18:**1040–1046.
37. **Kiehlbauch, J. A., B. D. Plikaytis, B. Swaminathan, D. N. Cameron, and I. K. Wachsmuth.** 1991. Restriction fragment length polymorphism in the ribosomal genes for species identification and subtyping of aerotolerant *Campylobacter* species. *J. Clin. Microbiol.* **29:**1670–1676.
38. **Kim, C., L. M. Graves, B. Swaminathan, L. W. Mayer, and R. E. Weaver.** 1991. Evaluation of hybridization characteristics of a cloned pRF106 probe for *Listeria monocytogenes* detection and development of a nonisotopic colony hybridization assay. *Appl. Environ. Microbiol.* **57:**289–294.
39. **Kostman, J. R., T. D. Edlind, J. J. LiPuma, and T. L. Stull.** 1992. Molecular epidemiology of *Pseudomonas cepacia* determined by polymerase chain reaction ribotyping. *J. Clin. Microbiol.* **30:**2084–2087.
40. **LiPuma, J. J., M. C. Fisher, S. E. Dasen, J. E. Mortensen, and T. L. Stull.** 1991. Ribotype stability of serial pulmonary isolates of *Pseudomonas cepacia. J. Infect. Dis.* **164:**133–136.
41. **Loutit, J. S., and L. S. Tompkins.** 1991. Restriction enzyme and Southern hybridization analyses of *Pseudomonas aeruginosa* strains from patients with cystic fibrosis. *J. Clin. Microbiol.* **29:**2897–2900.
42. **Luginbuhl, L. M., H. A. Rotbart, R. R. Facklam, M. H. Roe, and J. A. Elliot.** 1987. Neonatal enterococcal sepsis: case-control study and description of an outbreak. *Pediatr. Infect. Dis. J.* **6:**1022–1030.
43. **Maslow, J. N., S. Loring, A. Durbin, S. Brecher, and R. D. Arbeit.** 1991. *Klebsiella* isolates designated as *K. pneumoniae* and *K. oxytoca* by API and VITEK were identified as clonal by pulsed field gel electrophoresis, p. 352. *Program Abstr. 31st*

*Intersci. Conf. Antimicrob. Agents Chemother*. American Society for Microbiology, Washington, D.C.

44. **Matar, G. M., B. Swaminathan, L. N. Slater, and D. F. Welch.** 1992. Amplification and restriction endonuclease analysis of the ribosomal DNA 16S–23S spacer region from *Rochalimaea* species, abstr. C-463, p. 497. *Abstr. 92nd Gen. Meet. Am. Soc. Microbiol. 1992.* American Society for Microbiology, Washington, D.C.
45. **Mayer, L. W.** 1988. Use of plasmid profiles in epidemiologic surveillance of disease outbreaks and in tracing the transmission of antibiotic resistance. *Clin. Microbiol. Rev.* **1:**228–243.
46. **Mazurier, S., A. van de Giessen, K. Heuvelman, and K. Wernars.** 1992. RAPD analysis of *Campylobacter* isolates: DNA fingerprinting without the need to purify DNA. *Lett. Appl. Microbiol.* **14:**260–262.
47. **McLaughlin, G. L., D. K. Howe, D. R. Biggs, A. R. Smith, P. Ludwinski, B. C. Fox, D. N. Tripathy, C. E. Frasch, J. D. Wenger, R. B. Carey, M. Hassan-King, and M. H. Vodkin.** Amplification of rDNA loci to detect and type *Neisseria meningitidis* and other eubacteria. *Mol. Cell. Probes*, in press.
48. **McMillin, D. E., and L. L. Muldrow.** 1992. Typing of toxic strains of *Clostridium difficile* using DNA fingerprints generated with arbitrary polymerase chain reaction primers. *FEMS Microbiol. Lett.* **92:**5–10.
49. **Miranda, A. G., K. V. Singh, and B. E. Murray.** 1991. DNA fingerprinting of *Enterococcus faecium* by pulsed-field gel electrophoresis may be a useful epidemiologic tool. *J. Clin. Microbiol.* **29:**2752–2757.
50. **Mullis, K. B., and F. A. Faloona.** 1987. Specific synthesis of DNA *in vitro* via a polymerase-catalyzed chain reaction. *Methods Enzymol.* **155:**335–350.
51. **Murphy, T. F., J. M. Bernstein, D. M. Dryja, A. A. Campagnari, and M. A. Apicella.** 1987. Outer membrane protein and lipooligosaccharide analysis of paired nasopharyngeal and middle ear isolates in otitis media due to nontypable *Haemophilus influenzae*: pathogenetic and epidemiological observations. *J. Infect. Dis.* **156:**723–731.
52. **Nastasi, A., C. Mammina, and M. R. Villafrate.** 1991. rDNA fingerprinting as a tool in epidemiological analysis of *Salmonella typhi* infections. *Epidemiol. Infect.* **107:**565–576.
53. **Picard, B., J.-B. Denis, G. Decoux, and P. Goullet.** 1991. Computer-assisted statistical analyses of enzyme and ribosomal DNA electrophoretic polymorphism in *Yersinia. Electrophoresis* **12:**264–269.
54. **Plikaytis, B. D., G. M. Carlone, and B. B. Plikaytis.** 1986. Numerical analysis of normalized whole-cell protein profiles after sodium dodecyl sulphate-polyacrylamide gel electrophoresis. *J. Gen. Microbiol.* **132:**2653–2660.
55. **Poddar, S. K., and M. McClelland.** 1991. Restriction fragment fingerprint and genome sizes of *Staphylococcus* species using pulsed-field gel electrophoresis and infrequent cleaving enzymes. *DNA Cell Biol.* **10:**663–670.
56. **Press, W. H., B. P. Flannery, S. A. Teukolsky, and W. T. Vetterling.** 1986. *Numerical Recipes: The Art of Scientific Computing*. Cambridge University Press, Cambridge.
57. **Russell, P. J., J. M. Doenias, and S. J. Russell.** 1991. GELYMAC: a Macintosh application for calculating DNA fragment size from gel electrophoresis migration data. *CABIOS* **7:**265–266.
58. **Sadowski, P. L., B. C. Peterson, D. N. Gerding, and P. P. Cleary.** 1979. Physical characterization of ten R plasmids obtained from an outbreak of nosocomial *Klebsiella pneumoniae* infections. *Antimicrob. Agents Chemother.* **15:**616–624.
59. **Sambrook, J., E. F. Fritsch, and T. Maniatis.** 1989. *Molecular Cloning: A Laboratory Manual*, 2nd ed. Cold Spring Harbor Laboratory, Cold Spring Harbor, N.Y.
60. **Sayada, C., E. Denamur, J. Orfila, F. Catalan, and J. Elion.** 1991. Rapid genotyping of the *Chlamydia trachomatis* major outer membrane protein by the polymerase chain reaction. *FEMS Microbiol. Lett.* **83:**73–78.
61. **Schwartz, D. C., and C. R. Cantor.** 1984. Separation of yeast chromosome-sized DNAs by pulsed field gradient gel electrophoresis. *Cell* **37:**67–75.

62. **Selander, R. K., D. A. Caugant, H. Ochman, J. M. Musser, M. N. Gilmour, and T. S. Whittam.** 1986. Methods of multilocus enzyme electrophoresis for bacterial population genetics and systematics. *Appl. Environ. Microbiol.* **51:**873–884.
63. **Smith, C. L., T. Matsumoto, O. Niwa, S. Klco, J.-B. Fan, M. Yanagida, and C. R. Cantor.** 1987. An electrophoretic karyotype for *Schizosaccharomyces pombe* by pulsed field gel electrophoresis. *Nucleic Acids Res.* **15:**4481–4489.
64. **Sneath, P. H. A., and R. Sokal.** 1973. *Numerical Taxonomy: The Principles and Practice of Numerical Classification*. W. H. Freeman & Co., San Francisco.
65. **Soldati, L., and J. C. Piffaretti.** 1991. Molecular typing of *Shigella* strains using pulsed-field gel electrophoresis and genome hybridization with insertion sequences. *Res. Microbiol.* **142:**489–498.
66. **Stull, T. L., J. J. LiPuma, and T. D. Edlind.** 1988. A broad-spectrum probe for molecular epidemiology of bacteria: ribosomal RNA. *J. Infect. Dis.* **157:**280–286.
67. **Swofford, D. L., and G. J. Olsen.** 1990. Phylogeny reconstruction, p. 411–515. *In* D. M. Hillis and C. Moritz (ed.), *Molecular Systematics*. Sinauer Associates, Sunderland, Mass.
68. **Taylor, D. N., I. K. Wachsmuth, Y.-H. Shangkuan, E. V. Schmidt, T. J. Barrett, J. S. Schrader, C. S. Scherach, H. B. McGee, R. A. Feldman, and D. J. Brenner.** 1982. Salmonellosis associated with marijuana: a multistate outbreak traced by plasmid fingerprinting. *N. Engl. J. Med.* **306:**1249–1253.
69. **Tenover, F. C.** Personal communication.
70. **Tracy, T. E., and L. S. Mulcahy.** 1991. A simple method for direct automated sequencing of PCR fragments. *BioTechniques* **11:**68–75.
71. **Tram, C., M. Simonet, M.-H. Nicolas, C. Offredo, F. Grimont, M. Lefevre, E. Ageron, A. Debure, and P. A. D. Grimont.** 1990. Molecular typing of nosocomial isolates of *Legionella pneumophila* serogroup 3. *J. Clin. Microbiol.* **28:**242–245.
72. **van Soolingen, D., P. W. M. Hermans, P. E. W. de Haas, D. R. Soll, and J. D. A. van Embden.** 1991. Occurrence and stability of insertion sequences in *Mycobacterium tuberculosis* complex strains: evaluation of an insertion sequence-dependent DNA polymorphism as a tool in the epidemiology of tuberculosis. *J. Clin. Microbiol.* **29:**2578–2586.
73. **Versalovic, J., T. Koeuth, and J. R. Lupski.** 1991. Distribution of repetitive DNA sequences in eubacteria and application to fingerprinting of bacterial genomes. *Nucleic Acids Res.* **19:**6823–6831.
74. **Vines, A., M. W. Reeves, S. Hunter, and B. Swaminathan.** 1992. Restriction fragment length polymorphism in four virulence-associated genes of *Listeria monocytogenes*. *Res. Microbiol.* **143:**281–294.
75. **Vollrath, D., and R. W. Davis.** 1987. Resolution of DNA molecules greater than 5 megabases by contour-clamped homogeneous electric fields. *Nucleic Acids Res.* **15:**7865–7876.
76. **Wei, M.-Q., D. M. Groth, A. H. W. Mendis, J. Sampson, J. D. Wetherall, and W. B. Grubb.** 1992. Typing of methicillin-resistant *Staphylococcus aureus* with an M13 repeat probe. *J. Hosp. Infect.* **20:**233–245.
77. **Welch, D. F.** 1991. Applications of cellular fatty acid analysis. *Clin. Microbiol. Rev.* **4:**422–438.
78. **Welsh, J., and M. McClelland.** 1990. Fingerprinting genomes using PCR with arbitrary primers. *Nucleic Acids Res.* **18:**7213–7218.
79. **Wernars, K., and S. I. Mazurier.** 1992. Development of a rapid and powerful DNA fingerprinting technique for typing *Listeria* strains, p. 62–63. *In Listeria 1992. Proceedings of the Eleventh International Symposium on Problems of Listeriosis*. Statens Seruminstitut, Copenhagen, Denmark.
80. **Whittam, T. S., I. K. Wachsmuth, and R. A. Wilson.** 1988. Genetic evidence of clonal descent of *Escherichia coli* O157:H7 associated with hemorrhagic colitis and hemolytic uremic syndrome. *J. Infect. Dis.* **157:**1124–1133.
81. **Williams, J. G. K., A. R. Kubelik, K. J. Livak, J. A. Rafalski, and S. V. Tingey.** 1990.

DNA polymorphisms amplified by arbitrary primers are useful as genetic markers. *Nucleic Acids Res.* **18:**6531–6535.

82. **Woods, T. C., L. O. Helsel, B. Swaminathan, W. F. Bibb, R. W. Pinner, B. G. Gellin, S. F. Collin, S. H. Waterman, M. W. Reeves, D. J. Brenner, and C. V. Broome.** 1992. Characterization of *Neisseria meningitidis* serogroup C by multilocus enzyme electrophoresis and ribosomal DNA restriction profiles (ribotyping). *J. Clin. Microbiol.* **30:**132–137.
83. **Yan, W., N. Chang, and D. E. Taylor.** 1991. Pulsed-field gel electrophoresis of *Campylobacter jejuni* and *Campylobacter coli* genomic DNA and its epidemiologic application. *J. Infect. Dis.* **163:**1068–1072.
84. **Yogev, D., D. Halachmi, G. E. Kenny, and S. Razin.** 1988. Distinction of species and strains of mycoplasmas (Mollicutes) by genomic DNA fingerprints with an rRNA gene probe. *J. Clin. Microbiol.* **26:**1198–1201.
85. **Zervos, M. J., C. A. Kauffman, P. M. Therasse, A. G. Bergman, T. S. Mikesell, and D. R. Schaberg.** 1987. Nosocomial infection by gentamicin-resistant *Streptococcus faecalis*: an epidemiologic study. *Ann. Intern. Med.* **106:**687–691.

# In Vitro Nucleic Acid Amplification Techniques

# 3

David H. Persing

## Background

Only a decade ago, the prospect of producing billions of amplified copies of a specific nucleic acid sequence by performing successive rounds of in vitro nucleic acid replication would have been considered science fiction. Interestingly, the basic ingredients for an in vitro nucleic acid amplification method were described in a 1971 report by Kleppe et al. (84) in which the extensive synthesis of a tRNA gene by primer-directed DNA repair was postulated. However, this work apparently did not result in an exponential amplification process (the hallmark of a nucleic acid amplification method), and the concept remained dormant and largely unrecognized in the literature for many years (40).

In 1983, on a moonlit drive through the mountains of northern California, Cetus scientist Kary Mullis envisioned a process of in vitro nucleic acid amplification that would eventually become known as the polymerase chain reaction (PCR) (101). Buoyed by several technologies that matured in the 1980s, including automated oligonucleotide synthesis, PCR became a reality in a relatively short time, leading to the publication by Saiki et al. (136) of its first practical application in 1985. On the heels of PCR, several alternative methods for nucleic acid amplification have now been developed (12, 119) (Table 1).

The purpose of this chapter is to compile and review critically the methods available for in vitro nucleic acid amplification. The primary objective of such techniques is to improve the sensitivity of tests based on nucleic acids and to simplify them eventually through the use of automation and the incorporation of nonisotopic detection formats. The technology underlying in vitro amplification techniques is diverse and is in a state of constant transition. Consequently, it is important to provide a framework for understanding these techniques that is flexible enough to accommodate changes in this rapidly evolving field yet still functions as a valid classification system. One such classification system (119) (Table 1) divides nucleic acid amplification methods into three general categories: (i) target amplification systems, which use PCR, self-sustaining sequence replication (3SR), or the recently described strand displacement amplification (SDA); (ii) probe amplification systems, which include those involving Qβ replicase or thermostable DNA ligase chain reaction (LCR); and (iii) signal amplification, in which the signal generated from each probe molecule is increased by using compound probes or branched-probe technology. Current examples of each technology will be described in more detail later in this chapter.

**Table 1.** Nucleic acid amplification methods

| Amplification method | Key reference(s) | Enzymes used |
|---|---|---|
| Target | | |
| PCR | Saiki et al. (135, 136), Mullis and Faloona (102) | Thermophilic DNA polymerase |
| TAS | Kwoh et al. (87) | Reverse transcriptase, RNA polymerase |
| 3SR/nucleic acid sequence-based amplification | Guatelli et al. (63), Compton (34) | Reverse transcriptase, RNase H, RNA polymerase |
| SDA | Walker et al. (146) | Restriction endonucleases, DNA polymerase |
| Probe | | |
| LAR | Wu and Wallace (163) | DNA ligase |
| LCR | Barany (5) | Thermophilic DNA ligase |
| Qβ replicase-based amplification | Kramer and Lizardi (85), Lizardi et al. (91), Lomeli et al. (92) | Qβ replicase |
| Signal | | |
| Compound probes | Fahrlander and Klausner (47), Yang et al. (164) | None |
| bDNA probes | Urdea et al. (141), Sanchez-Pescador et al. (137) | None |

## Setting the Stage for Amplification Techniques: Nucleic Acid Sequence Information as a Commodity

The use of probes for the detection of a human pathogen depends on the identification of a specific nucleic acid sequence that can serve as a faithful surrogate for the identification of the organism as a whole. In the past, the availability of such sequence information was often limited. However, one of the practical benefits (and from the microbiologist's viewpoint, an unexpected one) of the "big science" of the Human Genome Initiative has been the tremendous advances in the technology of nucleic acid sequencing and in the computerized analysis of nucleotide sequence information (7). Already, nucleic acid sequencing protocols that are simple to perform and less prone to error have been developed (64); many such improvements have now been implemented in kit form. Partially automated systems for DNA sequencing are now installed in laboratories across the country, and fully automated systems that extract, amplify, and sequence DNA are now in development.

The acquisition of pathogen-specific DNA sequence information is now much more accessible to research laboratories. This has led to a virtual explosion in specific sequence data that are useful for designing diagnostic probes; much of this nucleotide sequence information can now be accessed through sequence data bases. If present trends continue, nucleic acid sequence information will soon become a commodity—a readily available resource that provides the basis for nearly unlimited applications.

## The PCR-PC Metaphor

The advent of nucleic acid amplification techniques in many ways resembles the introduction of the first personal computers. The original IBM personal computer (PC) was technologically unsophisticated by modern standards and could perform few tasks on its own. Nevertheless, the IBM PC gained widespread acceptance in the business community with the availability of computer software that performed useful functions, such as the Lotus 1-2-3 spreadsheet program. In time, technological improvements in the original IBM PC, many of which were introduced by companies other than IBM, drastically improved the computing power of the machine and made the instrument more user friendly. More importantly, software applications, mostly created by outside developers, greatly expanded the power of the PC technologic core to perform useful and even essential functions.

In some respects, nucleic acid amplification technology resembles that of a PC. Both technologies require programming to perform a specific application. The "software" of nucleic acid amplification is concoctions of oligonucleotides, enzymes, cosolvents, and temperature ramps that have been adapted to specific applications mostly by outside developers—in this case basic scientists, clinicians, clinical microbiologists, and corporate researchers. PCR has already undergone significant improvements since it was first introduced (see below). Other improvements of the core technology have occurred in the areas of contamination control (chapter 6), detection formats (chapter 6), and basic implementation. Moreover, the software library of PCR now includes a tremendous array of specific applications. Indeed, published reports of new applications for the diagnosis of infectious diseases, in the spirit of the technology they exploit, are accumulating at a seemingly exponential rate. However, like most analogies, the PCR/PC metaphor can be pushed too far; I have trouble envisioning a home PCR machine in the near future.

## The PCR-as-Petri-Dish Metaphor

The ability to isolate and cultivate microbial pathogens on artificial medium represented one of the great diagnostic triumphs of the last century. The refinement of in vitro culture techniques provided much of what is known about germ theory and the mechanisms and processes of infection. Indeed, cultivation of microbial pathogens from clinical specimens remains the mainstay of the clinical laboratory to this day. Primary isolation media serve to recover pathogens from normally sterile sources, and selective media are used to recover pathogens from polymicrobial sources or to further characterize primary isolates. In vitro cultivation also serves the function of a biologic amplification system, in which as many as a billion clones of a microorganism are grown from a single organism. Although it is impossible to measure the biochemical characteristics of a single microorganism by current methods, the collective measurement of the reactions of millions of clones of the same organism (a process we now take for granted in the clinical laboratory) has led to the establishment of useful markers for definitive identification and classification.

Despite its established value, however, in vitro cultivation has many drawbacks. The time required for growth of an organism in culture usually ranges from days to weeks, and many definitive tests can be performed only after subculture, which introduces additional delays. The ability to grow on artificial medium itself poses a significant selective pressure that not all microbial pathogens are able to overcome. A large number of known parasitic and viral pathogens lack any system for in vitro cultivation and therefore must be identified by direct visual examination, by antigen testing, or indirectly by serologic testing. Other organ-

Table 2. Amplification methods as analogs of artificial medium

| Amplification method | Specimen type | Amplification result | Analogous cultivation method |
|---|---|---|---|
| Universal | Polymicrobial | Polyclonal amplification | Primary isolation medium |
| Universal | Normally sterile, infection with single organism | Clonal amplification | Primary isolation medium |
| Universal | No organisms present | Negative | Primary isolation medium |
| Selective | Target organism present with or without other flora | Positive for target organism | Selective growth medium |
| Selective | Target organism absent | Negative | Selective growth medium |

isms, such as mycoplasmas, have such specialized growth requirements that routine screening is impractical and identification of specific disease associations is severely limited. Recent studies of the microbial composition of environmental samples revealed that only a small fraction of the total bacterial species present in such samples are capable of growth on artificial medium (1, 90, 133, 149). The human polymicrobial flora is probably no less complex than that of a hot spring; the microorganisms that we are able to identify by in vitro cultivation, whether classified as pathogens or members of the normal flora, probably represent only a fraction of what is actually present (42).

The promise of in vitro nucleic acid amplification techniques is the replacement of the process of biologic amplification—growth in culture—with the enzymatic duplication and amplification of specific nucleic acid sequences. This metaphor of PCR as petri dish may take several forms (Table 2). Some amplification protocols are selective. They are designed to amplify a nucleic acid target sequence from a particular organism or group of organisms. Thus, they are analogous to a selective medium, in which only the pathogen of interest is identified from a polymicrobial mixture. Other amplification protocols are purposely designed to be nonselective. These so-called universal systems are capable of amplifying nucleic acid sequences from a wide array of organisms. Therefore, they are analogous to a nonselective medium. In theory, a universal amplification system applied to a clinical specimen containing a single organism reflects clonality; specimens containing multiple microorganisms contain multiple variants of the amplified sequence. To produce definitive results, a universal amplification system normally relies on further characterization of the amplified sequence, such as a subsequent round of selective amplification or hybridization of the universal amplification product with a specific probe. This process is in turn analogous to plating on a selective medium after primary isolation. In nonselective detection, the PCR approach has a significant advantage over the petri dish: nucleic acid amplification techniques are not limited by the ability of an organism to grow in culture.

Despite the obvious power of nucleic acid amplification techniques, it would be a mistake to assume that they will trigger a wholesale replacement of conventional techniques in the near future. In fact, they may actually create many new problems of their own (chapter 4). Undoubtedly, the greatest initial impact of

**Figure 1.** Sensitivity of detection methods. The sensitivity of nucleic acid-based detection methods is expressed in moles and numbers of target molecules. Further description of this diagram is found in the text. (Adapted from reference 56.)

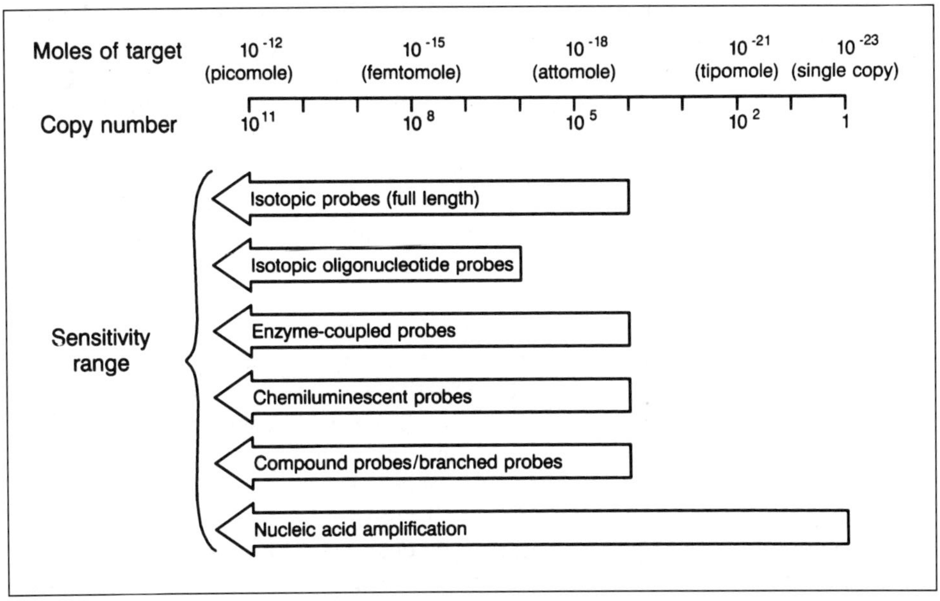


nucleic acid amplification technology will be in the detection of pathogens for which in vitro cultivation systems are labor intensive, lengthy, prohibitively expensive, or simply unavailable. As a result, the diagnostic repertoire of the clinical laboratory will expand significantly as new testing abilities are developed and new disease associations are identified (99, 128, 129).

## Analytical versus Clinical Sensitivity

Despite the auspicious introduction several years ago of nucleic acid probes for the detection and characterization of infectious agents, the average clinical laboratory has been relatively slow to accept tests based on this technology. Reasons for this include the technical difficulty of the procedures, their high cost relative to conventional methods, and, in some cases, the lack of a clear benefit with regard to clinical management. The analytical sensitivity of nucleic acid probes is at the core of many of these problems. The sensitivity of conventional (nonamplified) probes (under ideal circumstances, approximately $10^4$ to $10^5$ molecules of a homologous target) is not sufficiently high for them to be used widely for direct detection in specimens and thus to eliminate the need for prior in vitro cultivation (Fig. 1). However, there are several notable exceptions, some of which are reviewed in chapter 1.

Among the many touted advantages of in vitro amplification systems is their high analytical sensitivity. However, high analytical sensitivity does not necessarily translate into high clinical sensitivity. Clinical performance of an assay depends on various factors, including the method of sample preparation (to ensure that an adequate number of target molecules are present), the presence of inhibitors of the enzymatic processes of amplification, and the time at which the sample is collected (i.e., before, during, or after therapy). The volume of sample tested has a direct bearing on clinical performance if the target copy number is low. For example, for a system with an analytical sensitivity of one molecule, samples with an average of only one target copy give a positive result in less than 70% of all reactions performed at this concentration. In this hypothetical system, a sample would have to contain an average of 10 molecules to ensure that 99% of all specimens contain a single copy for detection. The effects of inhibitors are themselves amplified in a sample containing low copy numbers of the target sequence. An inhibitor that decreases the sensitivity of single-copy gene detection by 50% has a much greater effect on clinical performance if the sample contains a single copy than if the sample contains hundreds of target molecules. Thus, it is important to differentiate analytical sensitivity and clinical sensitivity in the assessment of an amplification technology for clinical use and to realize that an amplification system may not always provide adequate performance despite having high analytical sensitivity.

Sensitivity in practical terms also depends on the number of target molecules in the organism of interest. If many thousands of copies of a target sequence are present per organism, as is the case for repetitive genetic elements in mycobacteria (protocol 1.2) and leishmaniae (protocol 4.1) or rRNA sequences in *Borrelia burgdorferi* (protocol 1.5) (114, 115), a relatively insensitive amplification technique may provide adequate clinical performance. However, the advantages of a highly sensitive system are not lost in these settings, because the volume of sample required for reliable detection may be decreased (thus simplifying sample processing) or the effects of inhibitors may be dampened. For the detection of gene sequences in bacterial colonies or from viral cultures (in which biologic amplification has already occurred), a conventional nonamplified detection system may provide adequate performance. Thus, whether there is a need to amplify depends on many factors. The advantages of amplification must be weighed carefully against the need for extremely high levels of sensitivity.

## Target Amplification Methods

Target amplification systems are in vitro methods for enzymatic replication of a target molecule to levels at which it can be readily detected. To satisfy our definition of a target amplification technique, the process must incorporate target-specific sequence information into the amplification product. Thus, the final product contains sequence information, derived from the target, that is not present in the reagents at the start of the reaction. Target amplification systems are capable of simultaneously identifying a pathogen and providing a replica of the target sequence that can be further characterized. It is this property that distinguishes target amplification systems from all other amplification methods; this has certain advantages, which will be described later in this chapter.

### Polymerase Chain Reaction

The best-studied and most widely used target amplification technique is PCR (2, 43, 102, 119, 135, 136, 154, 160). In the 8 years since its first description by scientists at the Cetus Corporation, PCR has developed into a mainstay technique in many molecular biology laboratories. This method uses repeated cycles of oligonucleotide-directed DNA synthesis to perform in vitro replication of target nucleic acid sequences (Fig. 2). The oligonucleotides whose sequence is determined by the target nucleic acid are synthesized to be complementary to their annealing sites within the two different strands (the plus and minus strands) of a

**Figure 2.** PCR. In the first cycle, a double-stranded DNA target sequence is used as template, with the primer-binding sites indicated by hatched lines (A). These two strands are separated by heat denaturation, and two synthetic oligonucleotide primers (cross-hatched lines) anneal to their respective recognition sequences in the 5′-to-3′ orientation (B). Note that the 3′ ends of each primer are facing each other. *Taq* DNA polymerase initiates synthesis at the 3′ ends of each primer (C). Extension of the primer via DNA synthesis results in new primer-binding sites. The net result after one round of synthesis is two "ragged" copies of the original target DNA molecule. In the second cycle, each of the four DNA strands shown in panel C anneals to primers (present in excess) to initiate a new round of DNA synthesis (D). Of the eight single-stranded products, two are of a length defined by distance between and including the primer-annealing sites; this "short product" accumulates exponentially in subsequent cycles. (Reprinted from reference 118.)

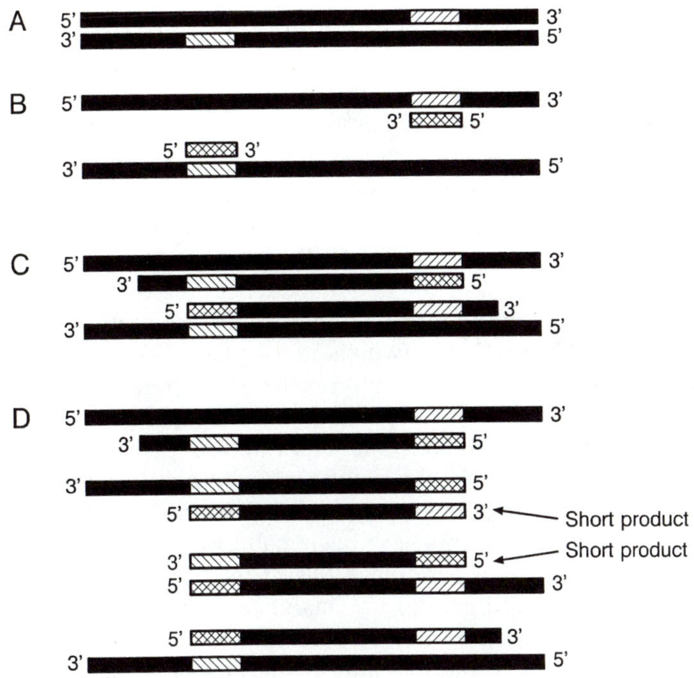


target DNA sequence. The distance between the primers is determined empirically and depends on many factors. The usual span between primers for diagnostic assays is 50 to 1,500 nucleotide bases.

In its rudimentary form, each cycle of PCR consists of three steps (Fig. 2): (i) a denaturing step, in which the target DNA is incubated at high temperature so that the target strands are melted apart and thus made accessible to hybridization by specific oligonucleotide primers; (ii) an annealing step, in which the reaction mixture is cooled to allow the primers to anneal to their complementary target sequences; and (iii) an extension reaction, usually done at an intermediate temperature, in which the primers are extended on the DNA template by a DNA polymerase. These three incubation steps are linked in what is referred to as a thermal cycle. A typical PCR protocol comprises 30 to 50 thermal cycles. Each time a cycle is completed, there is a theoretical doubling of the target sequence. Thus, repeating the thermal cycle results in a geometric accumulation of amplified target sequences.

During all of the thermal cycles of a PCR protocol, the individual primer extension reactions terminate at various distances from the primer, resulting in amplification products that contain opposite DNA strands of indeterminate lengths. However, after the second cycle, a specific amplification product begins to accumulate and rapidly becomes the predominant product. This so-called short product of PCR comprises the two primer-binding sites along with the stretch of intervening nucleic acids. It is identified by its precisely defined length and the presence of amplified internal target sequences. Nonspecific amplification products (that is, spuriously amplified sequences that do not contain the specific target sequence) are only rarely the same size as the target-specific product and do not contain internal sequences that are homologous to target-specific hybridization probes.

In conventional hybridization with a single oligonucleotide probe, imperfect pairing of bases allows the probe to anneal in a nonspecific manner so that it binds to multiple nontarget sites in the DNA mixture. Indeed, for low-abundance targets (especially if the hybridization is done at low stringency), the contribution of background hybridization may exceed that of the true target sequence. PCR overcomes much of this problem in part by requiring two specific hybridization events such that the two oligonucleotide primers each initiate DNA synthesis toward the annealing site for the primer on the opposite strand. Furthermore, to provide efficient amplification of the target sequence, the two priming sites must be within reasonable proximity of one another. Thus, the physical linkage of two priming sites by DNA synthesis (which itself places great demands on the stringency of primer-target hybridization) results in the primer-directed synthesis of new priming sites. This critical element of design gives PCR much of its sensitivity and specificity. In fact, the contribution of multiple independent specific hybridization events to the amplification process is the basis for the specificity of most of the amplification methods described in this section.

Many technical improvements have been made so that the PCR of today only remotely resembles, on an operational basis, its earlier counterpart. These improvements also enhanced greatly the analytical sensitivity and specificity of the technique. The first was the use of a thermostable DNA polymerase (*Taq* polymerase) isolated from a thermophilic bacterium, *Thermus aquaticus*, that grows in hot springs at temperatures of 70 to 75°C (135). The enzymatic activity of *Taq* polymerase has a half-life of approximately 40 min at 95°C and thus is able to withstand the repeated thermal cycling of PCR. Earlier versions of PCR required the addition of fresh polymerase to replenish the enzyme that had been denatured during the heating step. *Taq* polymerase is added once at the beginning of the reaction, thus greatly simplifying sample handling. In addition, conducting the annealing and extension reactions at higher temperatures significantly decreases nonspecific amplification. The annealing step, which is now performed at elevated temperatures, can be customized for each primer set so that conditions are unfavorable for the formation of imperfect base pair hybrids of primer and target (135).

A second innovation was the development of the programmable thermal cycler. The thermal cycler, essentially a programmable heating block, is capable of conducting successive heating and cooling cycles unattended, and it eliminates the tedious task of transferring reaction tubes between water baths or heating blocks set at the requisite temperatures (one of the least envied tasks of graduate students in the 1980s).

PCR is an extremely powerful technique for finding the nucleic acid equivalent of the proverbial needle in the haystack. PCR can be used to selectively amplify target sequences that are present in low abundance in a background of genomic DNA. This feature makes it potentially useful for the diagnosis of pathogens present in small numbers whose DNA (or RNA) copurifies with genomic DNA. One of the best-studied examples of this is the detection of proviral sequences for human immunodeficiency virus type 1 (HIV-1), in which the low prevalence in human mononuclear cells of virus-specific sequences precludes the use of conventional hybridization techniques (114) (protocol 2.1). The typical PCR-based HIV-1 detection scheme detects sequences located in highly conserved regions of the provirus such as the long terminal repeat or the *gag* and *pol* coding regions. The amplification products are then detected by hybridization to a probe that is homologous to internal sequences, that is, the amplified region bordered by the two amplification primers. The hybridization step serves to confirm the presence of the specific amplification products and to improve detection sensitivity. Several comparative studies of the PCR and conventional culture detection of HIV-1 in the peripheral blood of infected individuals confirmed the sensitivity of the nucleic acid-based technique. PCR may well become the preferred method for direct HIV detection, if only because it is faster, less costly, and potentially less hazardous than culture.

In the last few years, the availability of PCR reagents in kit form, the increased availability of oligonucleotide synthesis at low cost, and the tremendous increase in the availability of nucleotide sequence data have permitted the application of new technology to old problems in infectious-disease research laboratories. One of the most interesting applications of PCR has been the analysis of archived paraffin-embedded tissue specimens (60) and museum specimens (122). Such retrospective epidemiology studies add to our understanding of the incidence and manifestation of infectious diseases in previous years. Recently, PCR amplification in situ performed on fixed cells and tissues was described (31, 65, 110). This procedure may provide a sensitive means of determining the disposition of infectious organisms within tissues. The list of applications of PCR for the sensitive and specific detection of bacterial, fungal, and parasitic pathogens has grown immensely. The current applications of PCR to the detection of clinically relevant infectious microorganisms are given in Tables 3 through 6.

## Modifications to PCR

### *Multiplex PCR*

Many modifications to the basic PCR process, some of which are relevant to diagnostics, have been described. In multiplex PCR, multiple primer pairs specific for different targets are included in the same amplification reaction (29). Coamplification of different targets serves several purposes. (i) Large regions of the DNA sequence can be scanned for alterations. (ii) Unrelated segments of the target genome can be tested. (iii) Internal controls for the amplifiability of the sample can be included. (iv) Cost-effective panels of tests for multiple pathogens from a single specimen can be developed. Examples of multiplex PCR include detection of genes specifying toxins of *Staphylococcus aureus* (protocol 1.20) and *Clostridium difficile* (98) and coamplification of competitive internal controls (protocols 1.2, 1.13, and 2.6).

### *Point Mutation Detection*

The amplification refractory mutation system (ARMS) uses PCR primers with 3′ mismatches and a polymerase that lacks 3′ to 5′ exonuclease activity to detect specific alterations in a nucleotide sequence (106). The enzymatic extension of an

**Table 3.** Bacterial pathogens detected by nucleic acid amplification

| Bacterial pathogen | Reference or protocol |
|---|---|
| *Mycobacterium* sp. | 17, protocol 1.1 |
| *Mycobacterium tuberculosis* | 66, 76, protocol 1.2 |
| *Mycobacterium leprae* | 3, 67 |
| *Mycobacterium avium-intracellulare* | Protocol 1.1 |
| *Chlamydia trachomatis* | 32, 69, protocol 1.9 |
| *Borrelia burgdorferi* | 59, 94, 122, 123, 132, protocol 1.4 |
| *Mycoplasma pneumoniae* | Protocol 1.12 |
| *Mycoplasma* sp. | Protocol 1.13 |
| *Mycoplasma fermentans* | Protocol 5.4 |
| *Salmonella typhimurium* | 126 |
| *Bordetella pertussis* | 58, 71 |
| *Legionella pneumophila* | 74, protocol 1.14 |
| *Clostridium difficile* | 81, 98, protocol 1.19 |
| *Treponema pallidum* | Protocols 1.7 and 1.8 |
| Enterotoxigenic *Escherichia coli* | 33, 80, 111, protocol 1.16 |
| Enterohemorrhagic *Escherichia coli* | 21 |
| *Vibrio cholerae* | Protocol 1.15 |
| *Shigella* sp. | 33, protocol 1.17 |
| *Rickettsia* sp. | 4, 27, 150, 157, protocol 1.3 |
| *Ehrlichia* sp. | 16, protocol 1.3 |
| *Chlamydia pneumoniae* | Protocol 1.11 |
| *Staphylococcus aureus* | 20 |
| *Staphylococcus aureus* toxins | Protocol 1.20 |
| Bacterial meningitis pathogens | Protocol 1.21 |
| Whipple's disease-associated bacterium | 129, protocol 5.2 |
| *Yersinia pestis* | Protocol 1.6 |
| *Yersinia enterocolitica* | 75, 161 |
| *Helicobacter pylori* | Protocol 1.18 |
| *Aeromonas hydrophila* | 124 |
| Bacillary angiomatosis agent | 128 |
| *Listeria monocytogenes* | 22, 78, 107 |
| *Comamonas* sp. | 142 |
| *Bacillus anthracis* | 26 |
| Coliform bacteria | 13, 14 |
| *Ureaplasma urealyticum* | 156 |
| *Neisseria meningitidis* | 86 |
| *Neisseria gonorrhoeae* | 15, protocol 6.5 |
| *Mycoplasma genitalium* | 79 |
| *Campylobacter* sp. | 115 |
| *Corynebacterium diphtheriae* | 117 |
| *Actinobacillus pleuropneumoniae* | 140 |
| *Leptospira* sp. | 70, 143 |
| *Haemophilus influenzae* | 144 |

oligonucleotide primer depends on stable base pairing of the 3′ end of an oligonucleotide. When a mismatch occurs, primer extension (and the resultant PCR process) is often prevented. The ARMS approach uses selective amplification based on 3′ mismatches to identify specific mutations by demonstrating the presence of a nucleotide sequence that cannot be amplified with wild-type primers but can be amplified with primers containing the expected mutations. The ARMS has been used to detect point mutations associated with azidothymidine resistance

**Table 4.** Viral pathogens detected by nucleic acid amplification

| Viral pathogen[a] | Reference or protocol |
|---|---|
| *Herpesviridae* | |
| Herpes simplex virus types I and II | Protocols 2.5 and 2.6 |
| Epstein-Barr virus | Protocol 2.7 |
| Cytomegalovirus | 38, 165, protocol 2.8 |
| Varicella-zoster virus | Protocol 2.9 |
| Human herpesviruses 6 and 7 | Protocol 5.3 |
| *Papovaviridae* | |
| Human JC virus | Protocol 2.10 |
| Human papillomavirus | 11, 138, 165, protocol 2.17 |
| *Flaviviridae* | |
| Hepatitis C virus | 39, 155, 162, protocol 2.4 |
| Enteroviruses | Protocol 2.16 |
| *Hepadnaviridae* | |
| Hepatitis B virus | 68, 89, 113, protocol 2.3 |
| *Retroviridae* | |
| HIV-1 and HIV-2 | 62, 116, 127, protocol 2.1 |
| HTLV-I and HTLV-II | Protocol 2.2 |
| HBLV | 23 |
| *Orthomyxoviridae* | |
| Influenza virus | Protocol 2.12 |
| *Parvoviridae* | |
| Human parvovirus B19 | Protocol 2.11 |
| *Adenoviridae* | |
| Human adenovirus | Protocol 2.14 |
| *Togaviridae* | |
| Rubella virus | Protocol 2.15 |
| *Picornaviridae* | |
| Rhinovirus | 53 |
| Hepatitis A virus | 77 |

[a] HBLV, Human B-lymphotropic virus; HTLV, human T-cell leukemia virus.

in patients who test positive for HIV (protocol 6.2). The terms "competitive oligonucleotide priming" (54) and "mismatch amplification mutation assay" (28) have been used to describe similar PCR adaptations.

***Sequence Determination***

As our understanding of the roles of nucleotide sequence alterations and their associations with infectious disease increases, DNA sequencing will play a larger role in the clinical microbiology laboratory. Present examples include the detection of hepatitis B virus mutants that may be associated with progression to fulminant hepatic necrosis (113), development of rifampin resistance in mycobacteria (protocol 6.1), molecular epidemiology studies (77, 127), and microbial identification (10, 17, 112, 158, 159).

Direct nucleotide sequence determination will be especially useful for identifying the composition of highly variable target sequences that are amplified by using broad-range primers. Variations in the amplified sequence can be determined by several sequencing methods. To facilitate dideoxy-chain termination sequencing, single-stranded templates can be generated by asymmetric amplification or affinity capture of single-stranded DNA by using biotinylated primers and avidin-coated magnetic heads (protocol 1.1) (64). Recently, a powerful

**Table 5.** Fungal pathogens detected by nucleic acid amplification

| Fungal pathogen | Reference or protocol |
|---|---|
| *Blastomyces dermatitidis* | Protocol 3.1 |
| *Candida albicans* | 100, protocol 3.1 |
| *Coccidioides immitis* | Protocol 3.1 |
| *Cryptococcus neoformans* | 145, protocol 3.2 |
| *Histoplasma capsulatum* | Protocol 3.1 |
| *Pneumocystis carinii* | Protocol 3.3 |
| *Trichosporon beigelii* | 83 |

technique called cycle sequencing was described (134); in this technique, *Taq* polymerase is used in a thermal cycling protocol to perform dideoxy-chain termination sequencing or amplification and sequencing simultaneously. An optimized cycle-sequencing reaction requires only femtomole quantities (ca. $10^8$ copies) of target for the determination of stretches of nucleotides that are hundreds of bases long. This level of sensitivity approaches that of conventional (nonamplified) probes and may make it possible to determine pathogen-specific nucleotide sequences from clinical isolates or directly from clinical specimens.

***Nested Amplification***

A popular modification of PCR uses nested sets of PCR primers and is known as nested amplification (Fig. 3). In a typical nested-amplification protocol, a first round of amplification is performed with a single primer pair for 15 to 30 cycles. This step produces an amplification product, which is transferred to a new reaction tube for a second round of amplification by using a second primer pair that is specific for the internal sequence amplified by the first primer pair. The second amplification usually proceeds for an additional 15 to 30 cycles, after which the products are detected by gel electrophoresis.

Nested amplification offers both advantages and disadvantages. The sensitivity of most nested-amplification procedures is extremely high; often, a single copy of target can be detected without the need for hybridization with labeled probes. Reamplification with the second set of internal primers also serves to verify the specificity of the first-round product, in much the same way as a labeled probe does. In addition, the transfer of reaction products from the first reaction

**Table 6.** Protozoal pathogens detected by nucleic acid amplification

| Protozoal pathogen | Reference or protocol |
|---|---|
| *Babesia bigemina* | 51 |
| *Babesia bovis* | 46 |
| *Babesia microti* | 120, protocol 4.5 |
| *Entamoeba histolytica* | Protocol 4.4 |
| *Giardia lamblia* | Protocol 4.6 |
| *Leishmania* sp. | Protocol 4.2 |
| *Naegleria fowleri* | 97 |
| *Plasmodium falciparum* | 8, 168, protocol 4.3 |
| *Toxoplasma gondii* | 24, 61 |
| *Trichomonas vaginalis* | 131 |
| *Trypanosoma* sp. | Protocol 4.1 |

**Figure 3.** Nested amplification involving open transfer of the first-round amplification product, which usually leads to carryover contamination (A), and differential annealing of primers and products in a single-tube nested protocol (B). See text for additional description.

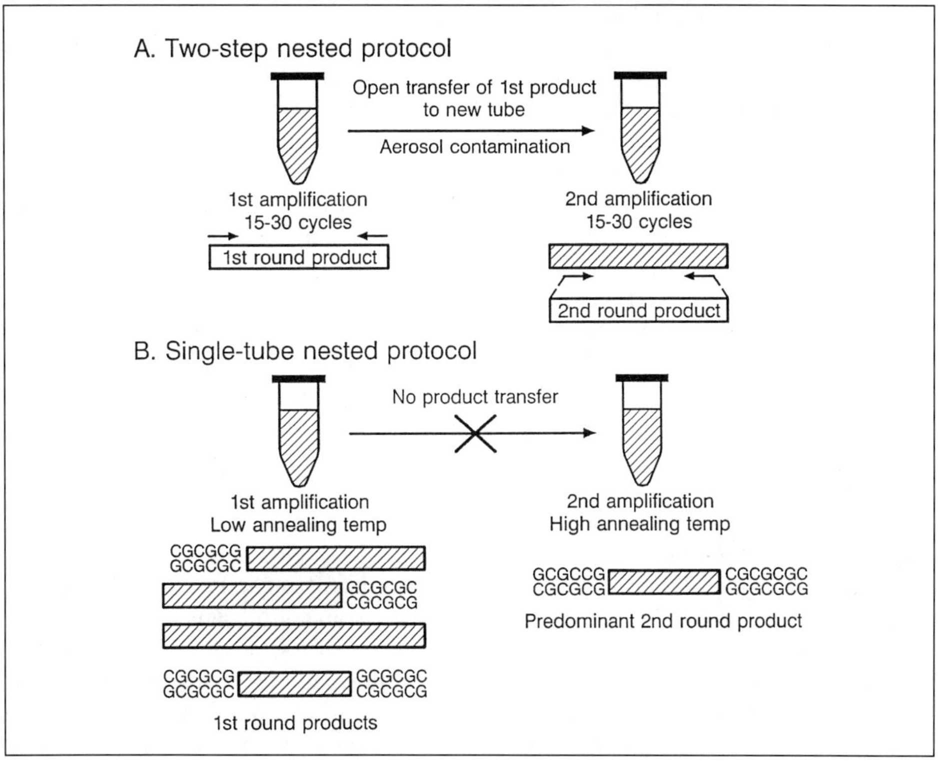


effectively serves to dilute out inhibitors that might be present in the sample initially. Thus, a conventional nonnested PCR protocol that might have progressed poorly because of suboptimal PCR conditions or inhibition can be salvaged, in many cases, by the use of a nested protocol.

Despite its advantages, however, nested amplification involving the open transfer of amplification products from tube to tube is not recommended for routine clinical laboratory use. This is because during the transfer of the reaction product from the first reaction into a new tube for reamplification, contamination by aerosolized amplified DNA inevitably occurs. The product of the first reaction cannot be inactivated by currently available protocols (chapter 5) because its activity as a template must remain intact for the nested protocol to function.

If a nested-PCR protocol must be used, several additional controls should be implemented, including interspersing multiple blank reactions at various intervals throughout the first and second rounds of amplification (chapter 9). In general, the incipient risk of eventual contamination outweighs the advantages of using a nested procedure. A well-optimized amplification protocol with a single primer pair can match the sensitivity of a nested reaction when hybridization is used to detect the product. Furthermore, reactions based on a single primer pair can be modified easily to prevent carryover contamination.

Several methods have been described recently for performing nested reactions without open transfer of amplification products between reaction tubes (44, 49, 167). These so-called single-tube nested protocols involve various approaches.

For the sensitive detection of hepatitis C virus in serum, Feray et al. (49) used a novel method to separate the first- and second-round reaction contents by placing the reaction mixture containing the internal primer within a thick layer of oil above the reaction mixture containing the first primer pair. After the first amplification, the reaction tubes are removed and spun in a microcentrifuge to mix the reaction components and dilute the first-round components, and then the contents are reamplified. This single-tube protocol has much less risk of contamination and is reported to be as sensitive as a two-tube nested protocol. Variations on this mechanical separation method have now been described (167).

Other methods for performing nested amplification in a single tube are based on differential annealing of outer and inner primer pairs and differential melting temperatures of the large and small amplification products (44). One approach investigated in our laboratory involves outer primers that are designed to anneal optimally at a lower temperature; the inner primers contain tails rich in G+C, thus elevating their optimal annealing temperatures during later cycles. Amplification is first performed at a low annealing temperature, which results in a mixture of products of the inner and outer primers. A switch to a higher temperature, however, favors amplification of the inner product. An example of this approach is shown in Fig. 3.

***Detection of RNA Targets***

The range of PCR applications to infectious diseases can be expanded greatly if RNA targets can be detected. For instance, some viral genomic targets are present only as RNA (hepatitis C virus, influenza virus, picornaviruses) and do not undergo reverse transcription into a DNA intermediate during their life cycle. Detection of RNA may also be useful for the identification of higher organisms such as bacteria and fungi, which may contain thousands of molecules of mRNA or rRNA. Furthermore, specific detection of RNA may prove to be useful for monitoring antimicrobial therapy (because of the more rapid turnover of RNA targets) or for studying pathogenic states that appear to be associated with specific gene expression, such as the link between genital neoplasia and expression of the E6 and E7 viral oncogenes of human papillomaviruses (103; see below).

RNA templates can be detected by PCR if the extracted RNA is first converted to cDNA by using a retroviral reverse transcriptase from avian myeloblastosis virus or Moloney murine leukemia virus. However, the use of these enzymes presents several difficulties (50). These enzymes are thermally labile and cannot be used at temperatures above about 42°C. Therefore, the stringency with which reverse transcription occurs is relatively low. This factor, coupled with the potential for single-stranded RNA templates to form stable secondary structures, leads to widely varying efficiencies of conversion of RNA to cDNA; in some protocols the efficiency is estimated to be as low as 5% (50). Thus, the relative inefficiency of reverse transcription enzymes represents a major obstacle to the sensitive and specific detection of RNA targets.

Recently, Myers and Gelfand (105) described the $Mn^{2+}$-dependent reverse transcriptase activity of a thermostable DNA polymerase isolated from *Thermus thermophilus*. This enzyme is capable of converting RNA to DNA at 72°C in the presence of $Mn^{2+}$; the addition of magnesium and ethylene glycol-bis(β-aminoethyl ether)-*N,N,N′,N′*-tetraacetic acid (EGTA) (to chelate the $Mn^{2+}$) to the reaction switches the template specificity of the *T. thermophilus* polymerase from RNA to DNA. Thus, a single enzyme is capable of converting RNA to cDNA and then amplifying the cDNA by PCR after a simple switch in buffer conditions. Our expectation is that the ability to conduct reverse transcription at 72°C will enhance the specificity of reverse transcription and will lead to greater conversion efficiency with concomitant increases in sensitivity for detection of RNA targets.

***Quantitative PCR***

A common assumption is that the exponential amplification of target nucleic acid sequences afforded by PCR might prevent it from being a quantitative procedure. However, there is now considerable evidence that under carefully controlled circumstances, levels of PCR product correlate directly with the number of input target molecules (50, 55, 104, 108, 148). The linear relationship between input copy number and level of amplification product usually extends across a range of 3 to 4 orders of magnitude (50). Factors affecting the quantitative nature of PCR include optimization of the reaction protocol for maximum efficiency, avoidance of excessive cycling, adequate processing of samples to eliminate inhibitors that may decrease the efficiency of amplification, and inclusion of internal coamplified standards. The last feature, in which a modified template added to PCRs serves as a control for amplification efficiency, was incorporated into several protocols in this book (protocols 1.2, 1.13, and 2.6). Ferre (50) has provided a recent review of quantitative PCR.

## Transcription-Based Amplification Systems: TAS and 3SR

The first non-PCR target amplification system was described in 1989 by Kwoh et al. (87). This ingenious system, known as the transcription-based amplification system (TAS), is based on amplification by in vitro transcription. Each cycle of TAS consists of two parts: synthesis of a DNA molecule that is complementary to the target nucleic acid, and in vitro transcription with the newly synthesized cDNA as a template. The synthesis of cDNA is primed by specially designed oligonucleotide primers. One end of the primer consists of a target-specific sequence, and the other end contains a promoter for phage T7 RNA polymerase. Thus, in the first step, one single-stranded DNA molecule is produced for each DNA or RNA target by the action of reverse transcriptase. After heat denaturation of RNA-DNA or DNA-DNA duplexes, a second strand is synthesized by newly added reverse transcriptase; the resulting double-stranded DNA copy of the target contains a T7 promoter at one or both ends. T7 RNA polymerase is then added to the cDNA mixture, and as many as 40 molecules of RNA are produced from each molecule of template. The action of T7 polymerase thus provides the amplification step. The newly synthesized cDNA serves as a template for the synthesis by T7 polymerase of a large molar excess of RNA, which then can be used as the substrate for the next cycle of TAS. Repeating this 40-fold amplification several times results in a several-millionfold amplification.

The utility of TAS as a target amplification method was demonstrated by Kwoh et al. (87) for the detection of the *vif* region of HIV-1 RNA. Fewer than one HIV-1-infected cell could be detected in a population of $10^6$ uninfected cells. After four cycles of TAS, target sequences were amplified $2 \times 10^6$- to $5 \times 10^6$-fold. The amplified RNA product was then detected by affinity capture via complementary oligonucleotides covalently attached to agarose beads, an approach that exploits the single-strandedness of the RNA product.

One of the drawbacks of TAS described by Kwoh et al. (87) was the need to heat denature the RNA-DNA duplexes during each amplification cycle. Because none of the enzymes used in TAS are thermophilic, they must be replenished at the end of each cycle. In exploring and modifying the reaction conditions of TAS,

Gingeras and colleagues (48, 57) and Compton (34) independently discovered that the process of heat denaturation of RNA-DNA duplexes could be replaced by the enzymatic activity of *Escherichia coli* RNase H. This led to the discovery of an isothermal amplification technique (that is, requiring no changes in the temperature of incubation) that exploited the simultaneous enzymatic activities of three enzymes—RNase H, avian myeloblastosis virus reverse transcriptase, and T7 polymerase—in a process called self-sustaining sequence replication (3SR) or nucleic acid sequence-based amplification (NASBA) (34, 56). In many respects, 3SR mimics in vitro the natural process of replication of the RNA genome of retroviruses (63). Figure 4 depicts the 3SR process.

In its initial description by Guatelli et al. (63), 3SR produced an average 10-fold amplification every 2.5 min for the first 15 min, leading to a $10^7$-fold amplification of an RNA target within 1 h. Subsequent developments, including optimization of substrate concentrations, ionic strength, pH, and temperature and addition of the cosolvents dimethyl sulfoxide and sorbitol, improved efficiency even further, so that $10^8$-fold amplification can now be achieved in 30 min (56). Furthermore, by exploiting the inherent RNase H activity of avian myeloblastosis virus reverse transcriptase, reaction conditions have now been established that allow 3SR to be performed with only two enzymes, avian myeloblastosis virus reverse transcriptase and T7 polymerase (56).

Although 3SR is an efficient and powerful target amplification system, in the 2 years after its description there was much less development of specific applications than for its older counterpart, PCR. However, a few 3SR applications did appear (57, 62) (protocols 1.10 and 2.18), and many are currently in development. 3SR may be well suited to the task of selective detection of RNA molecules. If the reactions are maintained under conditions in which target DNA is not denatured, the DNA remains inaccessible to the 3SR enzymatic machinery. Because RNA transcripts of the target are largely single stranded to begin with, they are accessible and can be amplified directly. In this respect, 3SR may have a slight edge over PCR, because in most protocols selective amplification of RNA templates by PCR requires selective isolation of RNA that is completely free of DNA or pretreatment of samples with RNase-free DNase (or both) (130). This incubation may be difficult to perform reliably and exposes RNA templates to possible degradation by endogenous RNase activity in the sample.

Possible applications of selective RNA amplification and detection in infectious-disease testing are many and varied. In detection of human papillomavirus DNA by PCR, many studies showed that the positive predictive value for identifying patients with cervical intracellular neoplasia is low because of the high prevalence of human papillomavirus infection in apparently disease-free individuals (11, 138). A 3SR-based technique that may provide a much-needed boost in diagnostic resolution is described in protocol 2.18. In that protocol, the RNA transcripts of the two viral early genes E6 and E7 are detected by selective 3SR amplification of cervical swabs and biopsy material. Expression of the E6 and E7 gene products and the subsequent inactivation of their cellular targets, the tumor suppressors p53 and RB (153), are both necessary and sufficient for neoplastic transformation in vitro (103); similar in vivo roles have been predicted. Thus, detection of the E6-E7 mRNA might have a better predictive value for determining which patients will require further testing or closer monitoring. Other applications of 3SR include detection of *Chlamydia trachomatis* (protocol 1.10), *Mycobacterium tuberculosis*, and azidothymidine resistance in HIV (57).

**Figure 4.** Nucleic acid amplification by 3SR. The 3SR reaction comprises continuous cycles of reverse transcriptase (RT) and RNA transcription to replicate a nucleic acid via a double-stranded cDNA intermediate. Oligonucleotides A and B prime DNA synthesis, producing a double-stranded cDNA containing a functional T7 promoter (steps 1 to 6). Complete cDNA synthesis is dependent on digestion of RNA in the intermediate RNA-DNA hybrid by RNase H (step 3). Transcription-competent cDNAs are used to produce multiple (50 to 1,000) copies of antisense RNA transcript of the original target (steps 7 and 8). These antisense transcripts are immediately converted to T7 promoter-containing double-stranded cDNA copies (steps 9 to 12) and used again as transcription templates. This process continues in a self-sustained cyclic fashion under isothermal conditions (42°C) until components in the reaction become limiting or inactivated (enzymes). · · · · , RNA; ——, DNA; ▬, T7 promoter sequence (see Fig. 4); ○, reverse transcriptase; ◇, T7 RNA polymerase; (TCS), target complementary sequence. (Reprinted with permission from reference 56.)

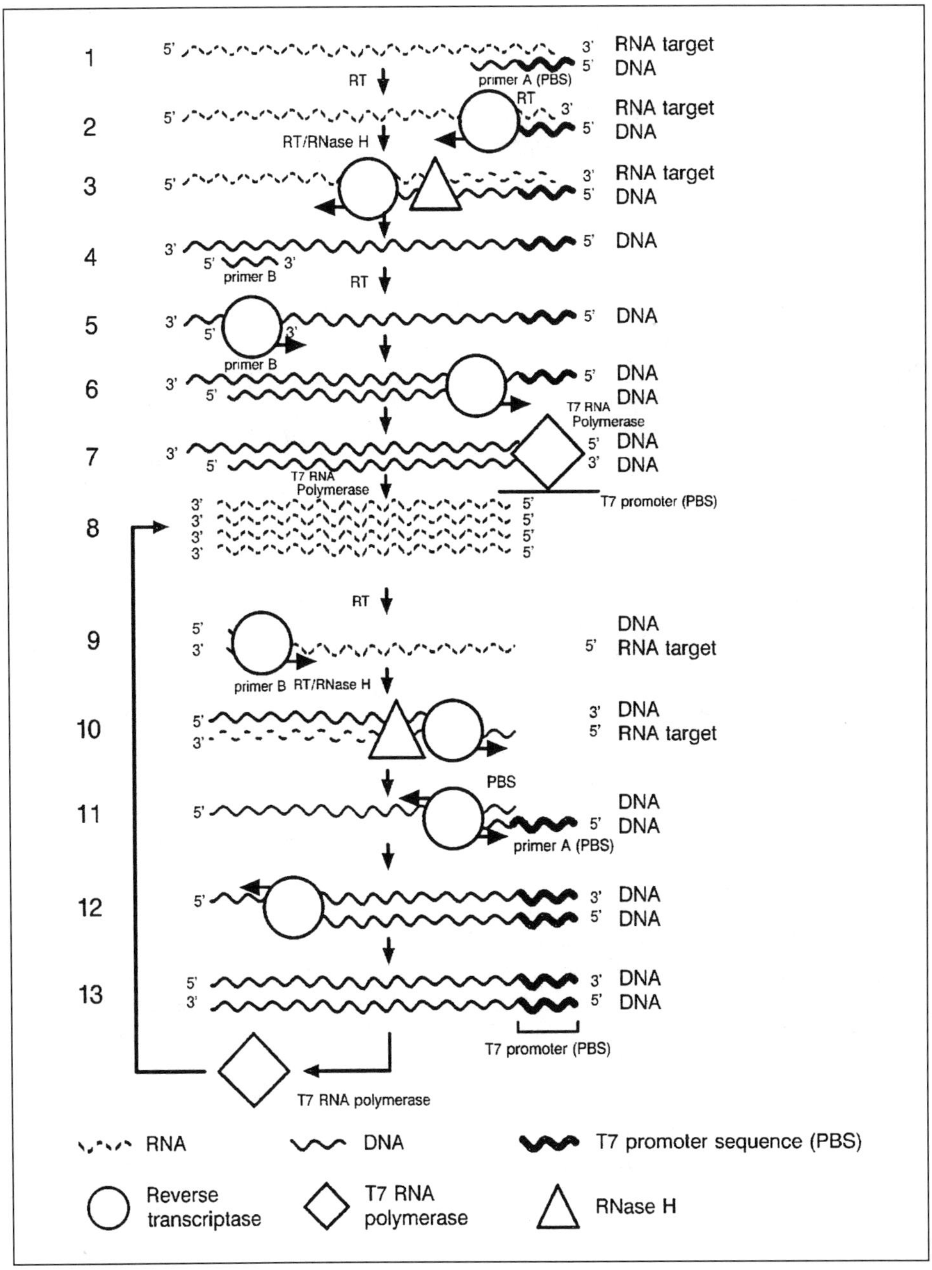

## Strand Displacement Amplification

The most recent addition to the litany of target amplification methods is known as strand displacement amplification (SDA) (146). SDA is based on the ability of DNA polymerases to initiate DNA synthesis at a single-stranded nick within a DNA target molecule and to displace the nicked strand during DNA synthesis. Displaced single-stranded molecules then serve as substrates for additional simultaneous nicking and displacement reactions, leading to an isothermal geometric accumulation of double- and single-stranded copies of the target sequence.

The key technology behind SDA is generation of site-specific nicks by a restriction endonuclease. Restriction endonucleases normally produce double-stranded DNA cleavage, which would not provide a proper template for strand displacement. However, if alpha-thio-substituted nucleotides are used to synthesize a double-stranded, hemiphosphorothioated DNA recognition site for endonuclease cleavage, the endonuclease is capable of cleaving (nicking) only one strand. The phosphorothioated strand remains unbroken and is capable of serving as a template for DNA synthesis at the site of the nick.

The primers used in SDA contain two functional components: a target-specific region of approximately 15 to 20 bp and a recognition sequence for a restriction endonuclease. After incubation with the target DNA, a 3′ to 5′ exonuclease-deficient polymerase is added along with the nucleotides dATP, dCTP, dTTP, and deoxyadenosine 5′-(alpha-thio)-triphosphate. Extensions of primer and target 3′ ends by the polymerase result in synthesis of an alpha-thiolated strand (with the primer as template) that is protected from cleavage by the restriction endonuclease. The primer, in addition to serving as a template for DNA synthesis in the opposite direction, initiates DNA synthesis on the target strand. Once the nick is created by the restriction enzyme, the 3′ end formed at the site of the nick initiates DNA synthesis, thereby displacing the newly created single strand. Displaced strands of both polarities then participate in a new round of nicking, polymerization, and displacement, resulting in a theoretical doubling of target sequences after each round of synthesis. The basic strategy behind SDA is shown in Fig. 5.

In the original description by Walker et al. (146), SDA was used to amplify a 47-bp target of the insertion sequence IS*6110* from *M. tuberculosis* and *M. bovis*. SDA reactions incubated for up to 5 h yielded a $10^6$-fold amplification, resulting in detection of as few as 100 copies of the target sequence or approximately 10 *M. tuberculosis* organisms (10 to 12 copies of the IS*6110* sequence are present in the *M. tuberculosis* genome).

Like 3SR, the reactions of SDA are isothermal and simultaneous. Temperature cycling is not required because the DNA strands are denatured enzymatically by strand displacement. However, the exploitation of this property of DNA polymerases exacts a price—the processivity of most DNA polymerases is lower for double-stranded templates (in which strand displacement is required) than for single-stranded templates (in which displacement is not required). The result is a limitation of the size of targets that can be amplified by SDA. However, this may not translate into a serious functional limitation, because for most infectious-disease applications, amplification of short sequences will be adequate for diagnostic purposes.

A second theoretical limitation of SDA is the need for restriction enzyme cleavage of the genomic DNA target prior to amplification. Such cleavage is

**Figure 5.** SDA reaction. After restriction enzyme digestion and denaturation of a double-stranded target fragment, primers $P_1$ and $P_2$ bind to single-strand target fragments $T_1$ and $T_2$. The primers contain recognition sites at their 5′ ends for a restriction enzyme. DNA replication with three deoxynucleoside triphosphates (dNTPs) and one dNTP[αS] produces double-stranded primer-target complexes with hemiphosphorothioate recognition sites, which are subsequently nicked by the appropriate restriction enzyme. A DNA polymerase lacking 5′-to-3′ exonuclease activity extends the 3′ end at the nick and displaces the downstream strand. Nicking and polymerization/displacement steps cycle continuously because of regeneration of a nickable recognition site. Exponential amplification occurs because strands displaced from the $P_1$-$T_1$ complex serve as target for primer $P_2$ and strands displaced from the $P_2$-$T_2$ complex serve as target for primer $P_1$. (Adapted from reference 146.)

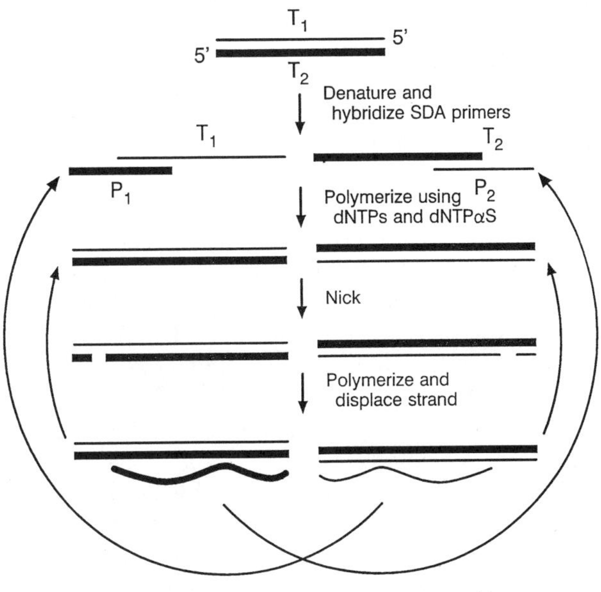


required to free the 3′ ends of the target sequence for the initiation of DNA synthesis. If one of the two restriction sites flanking the target is lost as a result of mutation, a linear, not geometric, accumulation of amplified target sequences results. This necessitates careful selection of target sequences to identify regions that are highly conserved within the primer-binding regions as well as in the upstream restriction endonuclease cleavage sites. An additional important consideration is that the target sequence must not contain a recognition site for the restriction endonuclease used in the nicking reaction. A recent modification of SDA eliminates the need for prior restriction enzyme digestion (147).

Future modifications to SDA are likely to include the use of thermophilic polymerases and eventually thermophilic restriction endonucleases. This will probably improve the analytical specificity of SDA amplification, just as the use of *Taq* polymerase improved the sensitivity and specificity of PCR. The high processivity of thermophilic DNA polymerases may also allow SDA amplification of large target sequences. Even in its present form, however, SDA is a unique and potentially powerful amplification method that may attain a prominent position among target amplification methods.

### Specialized Applications of Target Amplification

Target amplification methods are well suited to the task of detecting diverse forms of a given type of pathogen, in many ways fulfilling the role of a nonselective growth medium. Applications of broad-range or semiselective amplification have been used in the detection and characterization of slow-growing and uncultured pathogens (1, 17, 19, 30, 52, 90, 120, 149). Böttger et al. (17) were the first to apply universal primer amplification to the recovery of bacterial 16S rRNA genes, and they used the technique to recover and characterize such sequences in a wide range of mycobacteria. Using a system for genus-specific amplification of the same 16S gene target, this group recently identified a novel uncultured species, *Mycobacterium genovense* sp. nov., from the liver of AIDS patients with disseminated mycobacterial infection (protocol 1.1). By using a similar approach, Relman et al. recovered unique bacterial DNA sequences specific for the agents of bacillary angiomatosis (128) and Whipple's disease (129). Additional examples of the power of target amplification include replication of a portion of the HIV reverse transcriptase gene followed by allele-specific detection to identify multiple mutations associated with azidothymidine resistance (18, 57) (protocol 6.2) and the amplification of ribosomal DNA or intergenic spacer DNA to differentiate closely related bacterial species (151, 152; chapter 2).

A particularly promising application of target amplification is the harnessing of the identifying power of common sequence elements, such as those interspersed throughout the large- and small-subunit rRNA 23S and 16S genes of prokaryotes and eukaryotes, to develop microbial identification systems. By use of broad-range primers, such target sequences can be amplified from clinical isolates or directly from specimens (1, 17, 90, 120, 128, 129) (protocols 1.1, 1.21, 3.1, and 5.1). The nucleotide sequence flanked by the broad-range primers can then be determined directly to distinguish one organism from another (37, 42, 159). As a result of advances in DNA-sequencing technology, this approach may lead in the next several years to automated detection systems that use direct DNA sequencing (96, 134) or sensor-based sequence analysis (7, 45) followed by a computer-assisted search of a microbial sequence data base as a definitive method for detection and identification of bacterial, fungal, and parasitic pathogens (Fig. 6) (protocol 1.1).

## Amplification of the Probe Molecule

An alternative strategy for detecting rare target molecules is to amplify the probe molecule itself. Unlike target amplification methods, in which target sequences are copied into the molecule that is eventually detected, amplification of the probe molecule does not result in the incorporation of target information beyond what is present in the reaction at the start. Instead, the end product of the reaction is an amplified version of the original components used to detect the target. Such systems may be useful in settings in which the target-derived sequence information is not necessary, such as in the detection of a signature nucleotide sequence within a pathogen.

### Qβ Replicase Amplification of RNA Probe Molecules

Qβ replicase is a 215-kDa RNA-directed RNA polymerase that replicates the genomic RNA of bacteriophage Qβ. This unique multiple-subunit enzyme specifically recognizes a unique folded RNA structure (secondary structure) formed by

**Figure 6.** Nucleic acid sequence-based microbial identification. Broad-range (universal) primers are used in a target amplification protocol for amplification of 16S or 23S ribosomal DNA sequences. An automated sequencer determines the composition of the intervening nucleotides and directly enters sequence data into a computerized workstation. Data base analysis creates a hierarchical cluster of related sequences and then generates a list of related or identical organisms.

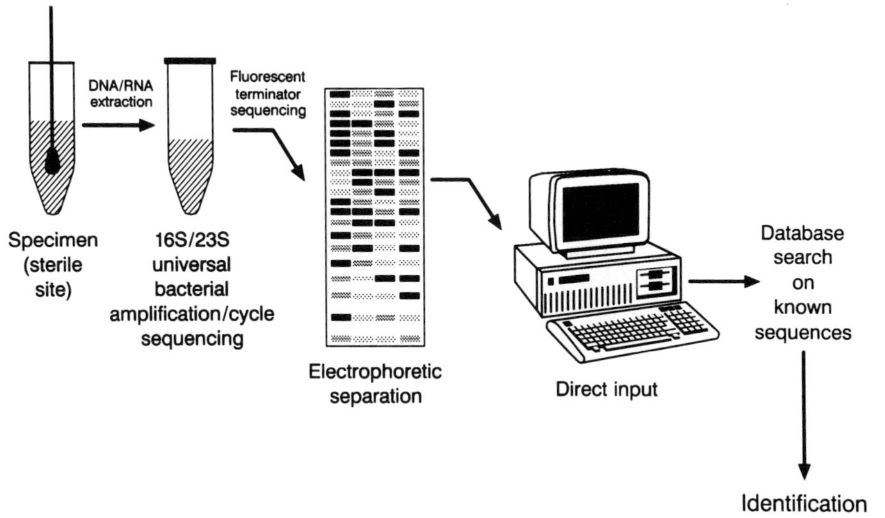


intramolecular base pairing of the Qβ RNA genome. Most other folded forms of RNA are not recognized as substrates (25, 92). Because some substrates of Qβ, such as the midvariant RNAs, can tolerate short probe inserts and still serve as a template for Qβ replicase, this system can be exploited to provide a means of amplifying hybridization signals by amplifying the probe molecule itself (85, 91, 92, 125).

The principle underlying Qβ-based amplification methods is straightforward. After the probe has annealed specifically to its target and the unbound or nonspecifically bound material (or both) is washed away, the probe is enzymatically replicated in vitro to levels that can be readily detected. The problem of decreasing the background binding may be formidable. Nonspecific hybridization of the probe molecule to sequences other than the intended target or the failure to remove it completely may result in amplification of background signals, potentially overwhelming the signal generated by a small number of target sequences. However, with the introduction of improvements such as molecular switches (85) or target cycling (in which the relative abundance of the target sequence is first increased by successive rounds of nucleic acid capture), background hybridization can be decreased markedly (73).

To construct a replicable reporter, RNA probes are produced by in vitro transcription of recombinant DNA templates. The probes have a secondary structure that is predicted to be similar to that of the wild-type substrate, except that they harbor target-specific extensions in one loop of the folded structure or at its 5′ or 3′ end. The inserts must be relatively short because of the presumed size constraints on the placement of unstructured sequences into the otherwise highly structured replication origin. The molecules are then incubated with purified Qβ replicase; despite the presence of the inserts, the replication of the chimeric molecules can be as efficient as that of the unmodified parent molecule, approaching $10^9$-fold in a 30-min incubation. Figure 7 depicts probe amplification with Qβ replicase.

After a certain level of amplification has been reached (approximately $10^7$-fold), Qβ replicase becomes saturated and the logarithmic reaction kinetics approach linearity. Thus, reactions involving more of the substrate molecules achieve saturation sooner, and further accumulation of replicated RNA becomes strictly time dependent. This property may make it possible to extend greatly the quantitative dynamic range of the assay because it allows a direct linear relationship between the number of input probe molecules and the amplified product, as determined by hybridization (91).

The major obstacle to development of a sensitive assay that uses this type of probe amplification is decreasing the levels of background amplification. One approach to decreasing background in Qβ replicase-based systems is to make the amplifiable probe conformation dependent; only probe molecules bound to the target would be recognized by the polymerase (85). One such molecular switch would be an RNase-sensitive site in a Qβ substrate molecule (Fig. 7) that is controlled by hybridization. When the probe is hybridized to the target, it is not recognized by RNase III because the double-stranded RNA substrate for RNase III cannot form. However, in the absence of target binding, the probe molecule forms a substrate for RNase III. Sequences unbound to the target are selectively cleaved, thus being rendered inactive as Qβ substrates. Hybridization-dependent switches, when coupled with target capture methods (166), may drastically decrease background problems associated with Qβ replicase-based amplification. A Qβ amplification system for detection of *Chlamydia trachomatis* has been described recently (69).

## DNA Ligase-Dependent Amplification

All amplification techniques described so far depend on polymerases to copy genetic information in vitro. An alternative probe amplification technique, based on target-dependent ligation of oligonucleotide probes, was described in 1989 by Wu and Wallace (163). This method uses DNA ligase to join two pairs of complementary oligonucleotide probes after they have bound to a target sequence in vitro. Successful ligation depends on the contiguous positioning and perfect base pairing of the 3′ and 5′ ends of oligonucleotides on a target DNA molecule. Once the two probes have been ligated, the ligation product, which now mimics one strand of the original target sequence, can serve as a template for ligation of complementary oligonucleotides. The components are then heated to denature the templates and allowed to anneal to new probes at a lower temperature. In the original procedure, fresh DNA ligase was added at the end of each round and the material was incubated for 30 min to 5 h (163). Repetition of this process resulted in a doubling of the ligation products, leading to a geometric accumulation (Fig. 8).

Potential problems that limited the usefulness of ligase-dependent amplification, as it was originally described, were high levels of background amplification due to blunt-end ligation of probe duplexes (9, 163), the need to add ligase after each amplification cycle, the relatively long reaction times (especially in the early cycles of amplification), carryover contamination, and relatively nonstringent annealing conditions, which compounded the problems of high background amplification. Barany (5) recently described the use of a thermophilic DNA ligase isolated from *T. aquaticus* that remains active during the thermal cycling steps

**Figure 7.** Qβ replicase-based probe amplification. A Qβ replicase substrate which contains both a target-specific probe region and an endonuclease cleavage site formed by intramolecular base pairing of the probe region is produced. When the molecule is annealed to the target sequence, it is RNase III resistant. Unbound probe is treated with RNase and then washed from the reaction mixture by target cycling (73). Qβ replicase is then added to the washed target-probe complex, and the reaction is incubated at 37°C, resulting in amplification of the probe.

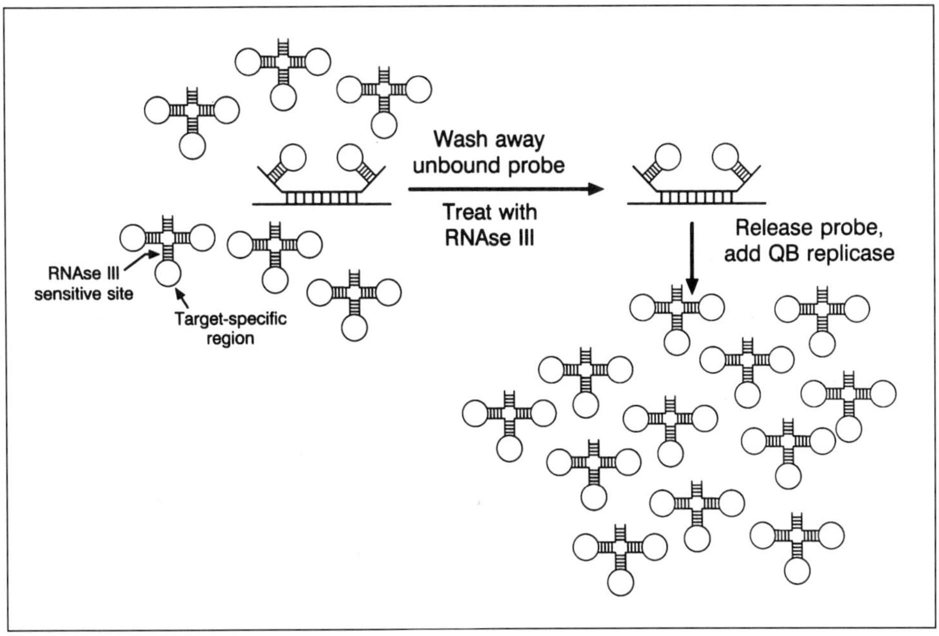


required for ligase-dependent amplification. This greatly simplified the LCR process because, like PCR, the enzyme can be added once at the beginning of the reaction. Furthermore, the problem of blunt-end ligation, precipitated largely by the low annealing temperature used, is substantially decreased because probe annealing can be performed at higher temperatures. These improvements in the basic methodology will greatly accelerate development of amplification-based tests based on ligation. Target-dependent ligation amplification is now commonly referred to as the ligase amplification reaction (163) or the ligase chain reaction (LCR) (6).

In the last 2 years, LCR was used for detection of specific pathogens: *Neisseria gonorrhoeae* (15), *B. burgdorferi* (72), and *Mycobacterium* species (76). LCR techniques do not copy new sequence information into the amplification products, so they are classified as probe amplification methods. However, many applications for detection of infectious agents do not require this capability; simple detection of a nucleotide signature sequence suffices in many cases. Although LCR-based methods cannot directly detect RNA targets, prior reverse transcription into cDNA results in a detectable target, thus further expanding the range of potential applications.

Target-directed ligation methods, either alone or in combination with a target amplification method, are potentially useful for the detection of point mutations. The oligonucleotides can be designed so that their point of ligation straddles the nucleotide position of a known mutation. Theoretically, efficient ligation and eventual detection of a ligation product occur only when the ends of the probes are perfectly base paired to the target sequence (5).

**Figure 8.** LCR. Oligonucleotide probes are annealed to template molecules in a head-to-tail fashion, with the 3′ end of one probe abutting the 5′ end of the second. DNA ligase joins the adjacent 3′ and 5′ ends to form a duplicate of one strand of the target. A second primer set, complementary to the first, then uses this duplicated strand (as well as the original target) as a template for ligation. Repeating the process results in a logarithmic accumulation of ligation products, which can be detected via the functional groups attached to the oligonucleotides.

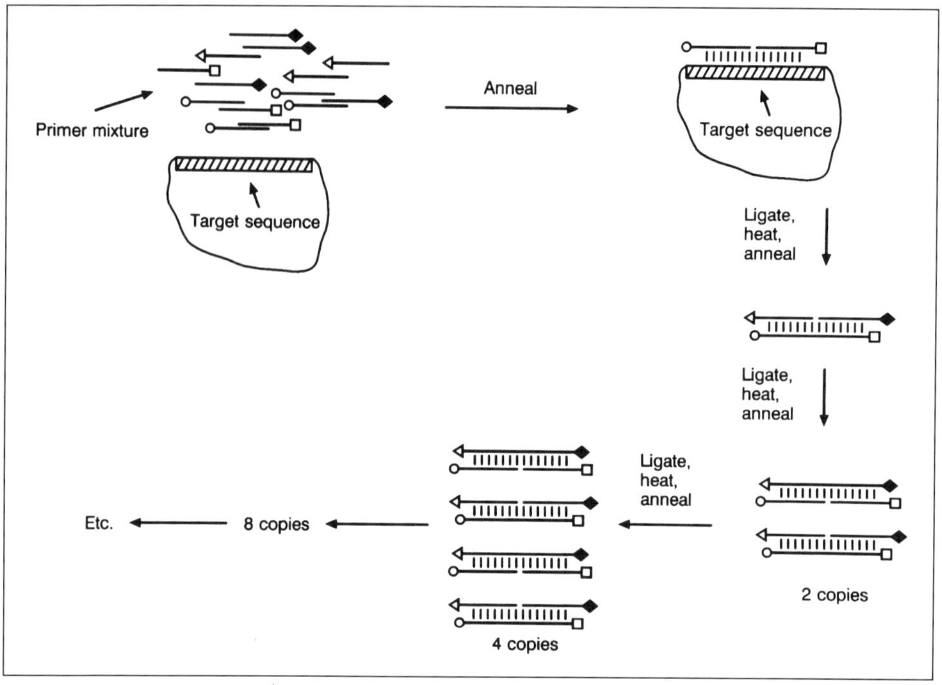


As with target amplification systems, the oligonucleotide probes used in ligation detection assays can be modified to aid in the detection of the amplification product. If one probe is synthesized with an affinity label such as biotin at its 5′ end and the other probe is labeled with a reporter group such as alkaline phosphatase or a fluorescent compound, only the ligation product incorporates both moieties. After the amplification reaction has been run over a streptavidin matrix and the mixture has been washed to eliminate unligated reporter groups, the bound material can be measured by assaying for enzyme activity or fluorescence. This approach, originally described by Landegren et al. (88) and termed the oligonucleotide ligation assay, may form the basis of an automated method for the detection of target sequences and nucleotide base changes.

## Signal Amplification Techniques

For the purpose of increasing the sensitivity of hybridization-based tools, an alternative to enzymatic duplication or ligation of nucleic acids is to increase the signal generated from the probe molecule itself. An increase in the signal generated from nucleic acid probes is theoretically possible if the reporter group can be attached in greater numbers to the probe molecule or if the signal intensity generated by each reporter is increased. Signal amplification methods, unlike probe or target amplification techniques, do not result in a replicable product and thus are less prone to product contamination (see below). However, the detection sensitivity ($10^4$ to $10^5$ target molecules) that can be achieved by signal amplification is generally lower than for enzymatic amplification methods, and thus the use of signal amplification may be limited in samples containing low target copy numbers.

## Compound Probes

Compound probes contain two functional components: a primary probe, recognizing the target sequence and containing multiple binding sites for secondary probes, and the secondary probes themselves, each of which contains a reporter group (47, 164). Incubation of these components with target DNA sequences results in the formation of a compound probe network, in which the primary probe is bound to its target via its target-specific region, leaving the remaining sequence available for binding by secondary probes that make up the reporter moiety. The probe trees that are generated can assume various linear or circular forms, but the net result is the same: the signal-generating capability of the probe is significantly increased. The practical outcome of this approach is increased sensitivity for detection of a target sequence. Of course, nonspecific hybridization signals are also increased in intensity, leading to higher levels of background. One approach to decreasing background would be to use a target capture method (73).

## Branched Probes

A second type of signal amplification system was developed by the Chiron Corporation and uses branched DNA (bDNA) to provide multiple hybridization sites for an enzyme-coupled reporter molecule (137, 141). The complex based on bDNA is referred to as an amplification multimer. The precursor molecules of amplification multimers are chemically synthesized. The backbone of their comb structure is built via a branching nucleoside analog that can be incorporated at regular intervals during oligonucleotide synthesis. Each branching site is then extended with a short oligonucleotide fragment, to which a longer target-specific region can be ligated.

Signal amplification by bDNA incorporates several simultaneous hybridization steps. After disruption of the target organism, target-specific oligonucleotide probes are used to capture the target nucleic acid to a solid support. A second set of target-specific probes (extenders) hybridize to adjoining sequences and also serve as binding sites (via 5′ extensions) for the branched amplification multimer. The branches of the latter then direct the binding of multiple alkaline phosphatase-labeled oligonucleotides. As many as 3,000 enzyme labels can be incorporated onto each target molecule in this manner. bDNA detection (Fig. 9) is very sensitive; several studies estimated the sensitivity of this system to be in the range of $10^3$ to $10^5$ target molecules.

The discontinuous design of capture probes and the target-specific extender probes makes the bDNA approach well suited to detecting a nucleic acid target with sequence heterogeneity, as is the case for hepatitis C virus and HIV. If one or two of the capture or extender probes fail to hybridize because of sequence variation, signal-generating capacity is not lost, because of the presence of several remaining probe complexes. As for the target amplification techniques, the requirement for multiple, independent probe-target hybridization events provides specificity. Nonspecific hybridization rarely allows hybridization of both capture and extender probes to the same nontarget sequence.

**Figure 9.** bDNA-based signal amplification. Target nucleic acid is released by disruption and captured to a solid surface via multiple contiguous capture probes. Contiguous extender probes hybridize with adjacent target sequences and contain additional sequences homologous to the branched amplification multimer. Enzyme-labeled oligonucleotides bind to the bDNA via homologous base pairing, and the enzyme-probe complex is measured by detection of chemiluminescence. All hybridization reactions occur simultaneously.

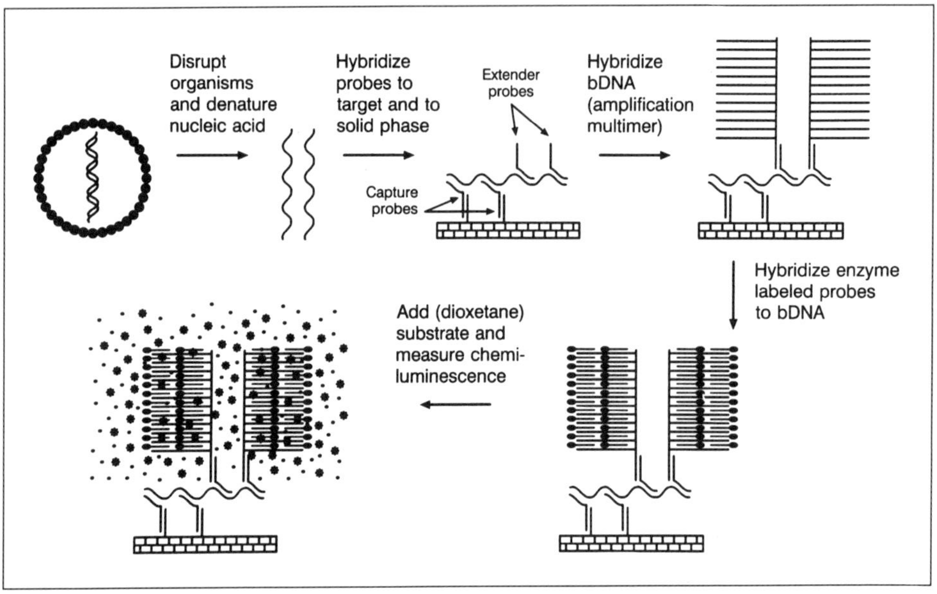


bDNA assays are being developed for hepatitis B and C viruses (39, 155), HIV-1 (116), and cytomegalovirus (139), for which specific antiviral therapies are available. Because bDNA probes provide quantitative detection over a range of several orders of magnitude, such assays may be useful for monitoring therapeutic responses to alpha interferon (hepatitis B and C viruses), azidothymidine (HIV), and ganciclovir (cytomegalovirus). The determination of responses to antiviral therapy may be useful in assessing the need for further therapy or for adjusting doses of antiviral agents.

## Potential Problems with Amplification Systems

The expanded roles of nucleic acid amplification technologies in the clinical laboratory will be accompanied by a new set of problems, many of which will result from the technologies themselves. Examples of such concerns include false positivity due to contaminating nucleic acids (chapter 5), the problem of oversensitivity (brought on by the extreme sensitivity of these techniques coupled with the presence of small numbers of a pathogenic organism in unaffected individuals) (121), potential problems in monitoring therapy (because of their potential ability to detect nonviable organisms) (93, 109), and inhibition by copurifying substances within clinical specimens of the complex enzymatic machinery that underlies many of the methods (35, 36, 82). Of all these problems, contamination by previously amplified nucleic acids is the most formidable (chapter 5). For diagnostic applications, product contamination represents the greatest impediment to routine implementation (118).

Each of the enzyme-based amplification methods (i.e., target and probe amplification methods) described in this chapter is subject to false-positive results due to amplification product carryover. In PCR, LCR, and SDA, DNA products are the main source of contamination, whereas in 3SR and Qβ replicase protocols, RNA products predominate. Fortunately, at least for PCR, several product inactivation methods that help circumvent this problem have been described (chapter 5). The application of some form of amplification product inactivation will probably be necessary for all amplification-based detection systems if quality control problems are to be avoided. However, even if amplification products can be contained successfully with these systems, the problem of false-positive results will probably persist. The exquisite sensitivity of amplification techniques makes them prone to false-positive results due to sample-to-sample carryover contamination and to stray nucleic acids circulating in areas where clinical samples are processed or aliquots are prepared. The best formula for dealing with this problem is first to realize that it exists and then to design assays and procedures such that problems can be quickly identified and remedied. Methods for controlling the build-up of amplification products are discussed in more detail in chapter 5.

## Future Challenges

The technology behind nucleic acid amplification is constantly changing. Consequently, it is difficult to predict with certainty what future roles these techniques will have in the clinical laboratory. When one considers further the need to control the rising costs of medical care, much of which is driven by technology, the future of these technologies seems even more tenuous. The initial use of nucleic acid amplification methods will be costly and limited to larger medical centers. However, the eventual broad-based acceptance of amplification techniques will decrease the associated costs via economies of scale, and increasing automation of sample extraction, amplification, and detection will greatly decrease the associated labor. Indeed, the implementation of such tests may actually save money. Rapid tests for identification of mycobacteria (protocols 1.1 and 1.2) may lead to lower expenditures by eliminating unnecessary respiratory isolation of patients. Sensitive detection of genes specifying methicillin resistance in staphylococci (protocol 6.4) may lead to treatment with less costly antibiotics for the majority of infections. Early recognition of herpes simplex virus DNA in cerebrospinal fluid obtained for patients with encephalopathy will provide an opportunity for early antiviral therapy and may eliminate the need for invasive diagnostic procedures (protocols 2.5 and 2.6). Molecular detection of mutations associated with multidrug-resistant tuberculosis (protocol 6.1) will very probably revolutionize the identification of such strains and lead to more timely and effective public health control measures.

With the increased implementation of nucleic acid amplification technology, an array of new problems will probably surface, including some already mentioned in this chapter. With the resolution of these new problems, however, old problems will probably resurface. Addressing the perennial problems associated with sample collection (chapter 9), transport, and preparation (82) (chapter 6) will be a major task of the clinical laboratory of the future, as will the establishment of guidelines for monitoring antimicrobial therapy.

The implementation of meaningful quality control programs will be a significant challenge. Because of the high sensitivity of these techniques, samples containing low copy numbers of target organisms should be included in quality control programs. Unfortunately, the preservation of small numbers of organisms in a quality control test sample will probably prove to be the most unreliable aspect of such a program, making difficult the large-scale administration of quality control programs via independent agencies. The next few years will also witness a growing need for laboratory standardization. Quality control problems can theoretically be kept to a minimum if procedural guidelines, such as those prepared by the National Committee for Clinical Laboratory Standards, can be developed for molecular diagnostic methods.

The greatest challenges ahead for the clinical laboratory are in the area of education. To remain at the forefront of technology, the clinical microbiology laboratory staff must rely to an increasing extent on an understanding of nucleic acid chemistry. Training programs for medical technologists must provide an appropriate theoretical framework and incorporate hands-on training in molecular methods. Fellowship and residency programs should devote substantial time to laboratory experience in molecular diagnostics, because those currently enrolled in these programs will be confronted with the new technology when they complete their training. Research training in molecular methods is helpful but not sufficient; amplification methods applied in a diagnostic laboratory are different in both style and substance from their research counterparts. Continuing-education programs that are tailored to the new technology should be developed and should provide continuous updates for this rapidly evolving field.

Despite many obstacles that are difficult to foresee, the prospects for nucleic acid amplification technology in the clinical laboratory remain excellent. The ability to provide more definitive diagnostic information and to greatly speed up the diagnosis of slow-growing or uncultured organisms ultimately will have a profound effect on patient management and on our understanding of infectious processes. The application of these techniques to the discovery of new organisms and then to the identification of their associations with disease will fuel investigations for decades to come. However, the use of these techniques probably will not remain limited to the detection of unusual pathogens. As nucleic acid amplification chemistry becomes more automated, accessible, and affordable, the eventual application of these methods to both esoteric and routine functions of the clinical microbiology laboratory will trigger the greatest expansion of our diagnostic repertoire in recent years.

## References


1. **Amann, R., N. Springer, W. Ludwig, H.-D. Görtz, and K.-H. Schleifer.** 1991. Identification in situ and phylogeny of uncultured bacterial endosymbionts. *Nature* (London) **351:**161–164.
2. **Arnheim, N., T. White, and W. E. Rainey.** 1990. Application of PCR: organismal and population biology. *BioScience* **40:**174–182.
3. **Arnoldi, J., C. Schlüter, M. Duchrow, L. Hübner, M. Ernst, A. Teske, H.-D. Flad, J. Gerdes, and E. C. Böttger.** Species-specific assessment of *Mycobacterium leprae* in skin biopsies by in situ hybridization and polymerase chain reaction. *Lab. Invest.* **66:**618–623.
4. **Azad, A. F., J. B. Sacci, Jr., W. M. Nelson, G. A. Dasch, E. T. Schmidtmann, and M. Carl.** 1992. Genetic characterization and transovarial transmission of a typhus-like rickettsia found in cat fleas. *Proc. Natl. Acad. Sci. USA* **89:**43–46.
5. **Barany, F.** 1991. Genetic disease detection and DNA amplification using cloned thermostable ligase. *Proc. Natl. Acad. Sci. USA* **88:**189–193.
6. **Barany, F.** 1991. The ligase chain reaction (LCR) in a PCR world. *PCR Methods Applic.* **1:**5–16.

7. **Barinaga, M.** 1991. Will "DNA chip" speed genome initiative? Science **253:**1489. (Editorial.)
8. **Barker, R. H., Jr., T. Banchongaksorn, J. M. Courval, W. Suwonkerd, K. Rimwungtragoon, and D. F. Wirth.** 1992. A simple method to detect *Plasmodium falciparum* directly from blood samples using the polymerase chain reaction. *Am. J. Trop. Med. Hyg.* **46:**416–426.
9. **Barringer, K. J., L. Orgel, G. Wahl, and T. R. Gingeras.** 1990. Blunt-end and single-strand ligations by *Escherichia coli* ligase: influence on an in vitro amplification scheme. *Gene* **89:**117–122.
10. **Barry, T., G. Colleran, M. Glennon, L. K. Dunican, and F. Gannon.** 1991. The 16S/23S ribosomal spacer region as a target for DNA probes to identify eubacteria. *PCR Methods Applic.* **1:**51–56.
11. **Bauer, H. M., Y. Ting, C. E. Greer, J. C. Chambers, C. J. Tashiro, J. Chimera, A. Reingold, and M. M. Manos.** 1991. Genital human papillomavirus infection in female university students as determined by a PCR-based method. *JAMA* **265:**472–477.
12. **Bej, A. K., M. H. Mahbubani, and R. M. Atlas.** 1991. Amplification of nucleic acids by polymerase chain reaction (PCR) and other methods and their applications. *Crit. Rev. Biochem. Mol. Biol.* **26:**301–334.
13. **Bej, A. K., S. C. McCarty, and R. M. Atlas.** 1991. Detection of coliform bacteria and *Escherichia coli* by multiplex polymerase chain reaction: comparison with defined substrate and plating methods for water quality monitoring. *Appl. Environ. Microbiol.* **57:**2429–2432.
14. **Bej, A. K., R. J. Steffan, J. DiCesare, L. Haff, and R. M. Atlas.** 1990. Detection of coliform bacteria in water by polymerase chain reaction and gene probes. *Appl. Environ. Microbiol.* **56:**307–314.
15. **Birkenmeyer, L., W. Bringer, and A. Armstrong.** 1992. Specific detection of *Neisseria gonorrhoeae* using the ligase chain reaction, p. 455. *Abstr. 92nd Gen. Meet. Am. Soc. Microbiol. 1992.* American Society for Microbiology, Washington, D.C.
16. **Biswas, B., D. Mukherjee, B. L. Mattingly-Napier, and S. K. Dutta.** 1991. Diagnostic application of polymerase chain reaction for detection of *Ehrlichia risticii* in equine monocytic ehrlichiosis (Potomac horse fever). *J. Clin. Microbiol.* **29:**2228–2233.
17. **Böttger, E. C.** 1989. Rapid determination of bacterial ribosomal RNA sequences by direct sequencing of enzymatically amplified DNA. *FEMS Microbiol. Lett.* **65:**171–176.
18. **Boucher, C. A. B., M. Tersmette, J. M. A. Lange, P. Kellam, R. E. Y. de Goede, J. W. Mulder, G. Darby, J. Goudsmit, and B. A. Larder.** 1990. Zidovudine sensitivity of human immunodeficiency viruses from high-risk symptom-free individuals during therapy. *Lancet* **336:**585–590.
19. **Bowman, B., J. W. Taylor, and T. J. White.** 1992. Molecular evolution of the fungi: human pathogens. *Mol. Biol. Evol.* **9:**893–904.
20. **Brakstad, O. G., K. Aasbakk, and J. A. Maeland.** 1992. Detection of *Staphylococcus aureus* by polymerase chain reaction amplification of the *nuc* gene. *J. Clin. Microbiol.* **30:**1654–1660.
21. **Brian, M. J., M. Frosolono, B. E. Murray, A. Miranda, E. L. Lopez, H. F. Gomez, and T. G. Cleary.** 1992. Polymerase chain reaction for diagnosis of enterohemorrhagic *Escherichia coli* infection and hemolytic-uremic syndrome. *J. Clin. Microbiol.* **30:**1801–1806.
22. **Bubert, A., S. Köhler, and W. Goebel.** 1992. The homologous and heterologous regions within the *iap* gene allow genus- and species-specific identification of *Listeria* spp. by polymerase chain reaction. *Appl. Environ. Microbiol.* **58:**2625–2632.
23. **Buchbinder, A., S. F. Josephs, D. Ablashi, S. Z. Salahuddin, M. E. Klotman, M. Manak, G. R. Krueger, F. Wong-Staal, and R. C. Gallo.** 1988. Polymerase chain reaction amplification and in situ hybridization for the detection of human B-lymphotropic virus. *J. Virol. Methods* **21:**191–197.
24. **Burg, J. L., C. M. Grover, P. Pouletty, and J. C. Boothroyd.** 1989. Direct and sensitive

detection of a pathogenic protozoan, *Toxoplasma gondii*, by polymerase chain reaction. *J. Clin. Microbiol.* **27:**1787–1792.

25. **Cahill, P., K. Foster, and D. E. Mahan.** 1991. Polymerase chain reaction and Qβ replicase amplification. *Clin. Chem.* **37:**1482–1485.
26. **Carl, M., R. Hawkins, N. Coulson, J. Lowe, D. L. Robertson, W. M. Nelson, R. W. Titball, and J. N. Woody.** 1992. Detection of spores of *Bacillus anthracis* using the polymerase chain reaction. *J. Infect. Dis.* **165:**1145–1148.
27. **Carl, M., C. W. Tibbs, M. E. Dobson, S. Paparello, and G. A. Dasch.** 1990. Diagnosis of acute typhus infection using the polymerase chain reaction. *J. Infect. Dis.* **161:**791–793.
28. **Cha, R. S., H. Zarbl, P. Keohavong, and W. G. Thilly.** 1992. Mismatch amplification mutation assay (MAMA): application to the c-H-ras gene. *PCR Methods Applic.* **2:**14–20.
29. **Chamberlain, J. S., R. A. Gibbs, J. E. Rainier, and C. T. Caskey.** 1990. Multiplex PCR for the diagnosis of Duchenne muscular dystrophy, p. 272–281. *In* M. A. Innis, D. H. Gelfand, J. J. Sninsky, and T. J. White (ed.), *PCR Protocols: A Guide to Methods and Applications*. Academic Press, Inc., San Diego, Calif.
30. **Chen, K., H. Neimark, P. Rumore, and C. R. Steinman.** 1989. Broad range DNA probes for detecting and amplifying eubacterial nucleic acids. *FEMS Microbiol. Lett.* **57:**19–24.
31. **Chiu, K. P., S. Cohen, D. Morris, and G. Jordan.** 1992. Intracellular amplification of proviral DNA in tissue sections using the polymerase chain reaction. *J. Histochem. Cytochem.* **40:**333–341.
32. **Claas, H. C. J., J. H. T. Wagenvoort, H. G. M. Niesters, T. T. Tio, J. H. Van Rijsoort-Vos, and W. G. V. Quint.** 1991. Diagnostic value of the polymerase chain reaction for *Chlamydia* detection as determined in a follow-up study. *J. Clin. Microbiol.* **29:**42–45.
33. **Cleuziat, P., and J. Robert-Baudouy.** 1990. Specific detection of *Escherichia coli* and *Shigella* species using fragments of genes coding for beta-glucuronidase. *FEMS Microbiol. Lett.* **60:**315–322.
34. **Compton, J.** 1991. Nucleic acid sequence-based amplification. *Nature* (London) **350:**91–92.
35. **de Franchis, R., N. C. P. Cross, N. S. Foulkes, and T. M. Cox.** 1988. A potent inhibitor of *Taq* polymerase copurifies with human genomic DNA. *Nucleic Acids Res.* **16:**10355.
36. **de Lomas, J. G., F. J. Sunzeri, and M. P. Busch.** 1992. False-negative results by polymerase chain reaction due to contamination by glove powder. *Transfusion* **32:**83–85.
37. **DeLong, E. F., G. S. Wickham, and N. R. Pace.** 1989. Phylogenetic stains: ribosomal RNA-based probes for the identification of single cells. *Science* **243:**1360–1363.
38. **Demmler, G. J., G. J. Buffone, C. M. Schimbor, and R. A. May.** 1988. Detection of cytomegalovirus in urine from newborns by using polymerase chain reaction DNA amplification. *J. Infect. Dis.* **158:**1177–1184.
39. **Douglas, D. D., J. Rakela, H. F. Taswell, D. Rabe, J. Detmer, W. Hunt, M. S. Urdea, and D. H. Persing.** 1992. Correlation of quantitative HCV RNA, anti-HCV IgM, and serum ALT with histologic presentation. American Association for the Study of Liver Diseases, Chicago, Ill.
40. **Dupont vs. Hoffman-La Roche.** 1991. San Francisco Federal Courts, cases 4683195 and 4683202. 1991.
41. **Eakin, A. E., J. Bouvier, J. A. Sakanari, C. S. Craik, and J. H. McKerrow.** 1990. Amplification and sequencing of genomic DNA fragments encoding cysteine proteases from protozoan parasites. *Mol. Biochem. Parasitol.* **43:**1–8.
42. **Eisenstein, B. I.** 1990. New opportunistic infections—more opportunities. *N. Engl. J. Med.* **323:**1625–1627. (Editorial.)

43. **Eisenstein, B. I.** 1990. The polymerase chain reaction: a new method of using molecular genetics for medical diagnosis. *N. Engl. J. Med.* **322:**178–183.
44. **Erlich, H. A., D. Gelfand, and J. J. Sninsky.** 1991. Recent advances in the polymerase chain reaction. *Science* **252:**1643–1651.
45. **Fägerstam, L. G., A. Frostell, R. Karlsson, M. Kullman, A. Larsson, M. Malmqvist, and H. Butt.** 1990. Detection of antigen-antibody interactions by surface plasmon resonance: application to epitope mapping. *J. Mol. Recogn.* **3:**208–214.
46. **Fahrimal, Y., W. L. Goff, and D. P. Jasmer.** 1992. Detection of *Babesia bovis* carrier cattle by using polymerase chain reaction amplification of parasite DNA. *J. Clin. Microbiol.* **30:**1374–1379.
47. **Fahrlander, P. D., and A. Klausner.** 1988. Amplifying DNA probe signals: a "Christmas tree" approach. *Bio/Technology* **6:**1165–1168.
48. **Fahy, E., D. Y. Kwoh, and T. R. Gingeras.** 1991. Self-sustained sequence replication (3SR): an isothermal transcription-based amplification system alternative to PCR. *PCR Methods Applic.* **1:**25–32.
49. **Feray, C., D. Samuel, V. Thiers, M. Gigou, F. Pichon, A. Bismuth, M. Reynes, P. Maisonneuve, H. Bismuth, and C. Bréchot.** 1992. Reinfection of liver graft by hepatitis C after liver transplantation. *J. Clin. Invest.* **89:**1361–1365.
50. **Ferre, F.** 1992. Quantitative or semi-quantitative PCR: reality versus myth. *PCR Methods Applic.* **2:**1–9.
51. **Figueroa, J. V., L. P. Chieves, G. S. Johnson, and G. M. Buening.** 1992. Detection of *Babesia bigemina*-infected carriers by polymerase chain reaction amplification. *J. Clin. Microbiol.* **30:**2576–2582.
52. **Frothingham, R., R. L. Allen, and K. H. Wilson.** 1991. Rapid 16S ribosomal DNA sequencing from a single colony without DNA extraction or purification. *BioTechniques* **11:**40–44.
53. **Gama, R. E., P. J. Hughes, C. B. Bruce, and G. Stanway.** 1988. Polymerase chain reaction amplification of rhinovirus nucleic acids from clinical material. *Nucleic Acids Res.* **16:**9346. (Abstract.)
54. **Gibbs, R. A., P.-N. Nguyen, and C. T. Caskey.** 1989. Detection of single DNA base differences by competitive oligonucleotide priming. *Nucleic Acids Res.* **17:**2437–2448.
55. **Gilliland, G., S. Perrin, K. Blanchard, and H. F. Bunn.** 1990. Analysis of cytokine mRNA and DNA: detection and quantitation by competitive polymerase chain reaction. *Proc. Natl. Acad. Sci. USA* **87:**2725–2729.
56. **Gingeras, T. R., and D. Y. Kwoh.** 1992. In vitro nucleic acid target amplification techniques: issues and benefits. *Praxis Biotechnol.* **4:**403–429.
57. **Gingeras, T. R., P. Prodanovich, T. Latimer, J. C. Guatelli, D. D. Richman, and K. J. Barringer.** 1991. Use of self-sustained sequence replication amplification reaction to analyze and detect mutations in zidovudine-resistant human immunodeficiency virus. *J. Infect. Dis.* **164:**1066–1074.
58. **Glare, E. M., J. C. Paton, R. R. Premier, A. J. Lawrence, and I. T. Nisbet.** 1990. Analysis of a repetitive DNA sequence from *Bordetella pertussis* and its application to the diagnosis of pertussis using the polymerase chain reaction. *J. Clin. Microbiol.* **28:**1982–1987.
59. **Goodman, J. L., P. Jurkovich, J. M. Kramber, and R. C. Johnson.** 1991. Molecular detection of persistent *Borrelia burgdorferi* in the urine of patients with active Lyme disease. *Infect. Immun.* **59:**269–278.
60. **Greer, C. E., S. L. Peterson, N. B. Kiviat, and M. M. Manos.** 1991. PCR amplification from paraffin-embedded tissues: effects of fixative and fixation time. *Am. J. Clin. Pathol.* **95:**117–124.
61. **Grover, C. M., P. Thulliez, J. S. Remington, and J. C. Boothroyd.** 1990. Rapid prenatal diagnosis of congenital *Toxoplasma* infection by using polymerase chain reaction and amniotic fluid. *J. Clin. Microbiol.* **28:**2297–2301.
62. **Guatelli, J. C., T. R. Gingeras, and D. D. Richman.** 1989. Nucleic acid amplification

in vitro: detection of sequences with low copy numbers and application to diagnosis of human immunodeficiency virus type 1 infection. *Clin. Microbiol. Rev.* **2:**217–226.

63. **Guatelli, J. C., K. M. Whitfield, D. Y. Kwoh, K. J. Barringer, D. D. Richman, and T. R. Gingeras.** 1990. Isothermal, *in vitro* amplification of nucleic acids by a multienzyme reaction modeled after retroviral replication. *Proc. Natl. Acad. Sci. USA* **87:**1874–1878.
64. **Gyllensten, U. B., and H. A. Erlich.** 1988. Generation of single-stranded DNA by the polymerase chain reaction and its application to direct sequencing of the *HLA-DQA* locus. *Proc. Natl. Acad. Sci. USA* **85:**7652–7656.
65. **Haase, A. T., E. Retzel, and K. Staskus.** 1990. Amplification and detection of lentiviral DNA inside cells. *Proc. Natl. Acad. Sci. USA* **87:**497–4975.
66. **Hance, A. J., B. Grandchamp, V. Lévy-Frébault, D. Lecossier, J. Rauzier, D. Bocart, and B. Gicquel.** 1989. Detection and identification of mycobacteria by amplification of mycobacterial DNA. *Mol. Microbiol.* **3:**843–849.
67. **Hartskeerl, R. A., M. Y. L. De Wit, and P. R. Klatser.** 1989. Polymerase chain reaction for the detection of *Mycobacterium leprae. J. Gen. Microbiol.* **135:**2357–2364.
68. **Hendricks, D. A., B. S. Hoo, B. J. Rathbun, M. Yano, and P. D. Neuwald.** 1992. Quantitative detection of HBV DNA in human sera, p. 406. *Abstr. 92nd Gen. Meet. Am. Soc. Microbiol. 1992.* American Society for Microbiology, Washington, D.C.
69. **Hewitt, P. L., L. Marselle, W. King, J. Shaw, A. Young, K. Borcherts, C. Williams, D. Mahan, and D. N. Halbert.** 1990. A prototype assay for the detection of *Chlamydia trachomatis* by nucleic acid hybridization and Q-beta amplification, p. 356. *Abstr. 90th Annu. Meet. Am. Soc. Microbiol. 1990.* American Society for Microbiology, Washington, D.C.
70. **Hookey, J. V.** 1992. Detection of *Leptospiraceae* by amplification of 16S ribosomal DNA. *FEMS Microbiol. Lett.* **69:**267–274.
71. **Houard, S., C. Hackel, A. Herzog, and A. Bollen.** 1989. Specific identification of *Bordetella pertussis* by the polymerase chain reaction. *Res. Microbiol.* **140:**477–487.
72. **Hu, H., K. Elmore, I. Facey, and D. Jenderzak.** 1991. Detection of *Borrelia burgdorferi* by ligase chain reaction, p. 79. *Abstr. 91st Gen. Meet. Am. Soc. Microbiol. 1991.* American Society for Microbiology, Washington, D.C.
73. **Hunsaker, W. R., H. Badri, M. Lombardo, and M. L. Collins.** 1989. Nucleic acid hybridization assays employing dA-tailed capture probes. II. Advanced multiple capture methods. *Anal. Biochem.* **181:**360–370.
74. **Hussong, D., R. R. Colwell, M. O'Brien, E. Weiss, A. D. Pearson, R. M. Weiner, and W. D. Burge.** 1987. Viable *Legionella pneumophila* not detectable by culture on agar media. *Bio/Technology* **5:**947–950.
75. **Ibrahim, A., W. Liesack, and E. Stackebrandt.** 1992. Polymerase chain reaction-gene probe detection system specific for pathogenic strains of *Yersinia enterocolitica. J. Clin. Microbiol.* **30:**1942–1947.
76. **Iovannisci, D. M., and E. S. Winn-Deen.** 1991. Ligation amplification and fluorescence detection of *Mycobacterium tuberculosis* DNA, p. 85. *Abstr. 91st Gen. Meet. Am. Soc. Microbiol. 1991.* American Society for Microbiology, Washington, D.C.
77. **Jansen, R. W., G. Siegl, and S. M. Lemon.** 1990. Molecular epidemiology of human hepatitis A virus defined by an antigen-capture polymerase chain reaction method. *Proc. Natl. Acad. Sci. USA* **87:**2867–2871.
78. **Jaton, K., R. Sahli, and J. Bille.** 1992. Development of polymerase chain reaction assays for detection of *Listeria monocytogenes* in clinical cerebrospinal fluid samples. *J. Clin. Microbiol.* **30:**1931–1936.
79. **Jensen, J. S., S. A. Uldum, J. Søndergåard-Andersen, J. Vuust, and K. Lind.** 1991. Polymerase chain reaction for detection of *Mycoplasma genitalium* in clinical samples. *J. Clin. Microbiol.* **29:**46–50.
80. **Karch, H., and T. Meyer.** 1989. Single primer pair for amplifying segments of distinct

Shiga-like-toxin genes by polymerase chain reaction. *J. Clin. Microbiol.* **27:**2751–2757.

81. **Kato, N., C.-Y. Ou, H. Kato, S. L. Bartley, V. K. Brown, V. R. Dowell, Jr., and K. Ueno.** 1991. Identification of toxigenic *Clostridium difficile* by the polymerase chain reaction. *J. Clin. Microbiol.* **29:**33–37.
82. **Kawasaki, E. S.** 1990. Sample preparation from blood, cells, and other fluids, p. 146–152. *In* M. A. Innis, D. H. Gelfand, J. J. Sninsky, and T. J. White (ed.), *PCR Protocols: A Guide to Methods and Applications*. Academic Press, Inc., San Diego, Calif.
83. **Kemker, B. J., P. F. Lehmann, J. W. Lee, and T. J. Walsh.** 1991. Distinction of deep versus superficial clinical and nonclinical isolates of *Trichosporon beigelii* by isoenzymes and restriction fragment length polymorphisms of rDNA generated by polymerase chain reaction. *J. Clin. Microbiol.* **29:**1677–1683.
84. **Kleppe, K., E. Ohtsuka, R. Kleppe, R. Molineux, and H. G. Khorana.** 1971. Studies on polynucleotides. XCVI. Repair replication of short synthetic DNA's as catalyzed by DNA polymerases. *J. Mol. Biol.* **56:**341–361.
85. **Kramer, F. R., and P. M. Lizardi.** 1989. Replicatable RNA reporters. *Nature* (London) **339:**401–402.
86. **Kristiansen, B.-E., E. Ask, A. Jenkins, C. Fermer, P. Rådstrøm, and O. Skøld.** 1991. Rapid diagnosis of meningococcal meningitis by polymerase chain reaction. *Lancet* **337:**1568–1569.
87. **Kwoh, D. Y., G. R. Davis, K. M. Whitfield, H. L. Chappelle, L. J. DiMichele, and T. R. Gingeras.** 1989. Transcription-based amplification system and detection of amplified human immunodeficiency virus type 1 with a bead-based sandwich hybridization format. *Proc. Natl. Acad. Sci. USA* **86:**1173–1177.
88. **Landegren, U., R. Kaiser, J. Sanders, and L. Hood.** 1988. A ligase-mediated gene detection technique. *Science* **241:**1077–1080.
89. **Larzul, D., F. Guigue, J. J. Sninsky, D. H. Mack, C. Bréchot, and J.-L. Guesdon.** 1988. Detection of hepatitis B virus sequences in serum by using in vitro enzymatic amplification. *J. Virol. Methods* **20:**227–237.
90. **Liesack, W., and E. Stackebrandt.** 1992. Occurrence of novel groups of the domain *Bacteria* as revealed by analysis of genetic material isolated from an Australian terrestrial environment. *J. Bacteriol.* **174:**5072–5078.
91. **Lizardi, P. M., C. E. Guerra, H. Lomeli, I. Tussie-Luna, and F. R. Kramer.** 1988. Exponential amplification of recombinant-RNA hybridization probes. *Bio/Technology* **6:**1197–1202.
92. **Lomeli, H., S. Tyagi, C. G. Pritchard, P. M. Lizardi, and F. R. Kramer.** 1989. Quantitative assays based on the use of replicatable hybridization probes. *Clin. Chem.* **35:**1826–1831.
93. **Malawista, S. E., R. T. Schoen, T. L. Moore, D. E. Dodge, T. J. White, and D. H. Persing.** 1992. Failure of multitarget detection of *Borrelia burgdorferi*-associated DNA sequences in synovial fluids of patients with juvenile rheumatoid arthritis: a cautionary note. *Arthritis Rheum.* **35:**246–247.
94. **Malloy, D. C., R. K. Nauman, and H. Paxton.** 1990. Detection of *Borrelia burgdorferi* using the polymerase chain reaction. *J. Clin. Microbiol.* **28:**1089–1093.
95. **Marshall, W. F., A. Telenti, J. Proper, A. J. Aksamit, and T. F. Smith.** 1991. Survey of urine from transplant recipients for polyomaviruses JC and BK using the polymerase chain reaction. *Mol. Cell. Probes* **5:**125–128.
96. **McBride, L. J., and M. D. O'Neill.** 1991. Automated analysis of mutations responsible for genetic diseases in humans. *Am. Lab.* **1991**(Nov.)**:**52–59.
97. **McLaughlin, G. L., M. H. Vodkin, and H. W. Huizinga.** 1991. Amplification of repetitive DNA for the specific detection of *Naegleria fowleri*. *J. Clin. Microbiol.* **29:**227–230.
98. **McMillin, D. E., L. L. Muldrow, and S. J. Laggette.** 1992. Simultaneous detection of

toxin A and toxin B genetic determinants of *Clostridium difficile* using the multiplex polymerase chain reaction. *Can. J. Microbiol.* **38:**81–83.
99. **Mills, J. A.** 1989. Do bacteria cause chronic polyarthritis? *N. Engl. J. Med.* **320:**245–246. (Editorial.)
100. **Miyakawa, Y., T. Mabuchi, K. Kagaya, and Y. Fukazawa.** 1992. Isolation and characterization of a species-specific DNA fragment for detection of *Candida albicans* by polymerase chain reaction. *J. Clin. Microbiol.* **30:**894–900.
101. **Mullis, K. B.** 1990. The unusual origin of the polymerase chain reaction. *Sci. Am.* **262**(Apr.)**:**56–65.
102. **Mullis, K. B., and F. A. Faloona.** 1987. Specific synthesis of DNA *in vitro* via a polymerase-catalyzed chain reaction. *Methods Enzymol.* **155:**335–350.
103. **Münger, K., W. C. Phelps, V. Bubb, P. M. Howley, and R. Schlegel.** 1989. The E6 and E7 genes of the human papillomavirus type 16 together are necessary and sufficient for transformation of primary human keratinocytes. *J. Virol.* **63:**4417–4421.
104. **Murphy, L. D., C. E. Herzog, J. B. Rudick, A. T. Fojo, and S. E. Bates.** 1990. Use of the polymerase chain reaction in the quantitation of *mdr*-1 gene expression. *Biochemistry* **29:**10351–10356.
105. **Myers, T. W., and D. H. Gelfand.** 1991. Reverse transcription and DNA amplification by a *Thermus thermophilus* DNA polymerase. *Biochemistry* **30:**7661–7666.
106. **Newton, C. R., A. Graham, L. E. Heptinstall, S. J. Powell, C. Summers, N. Kalsheker, J. C. Smith, and A. F. Markham.** 1989. Analysis of any point mutation in DNA: the amplification refractory mutation system (ARMS). *Nucleic Acids Res.* **17:**2503–2516.
107. **Niederhauser, C., U. Candrian, C. Höfelein, M. Jermini, H.-P. Bühler, and J. Lüthy.** 1992. Use of polymerase chain reaction for detection of *Listeria monocytogenes* in food. *Appl. Environ. Microbiol.* **58:**1564–1568.
108. **Noonan, K. E., C. Beck, T. A. Holzmayer, J. E. Chin, J. S. Wunder, I. L. Andrulis, A. F. Gazdar, C. L. Willman, B. Griffith, D. D. Von Hoff, and I. B. Roninson.** 1990. Quantitative analysis of MDR1 (multidrug resistance) gene expression in human tumors by polymerase chain reaction. *Proc. Natl. Acad. Sci. USA* **87:**7160–7164.
109. **Noordhoek, G. T., E. C. Wolters, M. E. J. de Jonge, and J. D. A. van Embden.** 1991. Detection by polymerase chain reaction of *Treponema pallidum* DNA in cerebrospinal fluid from neurosyphilis patients before and after antibiotic treatment. *J. Clin. Microbiol.* **29:**1976–1984.
110. **Nuovo, G. J., F. Gallery, P. MacConnel, J. Becker, and W. Bloch.** 1992. An improved technique for the *in situ* detection of DNA after polymerase chain reaction. *Pathology* **139:**1239–1244.
111. **Olive, D. M.** 1989. Detection of enterotoxigenic *Escherichia coli* after polymerase chain reaction amplification with a thermostable DNA polymerase. *J. Clin. Microbiol.* **27:**261–265.
112. **Olsen, G. J., D. J. Lane, S. J. Giovannoni, N. R. Pace, and D. A. Stahl.** 1986. Microbial ecology and evolution: a ribosomal RNA approach. *Annu. Rev. Microbiol.* **40:**337–365.
113. **Omata, M., T. Ehata, O. Yokosuka, K. Hosoda, and M. Ohto.** 1991. Mutations in the precore region of hepatitis B virus DNA in patients with fulminant and severe hepatitis. *N. Engl. J. Med.* **324:**1699–1704.
114. **Ou, C.-Y., S. Kwok, S. W. Mitchell, D. H. Mack, J. J. Sninsky, J. W. Krebs, P. Feorino, D. Warfield, and G. Schochetman.** 1988. DNA amplification for direct detection of HIV-1 in DNA of peripheral blood mononuclear cells. *Science* **239:**295–297.
115. **Oyofo, B. A., S. A. Thornton, D. H. Burr, T. J. Trust, O. R. Pavlovskis, and P. Guerry.** 1992. Specific detection of *Campylobacter jejuni* and *Campylobacter coli* by using polymerase chain reaction. *J. Clin. Microbiol.* **30:**2613–2619.
116. **Pachl, C. A., D. G. Kern, P. J. Sheridan, M. M. Stempien, J. A. Todd, Y. S. Zhu, Y. Gong, G. D. Cimino, J. C. Wilber, M. S. Urdea, and P. D. Neuwald.** 1992. Quantitative

detection of HIV RNA in plasma using a signal amplification probe assay. *Program Abstr. 32nd Intersci. Conf. Antimicrobial Agents Chemother.*, abstr. 1247.

117. **Pallen, M. J.** 1991. Rapid screening for toxigenic *Corynebacterium diphtheriae* by the polymerase chain reaction. *J. Clin. Pathol.* **44:**1025–1026.
118. **Persing, D. H.** 1991. Polymerase chain reaction: trenches to benches. *J. Clin. Microbiol.* **29:**1281–1285.
119. **Persing, D. H., and M. L. Landry.** 1989. *In vitro* amplification techniques for the detection of nucleic acids: new tools for the diagnostic laboratory. *Yale J. Biol. Med.* **62:**159–171.
120. **Persing, D. H., D. Mathiesen, W. F. Marshall, S. R. Telford, A. Spielman, J. W. Thomford, and P. A. Conrad.** 1992. Detection of *Babesia microti* by polymerase chain reaction. *J. Clin. Microbiol.* **30:**2097–2103.
121. **Persing, D. H., and J. Rakela.** 1992. Polymerase chain reaction for detection of hepatitis viruses: panacea or purgatory? *Gastroenterology* **103:**1098–1099.
122. **Persing, D. H., S. R. Telford III, P. N. Rys, D. E. Dodge, T. J. White, S. E. Malawista, and A. Spielman.** 1990. Detection of *Borrelia burgdorferi* DNA in museum specimens of *Ixodes dammini* ticks. *Science* **249:**1420–1423.
123. **Persing, D. H., S. R. Telford, III, A. Spielman, and S. W. Barthold.** 1990. Detection of *Borrelia burgdorferi* infection in *Ixodes dammini* ticks with the polymerase chain reaction. *J. Clin. Microbiol.* **28:**566–572.
124. **Pollard, D. R., W. M. Johnson, H. Lior, S. D. Tyler, and K. R. Rozee.** 1990. Detection of the aerolysin gene in *Aeromonas hydrophila* by the polymerase chain reaction. *J. Clin. Microbiol.* **28:**2477–2481.
125. **Pritchard, C. G., and J. E. Stefano.** 1990. Amplified detection of viral nucleic acid at subattomole levels using Q beta replicase. *Ann. Biol. Clin.* (Paris) **48:**492–497.
126. **Rahn, K., S. A. De Grandis, R. C. Clarke, S. A. McEwen, J. E. Galan, C. Ginocchio, R. Curtiss, and C. L. Gyles.** 1992. Amplification of an invA gene sequence of Salmonella typhimurium by polymerase chain reaction as a specific method of detection of Salmonella. *Mol. Cell. Probes* **4:**271–279.
127. **Rayfield, M., K. De Cock, W. Heyward, L. Goldstein, J. Krebs, S. Kwok, S. Lee, J. McCormick, J. M. Moreau, K. Odehouri, G. Schochetman, J. Sninsky, and C.-Y. Ou.** 1988. Mixed human immunodeficiency virus (HIV) infection in an individual: demonstration of both HIV type 1 and type 2 proviral sequences by using polymerase chain reaction. *J. Infect. Dis.* **158:**1170–1176.
128. **Relman, D. A., J. S. Loutit, T. M. Schmidt, S. Falkow, and L. S. Tompkins.** 1990. The agent of bacillary angiomatosis: an approach to the identification of uncultured pathogens. *N. Engl. J. Med.* **323:**1573–1580.
129. **Relman, D. A., T. M. Schmidt, R. P. MacDermott, and S. Falkow.** 1992. Identification of the uncultured bacillus of Whipple's disease. *N. Engl. J. Med.* **327:**293–301.
130. **Reyes, A., J. L. Dieguez, I. Garcia-Chacon, M. Morell, M. Ruiz, and J. C. Montilla.** 1992. Contaminating DNA in RNA amplification by polymerase chain reaction. *Clin. Chem.* **38:**1187.
131. **Riley, D. E., M. C. Roberts, T. Takayama, and J. N. Krieger.** 1992. Development of a polymerase chain reaction-based diagnosis of *Trichomonas vaginalis. J. Clin. Microbiol.* **30:**465–472.
132. **Rosa, P. A., and T. G. Schwan.** 1989. A specific and sensitive assay for the Lyme disease spirochete *Borrelia burgdorferi* using the polymerase chain reaction. *J. Infect. Dis.* **160:**1018–1029.
133. **Roszak, D. B., and R. R. Colwell.** 1987. Survival strategies of bacteria in the natural environment. *Microbiol. Rev.* **51:**365–379.
134. **Ruano, G., and K. K. Kidd.** 1991. Coupled amplification and sequencing of genomic DNA. *Proc. Natl. Acad. Sci. USA* **88:**2815–2819.
135. **Saiki, R. K., D. H. Gelfand, S. Stoffel, S. J. Scharf, R. Higuchi, G. T. Horn, K. B. Mullis, and H. A. Erlich.** 1988. Primer-directed enzymatic amplification of DNA with a thermostable DNA polymerase. *Science* **239:**487–491.

136. **Saiki, R. K., S. Scharf, F. Faloona, K. B. Mullis, G. T. Horn, H. A. Erlich, and N. Arnheim.** 1985. Enzymatic amplification of β-globin genomic sequences and restriction site analysis for the diagnosis of sickle-cell anemia. *Science* **230:**1350–1354.
137. **Sanchez-Pescador, R., M. S. Stempien, and M. S. Urdea.** 1988. Rapid chemiluminescent nucleic acid assays for detection of TEM-1 β-lactamase-mediated penicillin resistance in *Neisseria gonorrhoeae* and other bacteria. *J. Clin. Microbiol.* **26:**1934–1938.
138. **Schiffman, M. H., H. M. Bauer, A. T. Lorincz, M. M. Manos, J. C. Byrne, A. G. Glass, D. M. Cadell, and P. M. Howley.** 1991. Comparison of Southern blot hybridization and polymerase chain reaction methods for the detection of human papillomavirus DNA. *J. Clin. Microbiol.* **29:**573–577.
139. **Shen, L. P., J. A. Kolberg, R. R. Spaete, R. Miner, and W. L. Drew.** 1992. A quantitative method for detection of human cytomegalovirus DNA using a branched DNA enhanced label amplification assay, p. 408. *Abstr. 92nd Gen. Meet. Am. Soc. Microbiol. 1992*. American Society for Microbiology, Washington, D.C.
140. **Sirois, M., E. G. Lemire, and R. C. Levesque.** 1991. Construction of a DNA probe and detection of *Actinobacillus pleuropneumoniae* by using polymerase chain reaction. *J. Clin. Microbiol.* **29:**1183–1187.
141. **Urdea, M. S., T. Fultz, T. J. Anderson, M. Running, J. A. Hamren, S. Ahle, and C. A. Chang.** 1991. Branched amplification multimers for the sensitive, direct detection of human hepatitis viruses. *Nucleic Acids Symp. Ser.* **24:**197–200.
142. **Vaneechoutte, M., R. Rossau, P. DeVos, M. Gillis, D. Janssens, N. Paepe, A. DeRouck, T. Fiers, G. Claeys, and K. Kersters.** 1992. Rapid identification of bacteria of the *Comamonadaceae* with amplified ribosomal DNA-restriction analysis (ARDRA). *FEMS Microbiol. Lett.* **72:**227–234.
143. **Van Eys, G. J. J. M., C. Gravekamp, M. J. Gerritsen, W. Quint, M. T. E. Cornelissen, J. ter Schegget, W. J. Terpstra.** 1989. Detection of leptospires in urine by polymerase chain reaction. *J. Clin. Microbiol.* **27:**2258–2262.
144. **van Ketel, R. J., B. de Wever, and L. van Alphen.** 1990. Detection of Haemophilus influenzae in cerebrospinal fluid by polymerase chain reaction DNA amplification. *J. Med. Microbiol.* **33:**271–276.
145. **Vilgalys, R., and M. Hester.** 1990. Rapid genetic identification and mapping of enzymatically amplified ribosomal DNA from several *Cryptococcus* species. *J. Bacteriol.* **172:**4238–4246.
146. **Walker, G. T., M. C. Little, J. G. Nadeau, and D. D. Shank.** 1992. Isothermal *in vitro* amplification of DNA by a restriction enzyme/DNA polymerase system. *Proc. Natl. Acad. Sci. USA* **89:**392–396.
147. **Walker, G. T., M. S. Fraiser, J. L. Schram, M. C. Little, J. G. Nadeau, and D. P. Malinowski.** 1992. Strand displacement amplification—an isothermal, in vitro DNA amplification technique. *Nucleic Acids Res.* **20:**1691–1696.
148. **Wang, A. M., and D. F. Mark.** 1990. Quantitative PCR, p. 70–75. *In* M. A. Innis, D. H. Gelfand, J. J. Sninsky, and T. J. White (ed.), *PCR Protocols: A Guide to Methods and Applications*. Academic Press, Inc., San Diego, Calif.
149. **Ward, D. M., R. Weller, and M. M. Bateson.** 1990. 16S rRNA sequences reveal numerous uncultured microorganisms in a natural community. *Nature* (London) **345:**63–65.
150. **Webb, L., M. Carl, D. C. Malloy, G. A. Dasch, and A. F. Azad.** 1990. Detection of murine typhus infection in fleas by using the polymerase chain reaction. *J. Clin. Microbiol.* **28:**530–534.
151. **Welsh, J., and M. McClelland.** 1991. Genomic fingerprints produced by PCR with consensus tRNA gene primers. *Nucleic Acids Res.* **19:**861–866.
152. **Welsh, J., and M. McClelland.** 1992. PCR-amplified length polymorphisms in tRNA intergenic spacers for categorizing staphylococci. *Mol. Microbiol.* **6:**1673–1680.
153. **Werness, B. A., A. J. Levine, and P. M. Howley.** 1990. Association of human papillomavirus types 16 and 18 E6 proteins with p53. *Science* **248:**76–79.

154. **White, T. J., R. Madej, and D. H. Persing.** 1992. The polymerase chain reaction for the diagnosis of infectious diseases. *Adv. Clin. Chem.* **29:**161–196.

155. **Wilber, J. C., P. J. Johnson, P. J. Daley, C. S. Chan, R. Sánchez-Pescador, P. D. Neuwald, and M. S. Urdea.** 1992. Quantitation of hepatitis C viral RNA in serum or plasma, p. 406. *Abstr. 92nd Gen. Meet. Am. Soc. Microbiol. 1992.* American Society for Microbiology, Washington, D.C.

156. **Willoughby, J. J., W. C. Russell, D. Thirkell, and M. G. Burdon.** 1991. Isolation and detection of urease genes in *Ureaplasma urealyticum. Infect. Immun.* **59:**2463–2469.

157. **Wilson, K. H., R. Blitchington, P. Shah, G. McDonald, R. D. Gilmore, and L. P. Mallavia.** 1989. Probe directed at a segment of *Rickettsia rickettsii* rRNA amplified with polymerase chain reaction. *J. Clin. Microbiol.* **27:**2692–2696.

158. **Woese, C. R.** 1987. Bacterial evolution. *Microbiol. Rev.* **51:**221–271.

159. **Woese, C. R., E. Stackerbrant, T. J. Macke, and G. E. Fox.** 1985. A phylogenetic definition of the major eubacterial taxa. *Syst. Appl. Microbiol.* **6:**143–151.

160. **Wolcott, M. J.** 1992. Advances in nucleic acid-based detection methods. *Clin. Microbiol. Rev.* **5:**370–386.

161. **Wren, B. W., and S. Tabaqchali.** 1990. Detection of pathogenic *Yersinia enterocolitica* by a polymerase chain reaction. *Lancet* **336:**693. (Letter.)

162. **Wright, T. L., L. Ferrell, J. Lake, J. Roberts, N. Ascher, P. Sheridan, J. Detmer, W. Hunt, C. Chan, P. Neuwald, J. Wilber, M. Urdea, and R. Sanchez-Pescador.** 1992. Hepatitis C viral RNA in patients on interferon (IFN)-alpha for post-liver transplant hepatitis. Presented at the American Gastroenterological Association, San Francisco, 10 to 14 May 1992, and at the American Association for the Study of Liver Diseases, Chicago, 31 October to 3 November 1992.

163. **Wu, D. Y., and R. B. Wallace.** 1989. The ligase amplification reaction (LAR)—amplification of specific DNA sequences using sequential rounds of template-dependent ligation. *Genomics* **4:**560–569.

164. **Yang, J. Q., P. V. Tata, H. S. Park-Turkel, and H. W. Waksal.** 1991. The application of AmpliProbe in diagnostics. *BioTechniques* **11:**392–397.

165. **Yi, T., and M. M. Manos.** 1990. Detection and typing of genital human papillomaviruses, p. 356–367. *In* M. A. Innis, D. H. Gelfand, J. J. Sninsky, and T. J. White (ed.), *PCR Protocols: A Guide to Methods and Applications*. Academic Press, Inc., San Diego, Calif.

166. **Yolken, R. H., A. M. Sierra-Honigmann, and R. P. Viscidi.** 1991. Solid phase capture method for the specific amplification of microbial nucleic acids—avoidance of false-positive and false-negative reactions. *Mol. Cell. Probes* **5:**151–156.

167. **Yourno, J.** 1992. A method for nested PCR with single closed reaction tubes. *PCR Methods Applic.* **2:**60–65.

168. **Zolg, J. W., G.-X. Chen, and J. R. Plitt.** 1990. Detection of pyrimethamine resistance in *Plasmodium falciparum* by mutation-specific polymerase chain reaction. *Mol. Biochem. Parasitol.* **39:**257–265.

# Target Selection and Optimization of Amplification Reactions

# 4

DAVID H. PERSING

The term "signature sequence" was originated by bacterial geneticist Carl Woese in reference to unique sequences of ribosomal DNA (rDNA) or RNA (rRNA) that serve to distinguish one organism or group of organisms from another (49, 50). Because many nonribosomal sequences can also accomplish this purpose, the term might now be expanded to include any nucleic acid sequence that serves as a surrogate for the identification of an organism in a nucleic acid amplification protocol. The diversity of organisms in the microbial world (and the resultant diversity of the available nucleic acid sequence information) confounds attempts to produce hard and fast rules for selection of a genetic target for amplification. The decision of whether to select as a target sequence DNA or RNA, an open reading frame (ORF) or noncoding sequence, or a conserved or divergent region of the nucleic acid is interdependent upon many factors including the pathophysiology of the disease, the method of sample preparation (chapter 6), and the desired performance characteristics of the system. The purpose of this chapter is to provide guidelines for selection of genetic targets for use in amplification protocols and to outline a general approach to primer selection and optimization of amplification-based assays.

## Target Selection

Genetic stability of a target is probably the most important consideration in design of a protocol because loss or alteration of the target sequence may result in a loss of reactivity (Table 1). Target amplification procedures (including the polymerase chain reaction [PCR], self-sustaining sequence replication [3SR], and strand displacement amplification [chapter 3]) are susceptible to minor alterations in the target sequence, especially if those alterations occur near the 3′ positions of the primers used in the assays. This is due to the requirement by DNA polymerases that lack a 3′-to-5′ exonuclease activity for stable base pairing of the primer and template to initiate synthesis. Of course, this selectivity of DNA polymerase can be exploited when designing a system specifically for the detection of mutations, such as in the amplification refractory mutation system described in chapter 3. Probe amplification methods that use DNA ligase are also sensitive to mismatches near the point of ligation. For general detection of an organism or a class of organisms, high sequence conservation and stability are essential to good clinical performance.

A target sequence may provide adequate sensitivity and specificity even though it is present on a potentially unstable genetic element (i.e., a plasmid or

**Table 1.** Factors to consider in the selection of nucleic acid targets

| | |
|---|---|
| Target copy number | DNA versus RNA |
| Genetic linkage of target sequences | Sequence conservation |
| Sequence stability | Target size |

transposon sequence) if the presence of that element is associated with virulence. In fact, the detection of virulence-associated genetic alleles may be of great value in distinguishing pathogenic from nonpathogenic organisms that may be present in the same sample. Examples include the detection of exotoxin-producing strains of *Staphylococcus aureus* (protocol 1.20), verotoxin-producing strains of *Escherichia coli* (protocol 1.16), and the invasin gene of *Yersinia enterocolitica* (17). Since genetically unstable virulence determinants may be lost during primary isolation or serial passage, the optimal setting for the amplification of genetic tests for virulence determinants is probably soon after isolation or directly from the sample itself.

The sensitivity and reliability of an assay may be improved greatly if multiple target molecules are present in each microorganism. However, the mere presence of multiple copies of a target sequence does not necessarily translate into high sensitivity. Even if a target is present in only one copy per organism, adequate sensitivity may be achieved if many organisms are typically present in the sample being tested. Furthermore, if an organism has multiple, genetically linked copies of a target sequence, extraction of nucleic acids (with a minimum of shearing) would result in the preservation of the relative proximity of these targets, and the multiple targets would act as a single target for detection purposes. Although the sensitivity requirements for detection of this hypothetical organism would not have to be extremely high (because multiple targets are present whenever an organism is present), there would be a significant risk of missing the target in a sample with only a few organisms present; multiple analyses would then be necessary to identify the sequence. Dissociation of the target molecules may allow them to distribute uniformly throughout the sample, thus making the test less subject to sampling error. An example is shown in Fig. 1.

The application of amplification techniques to specimens with small numbers of organisms may require detection of unlinked target sequences or methods for concentrating the target nucleic acid from large volumes (chapter 6). One such example is in the detection of bacterial organisms in blood specimens from bacteremic patients, which may contain only a few organisms per milliliter. Processing of 0.1 ml of blood for amplification (a typical volume) could result in a false-negative result on purely statistical grounds. However, if rRNA (which may be present in a few thousand copies per bacterial cell) is isolated from a larger volume of blood and distributed throughout the extract, 0.1 ml of sample could contain hundreds to thousands of copies of the target. Examples provided in this book of detection of multicopy targets include PCR amplification of *Leishmania* nuclear DNA repeats (protocol 4.1), in which sensitivity is maximized by pretreating the DNA with DNase I to dissociate the repeat sequences, detection of *Borrelia burgdorferi* rRNA sequences by reverse transcription followed by PCR (protocol 1.5), and detection of mycobacterial insertion sequences (protocol 1.2).

## Use of DNA or RNA as Substrate

All of the nucleic acid amplification systems described in chapter 3 can use DNA and, with minor modifications, RNA as the substrate. PCR is designed to

**Figure 1.** Detection of dissociated and undissociated genetic targets. Genetic targets are more evenly distributed if their linkage is disrupted prior to detection by a sensitive method.

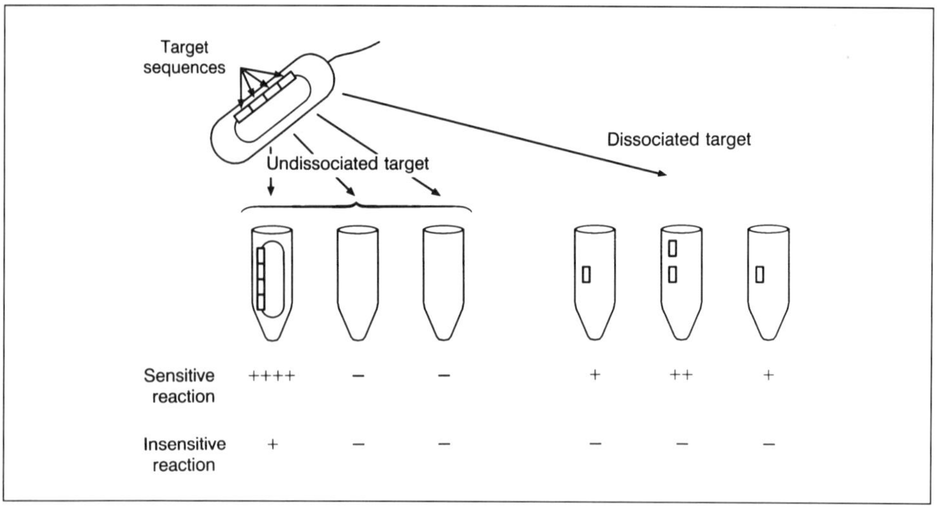


detect DNA sequences; if RNA detection is desired, a reverse transcription step to convert RNA into DNA must be incorporated. However, if selective detection of RNA is sought, primers must be designed to discriminate processed RNA from genomic sequences (13, 38) or the DNA template must first be eliminated by selective isolation or DNase treatment (9). In isothermal transcription-based amplification (3SR/NASBA; see chapter 3), reverse transcriptase activity is used from the start, so both DNA and RNA targets can be detected; selective RNA detection can be accomplished by avoiding denaturation of double-stranded DNA prior to the addition of reagents. Because the operating temperatures of 3SR are below the denaturation temperature of DNA, amplification of double-stranded DNA templates is avoided (chapter 3). Probe amplification methods such as ligase chain reaction (LCR) and those that use Qβ replicase can also use DNA or RNA targets; however, LCR, like PCR, requires conversion of RNA to DNA because RNA-DNA duplexes do not serve as efficient templates for ligation. Signal amplification techniques using branched DNA complexes are relatively unrestricted with regard to the detection of DNA or RNA.

Most of the amplification protocols described to date have used DNA targets. This is due in large part to the relative ease with which DNA can be isolated from clinical isolates or directly from specimens and the relative stability of DNA compared with RNA once isolated. Moreover, simple procedures such as boiling sufficiently eliminate DNase activity so that relatively crude specimens can be used in many cases, thus simplifying sample processing. For the majority of applications, DNA targets provide adequate sensitivity.

## Identification of Target Nucleic Acid Sequences

The optimal starting point in the process of identifying a target sequence for detection of an infectious organism is in the literature. The relative ease with which nucleic acids can now be cloned and sequenced has led to a virtual explosion in available sequence information, and much of this information is found in published reports. However, the combination of PCR amplification and

**Table 2.** Useful search strings for recovering rRNA

| Specificity | Search string | Mismatch tolerance | Reference |
|---|---|---|---|
| Eukaryotic nuclear small-subunit rDNA | 5′AACCTGGTTGATCCTGCCAGT3′<br>5′GATCCTTCTGCAGGTTCACCTAC3′ | 3 | 27 |
| Eukaryotic mitochondrial small-subunit rDNA | 5′CAGCAGTCAAGAATATTAGTCAATG3′ | 5 | 47 |
| Prokaryotic small-subunit rDNA | 5′CTGCTGCCTCCCGTAGGAGT3′ | 3 | 44 |
| Prokaryotic large-subunit rDNA | 5′GAACTGAAACATCTTAGT3′ | 3 | *[a] |
| Eukaryotic mitochondrial large-subunit rDNA | 5′GTACTTTTGCATAATGGGTCAGC3′ | 5 | 47 |

[a] *, our laboratory.

direct DNA sequencing can now generate considerable sequence data in only a few days; the mere presentation of sequence data is thus becoming less and less worthy of publication. This has led to a more recent trend of depositing sequence information directly into nucleic acid sequence data bases such as GenBank, Los Alamos, N.M., and the European Molecular Biology Laboratories (EMBL) data base, Heidelberg, Germany. Thus, whether or not relevant sequence information appears in the literature, it may be well worth the effort of searching these data bases by using appropriate keywords.

When searching nucleic acid sequence data bases, it is best to use a variety of search terms because sequences may be categorized in unexpected ways. In a hypothetical case, the gene encoding the 16S rRNA of *Treponema pallidum* may not be retrievable by searching for the strings "16S" or "ribosomal" because of misclassification of the sequence categorized as a "genomic DNA sequence." Rather, the search term "pallidum" would have to be used. The use of a genus name may also be misleading because genera are often indicated in abbreviated form. Thus, not all *T. pallidum* sequences would be retrieved when the search term "Treponema" is used.

In some cases, data base searches for a specific sequence motif will yield the best results. Table 2 lists some useful search strings for recovering eukaryotic and prokaryotic rDNA and rRNA sequences. String searches based on nucleotide sequence information can overcome difficulties associated with unusual naming conventions or even misspellings within sequence files. However, it should be realized that performing a string search on over 120 million nucleotides (the current number in the combined GenBank/EMBL data bases) is essentially the computerized equivalent of a hybridization assay. The sensitivity and specificity of "detecting" the desired sequences is highly dependent on the stringency of the search.

The relatively recent trend of patenting naturally occurring sequences in prokaryotic and eukaryotic organisms for diagnostic and therapeutic purposes has unquestionably led to delays in the dissemination of new sequence information and to interesting legal consequences for those wishing to develop diagnostic tests based on patented sequences. In infectious-disease testing, organisms presenting the greatest diagnostic challenges often have the least sequence information available in the public domain. For this reason, it may be necessary to isolate and characterize new diagnostic target sequences.

**Table 3.** Oligonucleotide design

| | |
|---|---|
| Oligonucleotide length | Primer-primer interactions |
| Melting temperature | Target length |
| Sequence composition | Disposition on the target sequence |
| Physical characteristics (self-annealing) | |

## Developing A New Target Sequence

If nucleotide sequence information is not available for the organism of interest, or if additional targets are sought for the purpose of confirming an assay, the task of isolating and characterizing a nucleic acid sequence is the task at hand. Although this daunting task would have been the domain of the experienced molecular biologist only a few years ago, several recent developments, including universal primer amplification and direct sequence analysis, have made the process more democratic. One of the most useful and reliable techniques for recovering a unique DNA sequence is by "universal" primer PCR amplification of rDNA sequences (27, 33, 36, 37, 45, 47). The principle of universal primer rDNA amplification is based on the finding that all nonviral organisms share homology within short segments of the rDNA genes. Some segments are found in all members of the bacterial kingdom; others are common to all nonviral life forms. This feature allows the design of synthetic oligonucleotide primers (universal primers) for the amplification and/or sequence determination of a wide variety of organisms. The universal sequences often flank regions of significant diversity that may be used as signature sequences.

Although universal primer amplification and analysis of the intervening sequences can be performed directly from clinical material (which may contain a mixture of members of the normal flora), extraction of genomic DNA from organisms in pure culture leads to less ambiguous results. Procedures for performing universal amplification of bacterial and fungal rDNA are provided in the second half of this book (protocols 1.1, 3.1, and 5.1). Once a segment of the rDNA has been amplified, it may be sequenced by using sequencing reagents in kit form. Protocols 1.1 and 5.1 provide detailed methods for sequence determination.

Diagnostic target sequences can be identified by other more labor-intensive means, such as differential screening of genomic clones or subtractive hybridization of cDNA (48). Once unique clones are identified, their nucleotide sequence should be determined and examined for inter- and intraspecies heterogeneity. Although the overall structure and nucleotide composition of a nucleic acid fragment may be conserved, there may exist regions of microheterogeneity that will make it difficult to design primers or probes that provide reliable results. Microheterogeneity is more difficult to predict for nonribosomal than for ribosomal sequences, because in the latter, selective pressure to maintain high sequence conservation is directed against the nucleotides themselves rather than indirectly against a protein product (see below).

## Primer Selection

The judicious choice of oligonucleotide primers is a critical element in determining the performance of a nucleic acid amplification-based detection system. Several considerations come into play when designing primers (Table 3): (i) oligonucleotide length, (ii) oligonucleotide melting temperature, (iii) sequence

**Figure 2.** Solid-support DNA synthesis. The oligonucleotide is synthesized on a solid support with the 3′ hydroxyl group attached to the support via an amide linkage. Successive rounds of deprotection, base addition, and capping lead to the synthesis of the oligonucleotide from the 3′ end. Thus, the last nucleotide to be added is at the 5′ end. The 5′-most nucleotide may be easily modified by using nucleotide analogs or derivatives in the last coupling step.

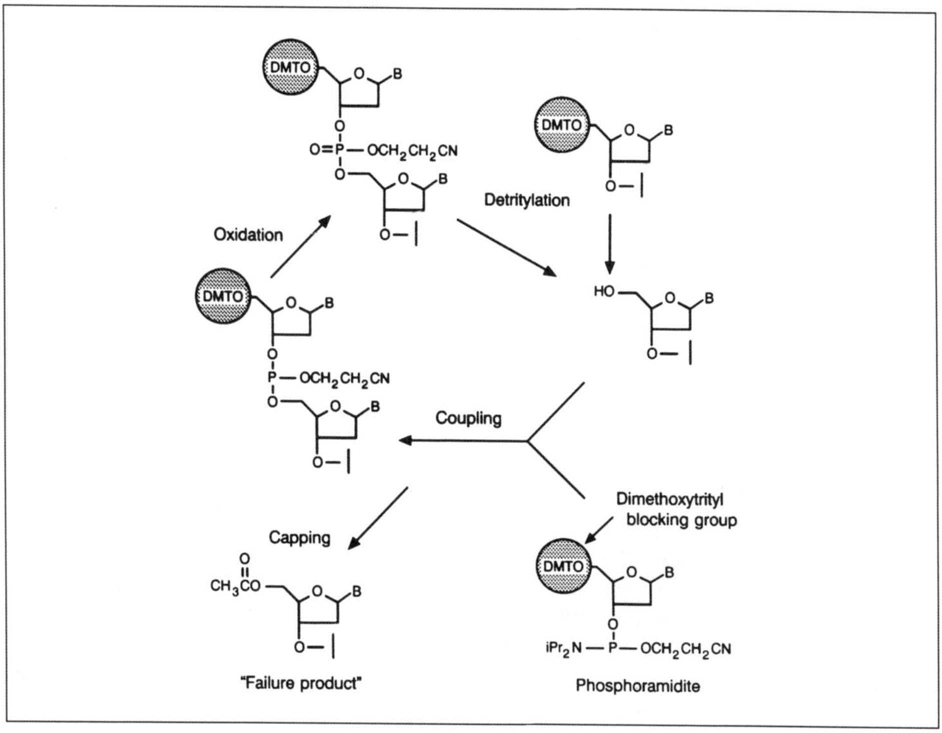


composition, (iv) physical characteristics, (v) primer-primer interactions, (vi) length of the amplified target, and (vii) location on the target sequence.

## Oligonucleotide Synthesis

Nucleic acid amplification techniques, regardless of the type, are driven by high concentrations of target-specific oligonucleotides. During the last decade, several types of instruments that automate the complex synthetic chemistry behind oligonucleotide synthesis have become available (2, 10, 24). These instruments are capable of producing large quantities of relatively pure oligonucleotides that can be used as primers or probes.

The chemistry of DNA synthesis is quite simple in concept. A reactive 3′ phosphorus of one nucleoside is chemically coupled to the 5′ hydroxyl moiety of a nucleoside bound to a solid support (1, 2, 19). Three other chemical reactions are then required to prepare the growing nucleotide for the next round of coupling: capping, oxidation, and deblocking (detritylation). Each of these steps is depicted in Fig. 2. The efficiency of oligonucleotide synthesis has improved greatly in recent years, so that the yield of the desired coupling is greater than 99% at each cycle. The purity of the finished product is thus dependent on the number of synthetic cycles; in other words, the longer the oligonucleotide, the lower the yield.

The synthetic process results in an oligonucleotide that grows from the 3′ end (the end coupled to a solid support) outward toward the 5′ end. Modifications to the 3′ end must therefore involve solid supports containing the modified base (such as inosine or a ribonucleoside). Modifications to the 5′ end, such as addition of an amino linker spacer for coupling of enzymatic reporters, are accomplished by incorporating the modified base in the last synthetic cycle.

### Oligonucleotide Length

The length of an oligonucleotide primer or probe for use in an amplification protocol is determined by the sequence conservation, size, and composition of the target sequence. If a target with uniformly high sequence conservation is chosen, the size of the oligonucleotide is less restricted. On the other hand, if a highly polymorphic sequence is selected for amplification, primer-binding sites displaying sufficient conservation (based on alignment of homologous sequences) may limit the size of the primer or the region within the sequence from which the primer can be selected.

For PCR protocols, the target-specific portion of primers used for amplification reactions usually ranges between 15 and 30 bp in length. Shorter primers may not provide adequate sequence specificity; longer primers usually do not add significant specificity and are more expensive to synthesize. Primer length also influences the ability of the system to tolerate mismatches (Table 3). The overall length of the primer is the sum of the target-specific portion of the primer plus any nontarget nucleotides (such as restriction endonuclease cleavage sites) that are added at the 5′ end; such additions add to the net length of the oligonucleotide without contributing to target specificity.

The addition of nontarget nucleotides to the 5′ end may alter significantly the hybridization characteristics of the primers and may ultimately affect performance. During the first few cycles, composite primers anneal to the target via the 3′-most (target-specific) region; the 5′ tail does not participate in hybridization. However, in the second and subsequent cycles, "short products" accumulate (chapter 3); these allow the entire length of the primer to bind. The net effect is an elevation in the melting temperature of the primer. Thus, for primers with 5′ extensions, a shift to higher annealing temperatures after the first 5 to 10 cycles may result in improved performance.

### Melting Temperature

The melting temperature ($T_m$) of an oligonucleotide bound to its target nucleic acid is the temperature at which 50% of the strands have become separated in solution. $T_m$ is dependent on the oligonucleotide concentration, the sequence composition, and the composition of the solvent (6, 12, 14, 20, 21, 26, 34, 35, 40, 46). $T_m$ is distinct from the dissociation temperature ($T_d$); the latter term relates to the retention of probes on blots or other solid supports during washing steps (46). Short oligonucleotides have lower $T_m$s than longer primers with the same base composition. However, as the primer length approaches 20 to 25 bp, the additive effect of length on $T_m$ is diminished, so that for primers commonly used in amplification protocols, base sequence composition plays the largest role.

There are several formulas for calculation of the $T_m$. The simplest method (28), based on assigning 2°C for every A or T and 4°C for every G or C, is the most widely used. Although reasonable approximations of $T_m$ can be achieved with this method, the accuracy of the formula suffers with increasing oligonucleotide length. For example, the calculated $T_m$ of a 20-mer with an equal A+T and G+C distribution would be 60°C, which is probably close to the actual $T_m$. However, the calculated $T_m$ for a 40-mer of the same sequence composition by this formula would be an absurd value of 120°C.

More accurate methods for calculating $T_m$ that are based on base-stacking interactions of "nearest neighbors" in the DNA duplex have recently been described (14, 21, 26, 40). Nearest-neighbor analysis assigns a temperature weight to different combinations of adjacent base pairs; the average of the temperature weight is what determines the $T_m$. Since the calculated value of $T_m$ is based on an average instead of a sum, nearest-neighbor analysis is less likely to be skewed by increased oligonucleotide length. Computerized oligonucleotide selection programs that eliminate the tedium of nearest-neighbor $T_m$ calculations are now available (23, 26, 30, 39). A useful formula for calculating an optimal annealing temperature ($T_a^{OPT}$) which is based on empirical evaluations of oligonucleotides is as follows (46):

$$T_a^{OPT} = 0.3T_m^{primer} + 0.7T_m^{product} - 14.9$$

in which $T_m^{primer}$ is the calculated $T_m$ of the less stable primer-template pair and $T_m^{product}$ is the $T_m$ of the PCR product.

The calculated $T_m$ values of both primers used in an amplification protocol should be as close as possible. This helps to avoid the problem of "asymmetric priming," in which one primer anneals to its target sequence with greater avidity than the other under the same hybridization conditions. With PCR, asymmetric priming is most commonly observed when the annealing temperature is set in between the $T_m$s of two primers with widely different $T_m$s.

In PCR protocols, it is generally advisable to select oligonucleotides with optimal annealing temperatures in the range of 55 to 65°C. Lower annealing temperatures may lead to poor target specificity and sensitivity (because of competition between nonspecific products and specific products for reaction components). Another consideration is the choice of an amplification product inactivation method; if the uracil *N*-glycosylase (UNG) inactivation protocol (chapter 5) is to be incorporated, annealing temperatures should not fall below 55°C, because UNG may become reactivated at lower temperatures. Reactivation of UNG may result in lower amplification efficiency because of the degradation of amplification products as soon as they are made (see chapter 5 for additional information).

## Sequence Composition

As mentioned in the previous section, the nucleotide composition of the oligonucleotide primer used in an amplification protocol will directly determine the annealing temperature to be used. A few additional points merit consideration. (i) Monotonous stretches of the same nucleotide should be avoided. Long tracts of a single nucleotide may decrease target specificity. (ii) The G+C content of the primers should approximate or slightly exceed the G+C content of the target. G+C-deficient primers flanking a G+C-rich target may not be able to compete effectively for binding, resulting in lower amplification efficiency. (iii) Special consideration should be given to the 3′ end of the primer (the end recognized by polymerases). A high proportion of G+C residues at the 3′ end may lead to lower target specificity, because of the tendency of this region to provide stable base pairing (via the G+C-rich region) to nontarget sequences. The selection of the

**Table 4.** PCR parameters for tolerance of primer mismatches

| Mismatch tolerance | Mismatch intolerance |
| --- | --- |
| Primer length 25–35 bases | Primer length 14–18 bases |
| High dNTP concentration (0.8–4.0 mM) | Low dNTP concentration (≤50 μM) |
| Lower annealing temperature | Higher annealing temperature |
| Mismatches reduced | Mismatches increased |
| More cycles | Fewer cycles |
| Mismatches located toward 5′ end | Mismatches located at or near 3′ end |
| 3′-terminal T residue | 3′ base pair: C · C, A · G, G · A (least tolerated) > A · A > > other mismatches (best tolerated) |

3′-most nucleotide is also critical to target specificity. Kwok et al. (22) have demonstrated that primers ending with a 3′ thymidine residue are more tolerant of mismatches than other nucleotides. Finally, for protein-coding regions, the position of the 3′ end of the primer with respect to the open reading frame of the target should be considered (see below). Parameters for mismatch tolerance and intolerance are listed in Table 4.

## Physical Characteristics

Self-annealing of primers (hairpin loop formation) should be avoided. The formation of hairpin structures in solution may reduce the effective concentration of a primer because of a competition between self-annealing and target annealing. In general, hairpin loop structures containing stems greater than 4 nucleotides in length with loops less than 5 nucleotides should be avoided. Self-complementarity at the 3′ end of a primer should be especially avoided, because it will reduce the effective concentration of the primer. Several commercially available computer programs are capable of predicting self-complementarity (23, 26, 30, 39).

## Primer Artifacts

The term "primer-dimer" has gradually crept into the lexicon of many laboratory investigators using PCR. It refers to an accumulated short PCR product that is roughly the size of the DNA duplex that would be created by two adjoining primers. The term primer-dimer itself is somewhat misleading; often, a few template-directed nucleotides are inserted between the primers. Primer-dimer artifacts can be generated by both primers, but structures produced from only one primer are theoretically possible. Small amounts of primer artifact are often generated in well-optimized reactions. However, large amounts can decrease the sensitivity of a reaction by competing for reaction components. Some steps toward avoidance of excess primer artifact can be taken by avoiding complementarity of the two primers used for amplification (especially if they are G · C residues) between the 3′ ends. Even single-nucleotide complementarity at the 3′ end should be avoided. Again, computer programs are available for predicting such interactions. However, a simple alignment by eye will often suffice.

## Target Length

Targets that can be amplified efficiently from clinical specimens usually range from 100 to 400 bp. The choice of an optimal size will depend on the desired features of the amplification protocol on the intended uses of the system and on the choice of an inactivation protocol. For example, if formalin-fixed tissue is to be tested, the size of the amplification product that can be produced will be quite limited because of the covalent modification of DNA by the fixative (15, 51). If fresh specimens are to be used, the size of the target is less limited but should generally stay below 700 to 1,000 bp. Although such large targets can often be amplified from clinical specimens, the process is less efficient because the processivity of the DNA polymerases is taxed. In addition, longer products may be more susceptible to low levels of inhibitors that affect polymerase processivity (8) (chapter 6). Amplification of long products will also increase the time necessary for amplification because of the longer extension times.

The choice of an inactivation protocol should be an important consideration in determining the size of the product. For the isopsoralen inactivation protocol (chapter 3), longer products are much more efficiently inactivated than short ones. This is especially relevant when the amplification products are G+C rich; short, G+C-rich products may be inactivated very poorly. Thus, if isopsoralen inactivation is to be used, products more than 200 to 250 bp long should be generated. Other methods for product inactivation are theoretically less subject to product size and G+C content than the photochemical method (chapter 5). However, recent evaluations of the photochemical and enzymatic inactivation methods have shown that similar length and nucleotide composition restrictions may apply to several of the inactivation protocols (43).

### Location on the Target Sequence

When protein-coding sequences are targeted for amplification, the position of the 3′ end of the primer relative to the ORF is an additional consideration. The selective pressure to maintain sequence conservation within ORFs is generally directed at the amino acid level. However, because of the degeneracy of the genetic code, multiple codons (nucleotide triplets) may encode the same amino acid; this may allow the nucleotide sequence to vary while preventing selective pressure at the level of protein structure from playing any part in sequence conservation. The nucleotide position in the codon with the greatest variability is in the third or "wobble" position of a codon. Thus, a reasonable approach is to design primers so that their 3′ nucleotides lie on the first or second base of a conserved codon.

For genetic targets that do not encode proteins, such as genetic regulatory regions and rDNA or rRNA, the selective pressure on sequence conservation is directed at the nucleotide level; primers designed to detect these targets do not have to consider their position relative to potential ORFs. Ideally, for both coding and noncoding regions, the choice of the 3′ and 5′ ends of a primer will be based on systematic comparison of related targets from different sources. The abundance of genetic information for small-subunit (16S-like) rDNA sequences lends itself naturally to such comparisons.

### Computerized Oligonucleotide Selection

Computer-assisted selection of oligonucleotides for amplification protocols or for nucleic acid sequencing has become popular in recent years. Such programs are available for both IBM PC-compatible and Macintosh (Apple Computer) platforms and are commercially available as stand-alone systems such as Oligo (IBM PC and Macintosh) (39) or Hyper-PCR (Macintosh) (30) or bundled with other software such as MacVector (IBI, Inc., New Haven, Conn.) for the

**Table 5.** Reaction optimization parameters for PCR

| | |
|---|---|
| Annealing temperature | Primer concentration |
| Magnesium ion concentration | Cosolvents |
| Nucleotide concentration | Product inactivation protocols |
| Enzyme concentration | "Hot-start" protocols |

Macintosh packages. Other oligonucleotide design programs are available as shareware (23, 26).

In general, these programs are very good at performing several important functions of oligonucleotide design, including determination of $T_m$ by accurate methods, and prediction of self-complementarity and interprimer annealing. However, none of these programs takes into consideration all of the details described in this section, and only one program identifies target regions based on multiple-sequence alignment (23). Some software allows for "automatic" selection of primers, but this selection process is based almost solely on the composition of the primers themselves. Thus, for the immediate future, primer design will remain an intelligent and intuitive process that cannot be duplicated by computer.

It is important to realize that the best use of oligonucleotide design software is to avoid poor choices, not to replace the empirical process of testing combinations of various primers to determine which provide the greatest analytical sensitivity and specificity. In this same vein, it is of little use to run string searches of oligonucleotide primers against sequence data bases to predict cross-reactivity. Only a tiny fraction of the total nucleotide sequences likely to be encountered in clinical specimens are known at present. Here again, empirical testing is more profitable (and less taxing of expensive computer resources) than digital deliberation.

## Reaction Optimization: The PCR Paradigm

Next to the selection of oligonucleotides, the optimization of amplification reactions is the greatest contributor to analytical sensitivity and specificity. Most of our experience to date has been with PCR. Factors to consider in PCR optimization include primer-annealing temperature, concentrations of divalent cations, enzyme concentration, primer concentrations, nucleotide concentration, reaction additives and cosolvents, and the use of thermal or enzymatic "hot-start" procedures (Table 5). Each of these will be discussed briefly below in roughly decreasing order of importance. However, because of the interdependent nature of PCR components, alteration of one promoter may have effects on some or all of the others. Thus, there is considerable controversy over which components of a reaction should be optimized first. For additional information on reaction optimization, the reader is referred to references 16 and 18.

### Annealing Temperatures and Thermal Cycling

As discussed in a previous section, primer annealing temperatures used in PCR protocols can be predicted by several formulae, but optimal temperatures should be determined empirically. Optimization of the annealing temperature becomes increasingly important as the genetic complexity of the sample increases. Higher complexity offers greater opportunities for nonspecific annealing, and under low-stringency conditions, poor specificity may result in nonspecific amplification products. Such products are often observed in agarose gels as ethidium bromide-stained bands that have higher or lower molecular weight than the expected product and that do not hybridize to internal probes.

In optimizing annealing temperatures, it is useful to perform reactions under conditions that will simulate the genetic composition of the sample to be tested. For example, in setting up a protocol for detection of an organism in human blood, in which human genomic DNA will copurify with that of the agent, dilutions of target DNA should be added to a constant amount of human DNA during optimization.

The optimal annealing temperature for a given PCR protocol is often higher than predicted by mathematical formulas. A good rule of thumb is to start at an annealing temperature predicted by nearest-neighbor analysis and then move up in increments of several degrees. At the first sign of a loss of sensitivity, a lower annealing temperature can be selected. Alternatively, a greater number of cycles in a two-step protocol (cycling between annealing and denaturation without stopping for extension) can be tested. However, our experience has been that two-step protocols generally work better for short targets (less than 300 to 400 nucleotides) and may not be applicable in all situations. Furthermore, the excessive cycling that is common in many two-step protocols may lead to lower specificity (3).

### Magnesium Ion Concentration

Increasing the concentration of magnesium ion has the net effect of decreasing the stringency of primer binding. Low magnesium ion concentrations may result in poor reaction efficiency; high concentrations may result in poor reaction specificity. The magnesium ion concentration should be calculated as a function of nucleotide concentration. For most PCR protocols, a range of magnesium concentrations between 0.5 and 3.0 mM above the total nucleotide concentration should be tested. Because of the tendency of $MgCl_2$ solutions to precipitate after freeze-thawing, stock solutions of $MgCl_2$ should be maintained in the refrigerator rather than the freezer and should be added to PCR master mixes separately prior to amplification.

### Deoxynucleoside Triphosphates

The optimal concentration of deoxynucleotide triphosphates (dNTPs) should be determined empirically and in conjunction with titration of $Mg^{2+}$ concentrations. Generally, concentrations of dNTPs will range from 20 and 200 mM. Concentrations that are too high may result in lower specificity and poor economy, and concentrations that are too low may result in a poor yield of amplification product.

As discussed in chapter 5, it may be prudent to consider replacing TTP partially or completely with dUTP in amplification reactions. Such a maneuver is compatible with a variety of inactivation protocols and provides the option of using UNG if amplification product contamination occurs (chapter 5). Replacement of TTP with dUTP does not result in a significant increase in cost and generally does not result in lower amplification efficiency. However, some investigators have observed that dUTP-substituted reactions have lower sensitivity (32). Additionally, dUTP substitution may have a negative effect on probe hybridization. Nevertheless, although establishing new reaction conditions for dUTP substitution represents a considerable effort, the potential long-term benefits are probably well worth it.

## Enzyme Choice and Concentration

The concentrations of polymerases used in PCR protocols vary significantly, but for *Taq* polymerase they usually range between 1 and 2.5 U/100-μl reaction. To optimize enzyme concentration, the protocol should be tested empirically at a variety of concentrations between 0.5 and 5 U/100-μl reaction (18). When enzyme concentrations are high, nonspecific priming will occur as determined by gel electrophoresis. Concentrations that are too low will result in poor amplification efficiency.

In recent years, a variety of DNA polymerases have become commercially available for use in PCR. The various polymerases may exhibit a range of thermal stabilities, exonuclease activities, processivities, incorporation error rates, and reverse transcriptase activities. However, very few independent evaluations have been performed to validate the manufacturer claims for these enzymes. In the absence of such comparisons, the best overall approach to choice of an enzyme is to test several different preparations empirically and then choose the enzyme that offers the best price/performance ratio.

## Primer Concentration

Primer concentrations that are low may result in poor reaction yields, and concentrations that are too high may result in lower specificity, as evidenced by an increase in nonspecific amplification products. For reasons of economy, the lowest primer concentrations that provide maximal sensitivity and specificity should be used.

Primer concentrations that are too high may also result in higher levels of primer-dimer artifact. For any PCR protocol, the production of large amounts of primer artifact is something to be avoided, because excess primer artifact reduces the amplification efficiency of the intended target (7, 29). An accumulation of primer artifact in the laboratory may also result in buildup of a highly competitive, reaction-specific inhibitor that may, over time, reduce the sensitivity of assays if contamination with primer artifact occurs. Such an accumulation of inhibitors may be one of the best arguments for using nucleic acid amplification prepackaged kits, which will presumably be free of contamination by either target nucleic acid or primer artifact.

## Additives and Cosolvents

The effects of a variety of additives and cosolvents on the properties of PCR and other amplification methods have been investigated in recent years (18, 42). Formamide and dimethyl sulfoxide can be tolerated at low concentrations in PCR and may help to amplify G+C-rich targets (25, 41). In this capacity as cosolvents, they probably facilitate strand separation of G+C-rich targets and thereby improve the accessibility of the template to primers and polymerase. The effects of glycerol on enhancement of PCR product yield are not well understood; presumably, glycerol helps to stabilize *Taq* polymerase and may also serve as a cosolvent. The addition of glycerol may be necessary if the photochemical inactivation (isopsoralen) method is used; inhibition of PCR by high concentrations of isopsoralens can be largely overcome by the addition of 10% glycerol (chapter 5).

Recently, it has been observed that the addition of single-strand DNA-binding proteins (SSBs) such as *E. coli* SSB and T4 phage gene 32 protein may reduce nonspecific amplification, improve product yield, and help overcome inhibition (16, 31). The mechanism underlying the action of SSBs is probably sequestration of primer during the reaction setup period, thereby reducing nonspecific primer events. The addition of SSBs is costly at present and is probably not justified for most reactions. However, if poor primer specificity cannot be overcome by adjusting reaction conditions, SSBs might be useful.

### Hot-Start Protocols

In a typical PCR reaction, all reaction components are mixed in a so-called master mix and held at 4 to 25°C before the components are dispensed into reaction tubes. Target DNA is then added to the reaction mixtures, and the tubes are placed in a thermal cycler. During the time that reactions are being prepared prior to placement into the thermal cycler and during the first heating cycle, *Taq* polymerase is capable of extending primer-template complexes under low-stringency hybridization conditions. Even though the subsequent temperatures may never drop below 55 to 60°C, the nonspecific extension products synthesized during the setup period may have deleterious effects on reaction specificity and sensitivity.

The hot-start technique (7, 15, 16, 28) avoids the formation of nonspecific extension products by keeping an essential component of the reaction (i.e., nucleotides, $Mg^{2+}$, primers, or polymerase) separate from the target during nonstringent temperatures. Once the reactions have reached an elevated temperature, the missing component is added. Reactions are then cooled to the appropriate annealing temperature, allowing the primer to "touch down" on the specific target sequence to initiate synthesis. The use of a hot start can lead to enhanced sensitivity for some systems, especially those for which nonspecific product formation is a problem (4, 7, 29).

Several methods can be used to achieve a hot start. One method involves wax beads, which form a thermally labile barrier to an essential reaction component (4). Alternatively, *Taq* polymerase can be withheld until reaction tubes are placed in the thermal cycler (16). Interestingly, the UNG method for product inactivation provides the enzymatic equivalent of a hot start (chapter 5). Nonspecific extension products are prevented from participating in amplification by the action of UNG, which remains active at temperatures up to 55°C (protocol 2.1). Recently, a *Taq* polymerase-specific monoclonal antibody was used to inhibit *Taq* activity until reaction components were heated to temperatures that denature the antibody (11).

## Conclusions

The purpose of this chapter has been to provide general guidelines for the selection of genetic targets and oligonucleotides for use in nucleic acid amplification protocols and to outline an approach for their optimization. Most of these guidelines pertain to the use of PCR, because PCR has been the most widely used to date. As other amplification methods become more common, specific recommendations pertinent to these systems will have to be developed. At present, however, it can be said that PCR has already taught us valuable lessons concerning target selection and reaction optimization, many of which can be generalized to other nucleic acid amplification protocols as they become more widely available.

## References


1. **Adams, S. P., and G. R. Galluppi.** 1986. DNA synthesis and applications to molecular biology. *Med. Res. Rev.* **6:**135–170.
2. **Atkinson, T., and M. Smith.** 1984. Solid phase synthesis of oligodeoxyribonucleotides by the phosphite-triester method, p. 35–81. *In* M. J. Gait (ed.), *Oligonucleotide Synthesis. A Practical Approach*. IRL Press, Oxford.
3. **Bell, D. A., and D. M. DeMarini.** 1991. Excessive cycling converts PCR products to random-length higher molecular weight fragments. *Nucleic Acids Res.* **19:**5079.
4. **Bloch, W.** 1992. Wax-mediated hot start PCR: AmpliWax™ PCR gems permit nonisotopic, unprobed detection of low-copy-number targets, p. 1–6. *In* L. S. Kelley (ed.), *Amplifications. A Forum for PCR Users*. Perkin-Elmer, Inc., Norwalk, Conn.
5. **Böttger, E. C.** 1989. Rapid determination of bacterial ribosomal RNA sequences by direct sequencing of enzymically amplified DNA. *FEMS Microbiol. Lett.* **65:**171–176.
6. **Breslauer, K. J., R. Frank, H. Blocker, and L. A. Marky.** 1986. Predicting DNA duplex stability from the base sequence. *Proc. Natl. Acad. Sci. USA* **83:**3746–3750.
7. **Chou, Q., M. Russell, D. E. Birch, J. Raymond, and W. Bloch.** 1992. Prevention of pre-PCR mis-priming and primer dimerization improves low-copy-number amplifications. *Nucleic Acids Res.* **20:**1717–1723.
8. **de Franchis, R., N. C. P. Cross, N. S. Foulkes, and T. M. Cox.** 1988. A potent inhibitor of *Taq* polymerase copurifies with human genomic DNA. *Nucleic Acids Res.* **16:**10355.
9. **Dilworth, D. D., and J. R. McCarrey.** 1992. Single-step elimination of contaminating DNA prior to reverse transcriptase-PCR. *PCR Methods Applic.* **1:**279–282.
10. **Efcavitch, J. W.** 1988. Automated system for the optimized chemical synthesis of oligodeoxyribonucleotides, p. 221–234. *In Macromolecular Sequencing and Synthesis. Selected Methods and Applications*. Alan R. Liss, Inc., New York.
11. **Findlay, J.** 1992. Development of PCR for in vitro diagnostics. *1992 San Diego Conference: Genetic Recognition*. San Diego Chapter of the American Association of Clinical Chemists, San Diego, Calif.
12. **Freier, S. M., R. Kierzek, J. A. Jaeger, N. Sugimoto, M. H. Caruthers, T. Neilson, and D. H. Turner.** 1986. Improved free-energy parameters for predictions of RNA duplex stability. *Proc. Natl. Acad. Sci. USA* **83:**9373.
13. **Gilliland, G., S. Perrin, K. Blanchard, and H. F. Bunn.** 1990. Analysis of cytokine mRNA and DNA: detection and quantitation by competitive polymerase chain reaction. *Proc. Natl. Acad. Sci. USA* **87:**2725–2729.
14. **Gotoh, O., and Y. Tagashira.** 1981. Stabilities of nearest-neighbor doublets in double-helical DNA determined by fitting calculated melting profiles to observed profiles. *Biopolymers* **20:**1033–1042.
15. **Greer, C. E., S. L. Peterson, N. B. Kiviat, and M. M. Manos.** 1991. PCR amplification from paraffin-embedded tissues. Effects of fixative and fixation time. *Am. J. Clin. Pathol.* **95:**117–124.
16. **Hosta, L., and P. Flick.** 1991/1992. Enhancement of specificity and yield in PCR, p. 1–5. *In Editorial Comments*, vol. 18. U.S. Biochemical Corp., Cleveland, Ohio.
17. **Ibrahim, A., W. Liesack, and E. Stackebrandt.** 1992. Polymerase chain reaction-gene probe detection system specific for pathogenic strains of *Yersinia enterocolitica*. *J. Clin. Microbiol.* **30:**1942–1947.
18. **Innis, M. A., and D. H. Gelfand.** 1990. Optimization of PCRs, p. 3–12. *In* M. A. Innis, D. H. Gelfand, J. J. Sninsky, and T. J. White (ed.), *PCR Protocols: A Guide to Methods and Applications*. Academic Press, Inc., San Diego, Calif.

19. **Itakura, K., J. S. Rossi, and R. B. Wallace.** 1984. Synthesis and use of synthetic oligonucleotides. *Annu. Rev. Biochem.* **53:**323–356.
20. **Jacobs, K. A., R. Rudersdorf, S. D. Neill, J. P. Dougherty, E. L. Brown, and E. F. Fritsch.** 1988. The thermal stability of oligonucleotide duplexes is sequence independent in tetra-alkylammonium salt solutions: application to identifying recombinant DNA clones. *Nucleic Acids Res.* **16:**4637–4650.
21. **Kierzek, R., M. H. Caruthers, C. E. Longfellow, D. Swinton, D. H. Turner, and S. M. Freier.** 1986. Polymer-supported RNA synthesis and its application to test the nearest-neighbor model for duplex stability. *Biochemistry* **25:**7840.
22. **Kwok, S., D. E. Kellogg, N. McKinney, D. Spasic, L. Coda, C. Levenson, and J. J. Sninsky.** 1990. Effects of primer-template mismatches on the polymerase chain reaction: human immunodeficiency virus type 1 model studies. *Nucleic Acids Res.* **18:**999–1005.
23. **Lucas, K., M. Busch, S. Mössinger, and J. A. Thompson.** 1991. An improved microcomputer program for finding gene- or gene family-specific oligonucleotides suitable as primers for polymerase chain reactions or as probes. *Comput. Appl. Biosci.* **7:**525–529.
24. **McBride, L. J., and M. H. Caruthers.** 1983. Phosphoramidites useful for synthetic deoxyoligonucleotides. *Tetrahedron Lett.* **24:**245–248.
25. **McConlogue, L., M. A. D. Brow, and M. A. Innis.** 1988. Structure-independent DNA amplification by PCR using 7-deaza-2′-deoxyguanosine. *Nucleic Acids Res.* **16:**9869.
26. **McGraw, R. A., E. K. Steffe, and S. M. Baxter.** 1990. Sequence-dependent oligonucleotide-target duplex stabilities: rules from empirical studies with a set of twenty-mers. *BioTechniques* **8:**674–678.
27. **Medlin, L., H. J. Elwood, S. Stickel, and M. L. Sogin.** 1988. The characterization of enzymatically amplified eukaryotic 16S-like rRNA-coding regions. *Gene* **71:**491–499.
28. **Meinkoth, J., and G. Wahl.** 1984. Hybridization of nucleic acids immobilized on solid supports. *Anal. Biochem.* **138:**267–284.
29. **Mullis, K.** 1991. The polymerase chain reaction in an anemic mode: how to avoid oligodeoxyribonuclear fusion. *PCR Methods Applic.* **1:**1–4.
30. **Osborne, B. I.** 1992. HyperPCR: a Macintosh HyperCard program for the determination of optimal PCR annealing temperature. *Comput. Appl. Biosci.* **8:**83.
31. **Panaccio, M., and A. Lew.** 1991. PCR based diagnosis in the presence of 8% (v/v) blood. *Nucleic Acids Res.* **19:**1151.
32. **Pang, J., J. Modlin, and R. Yolken.** 1992. Use of modified nucleotides and uracil-DNA glycosylase (UNG) for the control of contamination in the PCR-based amplification of RNA. *Mol. Cell. Probes* **6:**251–256.
33. **Persing, D. H., D. Mathiesen, W. F. Marshall, S. R. Telford, A. Spielman, J. W. Thomford, and P. A. Conrad.** 1992. Detection of *Babesia microti* by polymerase chain reaction. *J. Clin. Microbiol.* **30:**2097–2103.
34. **Powell, J. R., and A. Caccone.** 1990. The TEACL method of DNA-DNA hybridization: technical considerations. *J. Mol. Evol.* **30:**267.
35. **Quartin, R. S., and J. G. Wetmur.** 1989. The effect of ionic strength on the hybridization of oligodeoxynucleotides with reduced charge due to methylphosphonate linkages to unmodified oligodeoxynucleotides containing the complementary sequence. *Biochemistry* **28:**1040.
36. **Relman, D. A., J. S. Loutit, T. M. Schmidt, S. Falkow, and L. S. Tompkins.** 1990. The agent of bacillary angiomatosis: an approach to the identification of uncultured pathogens. *N. Engl. J. Med.* **323:**1573–1580.
37. **Relman, D. A., T. M. Schmidt, R. P. MacDermott, and S. Falkow.** 1992. Identification of the uncultured bacillus of Whipple's disease. *N. Engl. J. Med.* **327:**293–301.
38. **Reyes, A., J. L. Dieguez, I. Garcia-Chacon, M. Morell, M. Ruiz, and J. C. Montilla.** 1992. Contaminating DNA in RNA amplification by polymerase chain reaction. *Clin. Chem.* **38:**1187.
39. **Rychlik, W., and R. E. Rhoades.** 1989. A computer program for choosing optimal

oligonucleotides for filter hybridization, sequencing and *in vitro* amplification of DNA. *Nucleic Acids Res.* **17:**8543–8551.
40. **Rychlik, W., W. J. Spencer, and R. E. Rhoads.** 1990. Optimization of the annealing temperature for DNA amplification *in vitro. Nucleic Acids Res.* **18:**6409–6412.
41. **Sarkar, G., J. Cassady, C. D. Bottema, and S. S. Sommer.** 1990. Characterization of polymerase chain reaction amplification of specific alleles. *Anal. Biochem.* **186:**64.
42. **Sarkar, G., and S. S. Sommer.** 1990. Formamide can dramatically improve the specificity of PCR. *Nucleic Acids Res.* **18:**7465.
43. **Smith, T. J., M. Espy, and D. H. Persing.** Submitted for publication.
44. **Weisburg, W.** Personal communication.
45. **Weisburg, W., S. M. Barns, D. A. Pelletier, D. J. Lane.** DNA amplification for phylogenetic study. *J. Bacteriol.* **173:**697–703.
46. **Wetmur, J. G.** 1991. DNA probes: applications of the principles of nucleic acid hybridization. *Crit. Rev. Biochem. Mol. Biol.* **26:**227–259.
47. **White, T. J., T. Bruns, S. Lee, and J. Taylor.** 1990. Amplification and direct sequencing of fungal ribosomal RNA genes for phylogenetics, p. 315–322. *In* M. A. Innis, D. H. Gelfand, J. J. Sninsky, and T. J. White (ed.), *PCR Protocols: A Guide to Methods and Applications*. Academic Press, Inc., San Diego, Calif.
48. **Wieland, I., G. Bolger, G. Asouline, and M. Wigler.** 1990. A method for difference cloning: gene amplification following subtractive hybridization. *Proc. Natl. Acad. Sci. USA* **87:**2720–2724.
49. **Woese, C. R.** 1987. Bacterial evolution. *Microbiol. Rev.* **51:**221–271.
50. **Woese, C. R., E. Stackerbrant, T. J. Macke, and G. E. Fox.** 1985. A phylogenetic definition of the major eubacterial taxa. *Syst. Appl. Microbiol.* **6:**143–151.
51. **Wright, D. K., and M. M. Manos.** 1992. Sample preparation from paraffin-embedded tissues, p. 153–158. *In* M. A. Innis, D. H. Gelfand, J. J. Sninsky, and T. J. White (ed.), *PCR Protocols: A Guide to Methods and Applications*. Academic Press, Inc., San Diego, Calif.

# Amplification Product Inactivation Methods

# 5

David H. Persing and George D. Cimino

The expanding role of nucleic acid amplification technology in the clinical laboratory will be accompanied by new types of problems, many of which are precipitated by the technologies themselves. Ironically, the very attribute that makes amplification techniques so powerful—extreme sensitivity—also represents their Achilles' heel; contamination of samples by tiny amounts of DNA or RNA targets (amounts that would be undetectable by conventional methods) may lead to false-positive results. Nucleic acid contamination may occur from a variety of sources: (i) clinical specimens containing large numbers of target molecules, resulting in crossover contamination between specimens (this type of contamination is already well known to laboratory personnel) (38); (ii) nucleic acids from organisms grown in culture, or cloned DNA from such organisms (DNA clones derived from the organism of interest represent a particularly insidious problem because widespread contamination of laboratory equipment and reagents may occur); and (iii) accumulation of amplification products by repeated amplification of the same target sequence. The first two forms of contamination can be relatively easily avoided by using careful techniques, observing stringent quality control practices, and avoiding areas of the laboratory used for culture or DNA cloning. However, contamination by previously amplified nucleic acids can occur in any laboratory and may be extremely difficult to overcome, despite strict precautions (22, 23, 25, 32).

Nucleic acid contamination was first identified in routine applications of the polymerase chain reaction (PCR) in research laboratories (22, 37). A single PCR may contain as many as a trillion copies of an amplified target sequence, each of which is a proven substrate for further amplification. This is a staggering number of molecules—enough to cover every square mile of land and water on the surface of the earth with 5,000 copies. When the contents of the same reaction are distributed across the modest square footage of a clinical laboratory, serious problems of false positivity may result. The seemingly inexorable accumulation of amplified products may lead to contamination of reagents, working surfaces, air supply ducts, and the skin and hair of laboratory workers (22).

Each of the enzyme-catalyzed amplification methods described in chapter 3 (and virtually all such methods described in the future) are vulnerable to the problem of amplification product contamination. In PCR, ligase chain reaction, and strand displacement amplification, DNA products will accumulate, whereas in protocols involving self-sustaining sequence replication (3SR) or Qβ replicase, RNA products will predominate. Fortunately, several amplification product inactivation methods have now been described for use with PCR; many of these

**Figure 1.** Temporal steps in a diagnostic assay that involves PCR amplification. The stippled boxes indicate time points where inactivation protocols can be implemented.

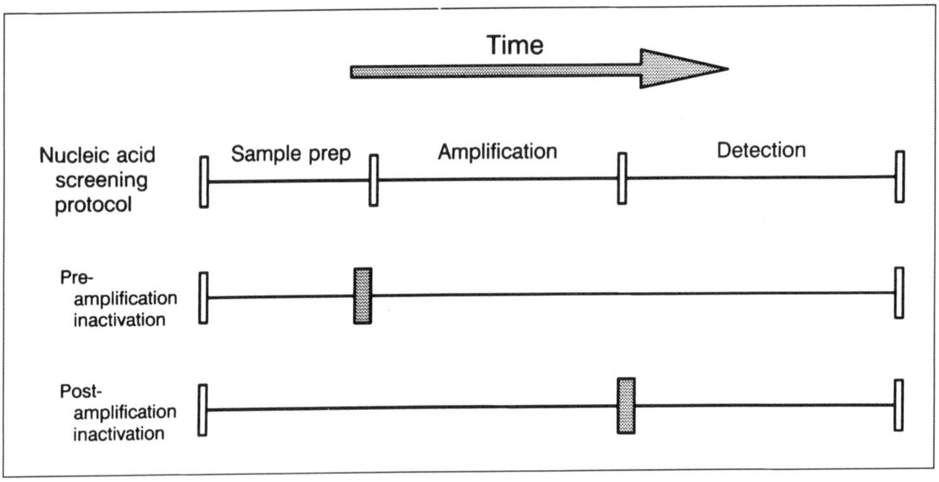


may also be useful for inactivating the products of other amplification methods. This chapter will describe these techniques in detail and will provide practical suggestions for their implementation.

## Inactivation of Amplified Nucleic Acids: Theoretical and Temporal Considerations

The numerous techniques available for sterilizing biological organisms either are not directly compatible with the amplification and detection procedures or are simply inappropriate. Nucleic acid denaturation by thermal or chemical means is ineffective since a single-stranded nucleic acid sequence is readily amplifiable. A major mode by which carryover is spread is through aerosols created by handling PCR products (amplicons). Initial attempts to control contamination in the research laboratory included physical separation of pre- and postamplification steps (23). However, despite these containment procedures, in the practical experience of the research laboratory, carryover contamination from a few to a few hundred copies of amplicons is still observed. Several sophisticated techniques are now available for inactivation of amplification products. These techniques include photochemical, enzymatic, and chemical modifications of the nucleotide bases or sugars of the amplification products, thus preventing the altered strands from acting as templates for *Taq* DNA polymerase extension reactions.

A diagnostic assay based on nucleic acid amplification generally consists of the three phases illustrated in Fig. 1. Nucleic acid from a clinical sample is prepared, mixed with amplification reagents, and placed in a reaction vessel. The reaction vessel is closed and incubated at a set temperature or temperatures at which amplification occurs through a cascading series of polymerase extension or ligation reactions. At the end of the incubation, the reaction vessel is opened and a sample of the reaction products is removed for detection. Two criteria must be met by amplification product control procedures to ensure that future diagnostic assays are not compromised by a carryover event. The first is that the reaction products are not exposed to the environment until they are modified in a specific manner that distinguishes them from true target sequences in subsequent amplifications. This requires that the distinguishing modification occur before opening the reaction vessel for the detection phase of the diagnostic assay. The second criterion is that the modification(s) must not interfere with amplification or subsequent detection steps. Sensitivity and specificity must be maintained.

An examination of the sequence of events in an amplification protocol leads to different philosophical approaches to inactivation. In one approach, inactivation occurs prior to amplification (preamplification). The basic principle behind preamplification inactivation is to eliminate contaminating nucleic acids by selective modification or degradation prior to initiation of amplification. This approach acknowledges that contamination will occur but provides an effective remedy in the form of a preamplification "clean-up" step.

In the postamplification approach, the amplified nucleic acid is modified after amplification. Two types of postamplification methods will be discussed in this chapter: chemical and photochemical. Chemical inactivation is performed by the addition of hydroxylamine or sodium hydroxide to the reaction vessels after amplification. For photochemical inactivation, chemical reagents (furocoumarins) which tolerate the temperature conditions used for amplification are added to the reaction mixtures prior to amplification (8); after amplification, but before the reaction vessels are opened, the furocoumarins are covalently linked to the nucleic acids by exposure to long-wavelength UV light. The philosophy of the chemical and photochemical approaches is to create inactivated amplification products that will meet the detection criteria of the assay yet will not function as a template if carried over into a subsequent reaction.

## Preamplification UV Light Irradiation

The first method described for inactivation of amplification products was preamplification irradiation of PCR components by short-wavelength UV light (7, 14, 30, 39–41). This method is based on the well-known property of short-wavelength UV light to create thymidine dimers and other covalent modifications of DNA, thus rendering the contaminating nucleic acid inactive as a template for further amplification. The method offers several advantages—it is simple to perform, is inexpensive, and does not require modification of existing protocols. However, widely varying efficiencies have been reported, probably because the effectiveness of UV light for inactivation of a template is highly dependent on the length and nucleotide composition of the template (7, 14, 30). Short templates (less than ca. 300 bp) are incompletely inactivated, and the efficiency of the process is decreased further for G+C-rich templates (7, 12). Since many amplification protocols detect and amplify templates that are less than 300 nucleotides, the usefulness of this technique may be limited. Furthermore, short-wavelength UV irradiation may have deleterious effects on *Taq* polymerase and oligonucleotides, which may reduce amplification efficiency (30). Recent studies have indicated that free deoxyribonucleoside triphosphates (dNTPs) present in amplification mixtures may exert a powerful shielding effect that protects against UV-induced damage of DNA templates (16). Thus, the elimination of dNTPs from reaction mixtures may increase the effectiveness of pre-PCR UV irradiation. Gamma irradiation has recently been used for a similar purpose (11).

Pre-PCR UV irradiation may be effective for reducing the template activity of contaminating bacterial DNA within *Taq* polymerase (most probably introduced during the commercial enzyme purification procedures) (3, 36, 42) or PCR

**Figure 2.** Steps of the photochemical post-PCR inactivation procedure.

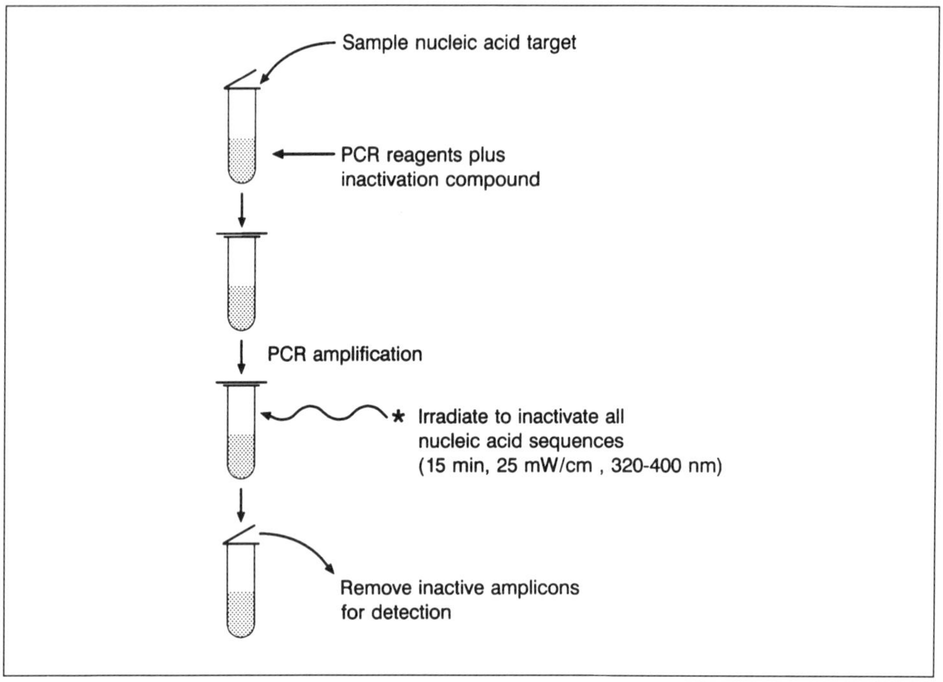


reagents prior to amplification with broad-range (universal bacterial) primers (chapter 3; protocols 1.1, 1.21, and 5.1). Elimination of exogenous sources of bacterial DNA is especially important when attempting to amplify small numbers of bacterial genomes from clinical specimens by this approach. In general, the targets selected for amplification in this setting are large (500 to 1,600 bp) and are effectively eliminated. However, Meier et al. (27) found UV irradiation to be ineffective for elimination of a contaminating 920-bp 16S bacterial ribosomal DNA target in reactions containing commercially prepared *Taq* polymerase. These investigators employed 8-methoxypsoralen followed by long-wavelength UV irradiation to eliminate contaminating DNA; this simple pretreatment enabled the identification of a novel uncultured mycobacterial species (21, 27).

## Photochemical Inactivation with Furocoumarins

The conditions and requirements for postamplification photochemical inactivation of PCR amplification products have been described recently (8, 20). The procedure is illustrated schematically in Fig. 2. At the time of preparation, the target nucleic acid and PCR reagents are placed in a standard microcentrifuge tube along with a photochemical reagent. The tube is closed and subjected to the amplification conditions. After amplification, but before the tube is opened and the amplification products are exposed to the atmosphere, the photochemical reagent is activated by briefly subjecting the tube to light at 300 to 400 nm. The polypropylene tubes that are used in PCR thermal cyclers are adequately transparent to these wavelengths to permit photoactivation. After the photochemical process is complete, the tube is opened and the modified PCR products are

**Figure 3.** Structure and reactive sites of psoralens and isopsoralens in double-stranded nucleic acids. (A) A psoralen molecule is shown intercalated between two strands of a DNA double helix. The intercalation site shown is created by adjacent pyrimidines on opposite strands of DNA. Two types of monoadducts can exist at this site when a cyclobutane ring is formed between either the 4′,5′ position of the psoralen and the 5,6 position of the pyrimidine above the intercalated psoralen molecule (a furan-side monoadduct), or the 3,4 position of the psoralen and the 5,6 position of the pyrimidine below the psoralen molecule (a pyrone-side monoadduct). The geometric orientation of the two reactive bonds (3,4 and 4′,5′) of the psoralen permits both adducts to occur simultaneously, forming a covalent cross-link between the two DNA strands. (B) An isopsoralen molecule is shown intercalated between the same two strands of a DNA helix. In analogy with the psoralens, two types of monoadducts can occur on reaction with the pyrimidines at the intercalation site. However, the alignment of the two reactive positions excludes the simultaneous occurrence of both the furan-side and pyrone-side adducts in a DNA helix. Only monoadducts are formed with this class of compounds.

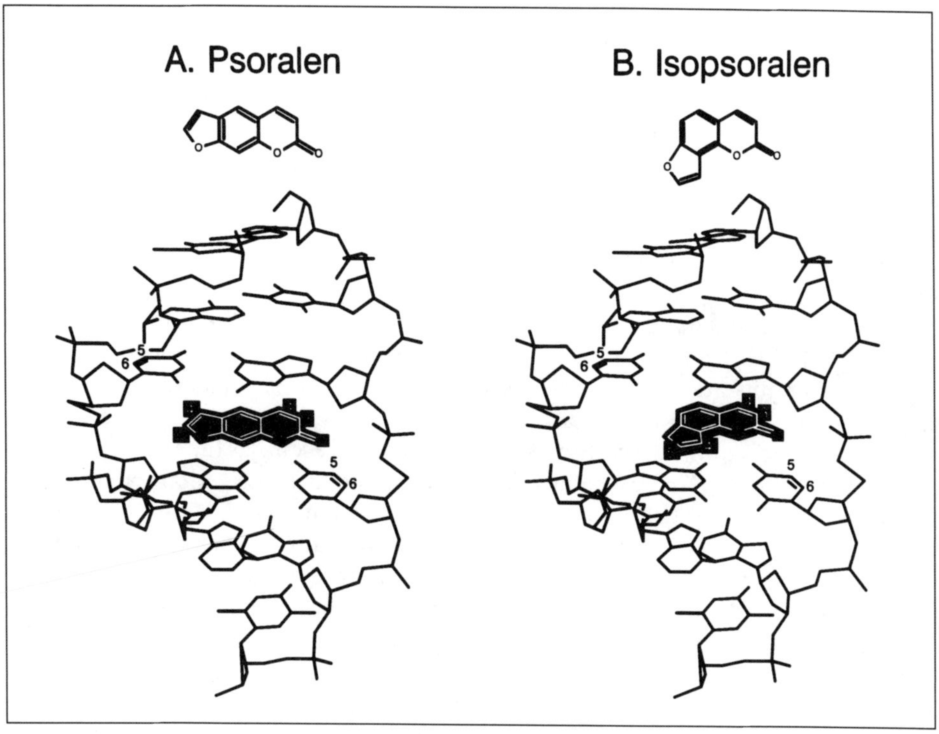


detected in the usual fashion as either double-stranded DNA on an agarose gel or single-stranded DNA with hybridization to a complementary probe.

Derivatives of both psoralens and isopsoralens have been evaluated as photochemical reagents for this inactivation process. These furocoumarin compounds represent a class of planar tricyclic reagents that are known to intercalate between base pairs of nucleic acids as illustrated in Fig. 3 (6). These compounds contain two reactive double bonds which, when excited, covalently bind to nucleic acid polymers by forming cyclobutane adducts with pyrimidine bases. Modified amplicons that are carried over into new PCRs are refractory to a new round of amplification because the furocoumarin-pyrimidine adducts on the carryover molecules serve to block the template-dependent extension reactions carried out by *Taq* DNA polymerase (8). The wavelengths used for this inactivation process are in the range of 300 to 400 nm. These wavelengths minimize direct UV damage to nucleotides contained within the amplicons yet are suitable for use in the laboratory without the need for special lighting conditions that prevent unwanted activation of the furocoumarin compounds on the benchtop. In laboratories that have direct exposure to sunlight, caution is warranted because there may be sufficient light to activate furocoumarins on the bench.

The photochemical inactivation procedure is effective with either psoralen or isopsoralen derivatives. The psoralen reagents are capable of forming covalent cross-links between the complementary strands of a PCR product (Fig. 3A). This mode of inactivation limits the detection of the amplification products to formats that rely on double-stranded DNA products only. Although this is possible in some instances, the usual detection format used in PCR protocols relies on hybridization of one strand of the product to a complementary internal probe. Isopsoralen compounds (Fig. 3B) should be used when hybridization-based detection methods are employed since they do not cross-link complementary strands and the modified amplicons are themselves capable of efficient and specific hybridization (3, 9, 20, 43).

Practical application of the photochemical inactivation technique required the synthesis of novel amino derivatives of isopsoralen. The available natural compounds did not have sufficient reactivity with polynucleotides; added amino groups were required to solubilize the isopsoralens for use at concentrations that are effective. Additionally, the positive charge of the new derivatives enhanced binding to the negatively charged single strand of the amplicon sequences. In a previous report (8), identical copy numbers of each of the single strands of a 115-mer PCR product were inactivated with 6-aminomethyl-4,5′-dimethylisopsoralen (6-AMDMIP) either as single-stranded oligomers or as a complementary duplex. Reamplification of known copy numbers of these molecules illustrated that the single-stranded oligomers were as effectively inactivated as the duplex hybrid with the amino-derivativized isopsoralen. Presently, there is extensive experience with three isopsoralens for their amplicon inactivation capabilities: 4′-AMDMIP, 6-AMDMIP, and IP-10 (a proprietary amino derivative of isopsoralen available from HRI Associates, Inc., Concord, Calif.). At similar concentrations, different levels of inactivation efficiency of these compounds are observed with different PCR products. These differences can be attributed to preferential reactivities of these compounds with different sequences. However, at this time there are not enough data to determine the sequence preferences of these compounds; one must rely on empirical results to determine the effectiveness of the reagents with a particular amplicon sequence.

## Choice of Primers, Isopsoralen Concentration, and Cosolvents for Effective Inactivation

Photochemical inactivation can be designed so that virtually all amplicons are inactivated. However, the inactivation process is statistical and partial inactivation can occur when some of the amplicon strands escape the modification process. The absolute number of adducts per amplicon strand is not a constant value for each strand but is a variable number and is statistically distributed throughout the entire population of the photochemically altered molecules. The fraction of strands that escape the photochemical modification, and are therefore not inactivated, is determined by the length and sequence of amplicon and the concentration of the inactivation reagent. Figure 4 illustrates the length effect of this process. A 115-nucleotide PCR product is much more difficult to inactivate than a 500-nucleotide product under identical conditions. Since A+T-rich se-

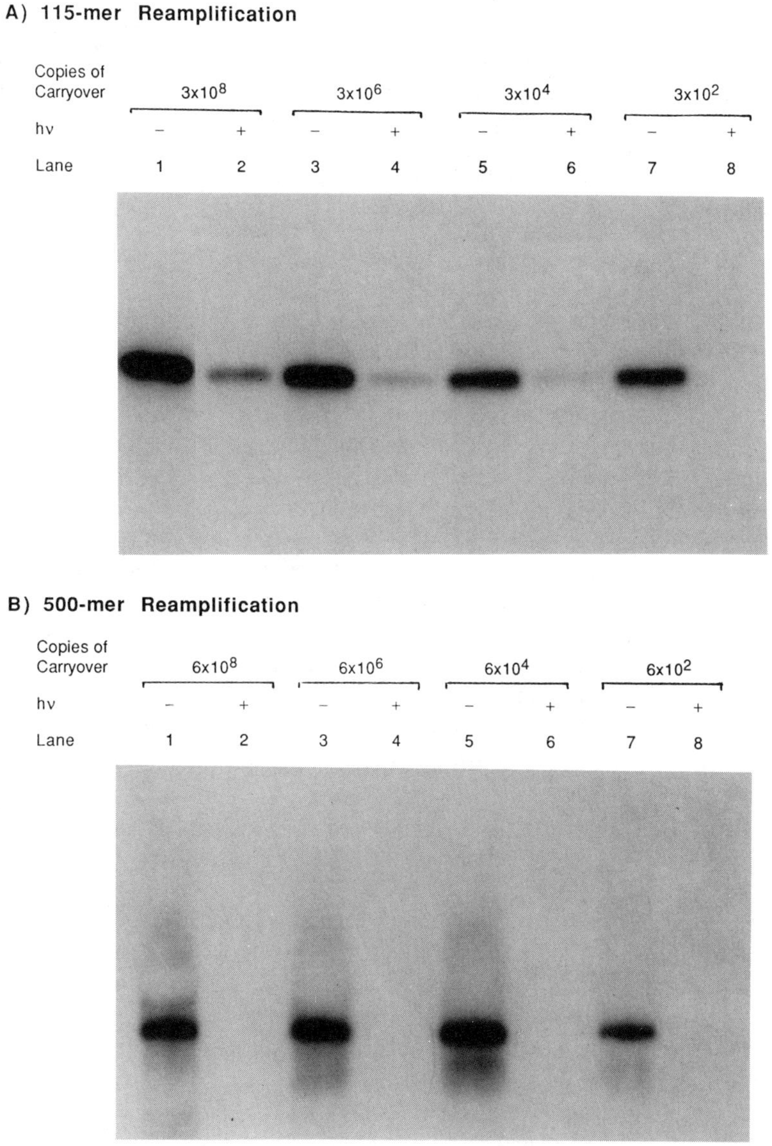


**Figure 4.** Comparison of 4′-AMDMIP inactivation of a 115-mer amplicon with inactivation of a 500-mer amplicon. The reamplification results are shown in the autoradiograph. Panel A corresponds to reamplification of a 115-mer amplicon and panel B corresponds to reamplification of a 500-mer amplicon. Lanes 1, 3, 5, and 7 are reamplifications of the nonirradiated controls. Lanes 2, 4, 6, and 8 are reamplifications from inactivated amplicons. (Reprinted from reference 8.)

quences are more reactive with psoralens, a judicious choice of primers that generate long (>200 to 300 bp, depending on the G+C content) or A+T-rich products will facilitate the inactivation process (see chapter 4). Inactivation efficiency is also improved by increasing the isopsoralen concentration. The most useful range of concentrations is 25 to 200 μg/ml. Higher concentrations of the positively charged reagents compromise detection sensitivity by either inhibiting the amplification process or reducing the hybridization efficiency of the inactivated products. The inhibition of PCR amplification is partially reduced by the addition of up to 10% glycerol as a cosolvent in the amplification mixture (20).

**Figure 5.** Effects of isopsoralen cross-linking on amplification efficiency, gel mobility, and hybridization. A 156-bp *B. burgdorferi* DNA target was amplified in the presence of various concentrations of isopsoralen IP-10. Concentrations of IP-10 are listed at the top of each lane. See text for additional details. (Reprinted from reference 34.)

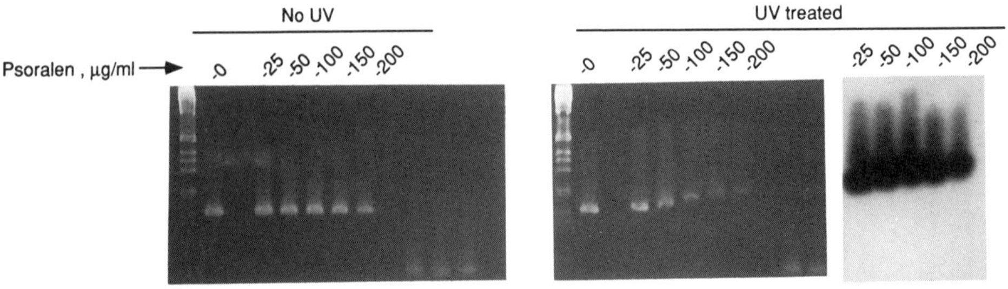


The advantage afforded by the isopsoralen-modified PCR products is their ability to retain specific and efficient hybridization properties. This attribute defines the upper concentration limit for the isopsoralen reagent in an inactivation protocol. An extreme carryover event can lead to a positive signal by direct detection of the carryover molecules themselves. The number of carryover molecules that are directly detected in the absence of amplification depends on the detection format. For a hybridization format in which a single $^{32}$P-radiolabeled oligonucleotide probe is used, $10^7$ to $10^8$ molecules are sufficient to be directly detected. It is important to identify the direct detection level because it also represents the level of inactivation that should be strived for when designing a postamplification inactivation protocol. Levels sufficient to produce maximal inactivation without affecting sensitivity should be chosen to help prevent "amplicon buildup" (32). However, use of an extremely high isopsoralen concentration that inactivates beyond the direct detection level is not required and may result in unnecessary sensitivity losses.

## Modification of Existing PCR Protocols to Incorporate Isopsoralen Inactivation

The use of isopsoralens as photochemical product inactivation reagents requires that the statistical nature of the modification process be matched with the level of amplification and the length and sequence of the amplicon. To incorporate the isopsoralen method into an existing PCR assay, the design procedure should begin with an assessment of the tolerable level of carryover so that the length and sequence of an amplicon can be matched with a practical level of isopsoralen concentration. At the direct detection level of $10^8$ molecules, 20 effective adducts per PCR strand are necessary to be confident that the carryover event does not contain an active template. For genomic DNA with a 50% G+C content, modification densities from 1 adduct per 100 bases to 1 adduct per 11 bases can be achieved by varying the isopsoralen (4′-AMDMIP) concentration range from 10 to 400 µg/ml. These considerations will determine a range of isopsoralen concentrations that should be appraised for their ability to inactivate and for their effect on detection sensitivity.

Amplification efficiency and hybridization efficiency both contribute to the detection sensitivity of the assay. Amplification efficiency in the presence of isopsoralen should be evaluated under conditions that are far away from the PCR plateau so that inhibitory effects can be seen. By using tolerable concentrations of isopsoralen, the hybridization efficiency of the modified amplicons should be determined. Finally, the number of PCR cycles should be adjusted to bring the detection level back to the level that exists prior to implementation of the inactivation procedure. Typically, two or three additional cycles of amplification are necessary for protocols in which high concentrations of isopsoralen are used. By following these steps, an effective and practical post-PCR inactivation protocol that helps to eliminate the carryover problem associated with the PCR will emerge.

The effects of isopsoralen concentration on the mobility and hybridization of a 156-bp *Borrelia burgdorferi* target are shown in Fig. 5. At various isopsoralen concentrations without UV exposure, there is a slight inhibitory effect on amplification efficiency at the highest isopsoralen concentration (left panel). With UV exposure, there is a significant, concentration-dependent increase in the apparent molecular weight of the products (center panel). In addition, there is a small concentration-dependent decrease in hybridization efficiency that is most pronounced at the higher isopsoralen concentrations (right panel).

## Enzymatic Inactivation

### Inactivation by Uracil-*N*-Glycosylase

Bacteria have evolved sophisticated systems for DNA repair following DNA damage, misincorporation of nucleotides by polymerases, or chemical modification of nucleotide residues. Uracil-*N*-glycosylase (UNG) is a DNA repair enzyme found in a wide variety of bacterial species (24, 44). The normal function of this enzyme is to remove uracil residues created by the spontaneous deamination of cytosine residues in double-helical DNA; failure of the organism to recognize and remove uracil residues would otherwise result in a high frequency of C-to-T transition mutations as a result of misincorporation of A residues on deoxyuracil-containing templates. The function of UNG is thus to recognize and cleave uracil residues from the DNA backbone as a component of excision repair.

In 1990, Longo et al. reported on the use of UNG to inactivate PCR amplification products (26). The principle of UNG inactivation is illustrated in Fig. 6. During amplification, TTP is replaced with dUTP, which serves as a substrate for *Taq* polymerase in place of TTP. The amplification products thus contain deoxyuracil residues. This serves to distinguish PCR products from starting template on a chemical basis; only newly created products (in contrast to the starting template) will contain deoxyuracil. If dUTP-containing amplification products are present as contaminants, the addition of UNG to reaction mixtures results in cleavage of deoxyuracil residues from the DNA backbone. The resulting abasic site is highly susceptible to alkaline hydrolysis at elevated temperatures and subsequent strand scission of the template. Since PCR is normally carried out under conditions of slightly alkaline pH and high temperature, the template activity of UNG-treated PCR products is eliminated (Fig. 6).

To serve as a reagent in an amplification product inactivation method, UNG is added to reactions prior to amplification. During a short incubation prior to amplification, contaminating (dU-containing) PCR products that may be present are cleaved by UNG, inactivating them as templates. In the initial cycles of PCR, primer-directed DNA synthesis thus occurs exclusively from the input template, which is normally spared the action of UNG. The UNG inactivation protocol is shown schematically in Fig. 7.

**Figure 6.** Inactivation of previously amplified DNA by selective action of UNG. UNG is added to reactions prior to amplification. Selective cleavage of uracil residues from any previously amplified DNA present as a contaminant results in elimination of the template activity of contaminants.

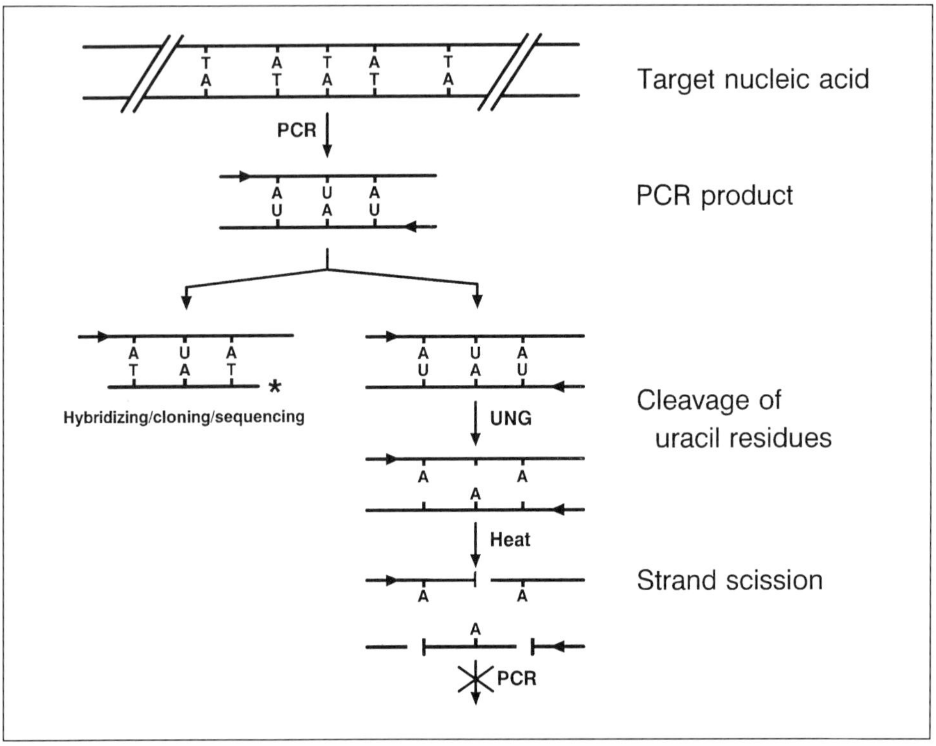


## Capacity of UNG Inactivation

In the original report of Longo et al. (26), UNG inactivation was capable of eliminating up to 0.1 fg of dU-containing PCR product, depending on whether dU was present in the primers used for PCR or incorporated into the product by substitution of dUTP for TTP in the reaction mixtures. In our experience (DHP) the UNG protocol has been capable of eliminating up to $10^{10}$ copies of a 156-bp A+T-rich amplification product (34). Like the photochemical inactivation method, the efficiency of inactivation may depend on the length and G+C content of the product (with G+C-rich products being more difficult to inactivate). However, because every thymidine residue in the product can theoretically be replaced by uracil, the UNG protocol may be more effective for short G+C-rich amplicons than UV irradiation or photochemical treatment.

## Enzymatic "Hot Start"

In a typical PCR protocol, reaction components and template are mixed at ambient temperatures and then placed into a thermal cycler in which higher temperatures are maintained. However, during the setup period prior to thermal cycling, *Taq* polymerase is active and can extend primer-target duplexes at temperatures as low as 4°C. Because the primer annealing conditions at these temperatures are nonstringent, numerous nonspecific primer extension products are formed, which may result in poor reaction specificity. Poor specificity may

**Figure 7.** Protocol for the use of UNG inactivation. See the text for additional details.

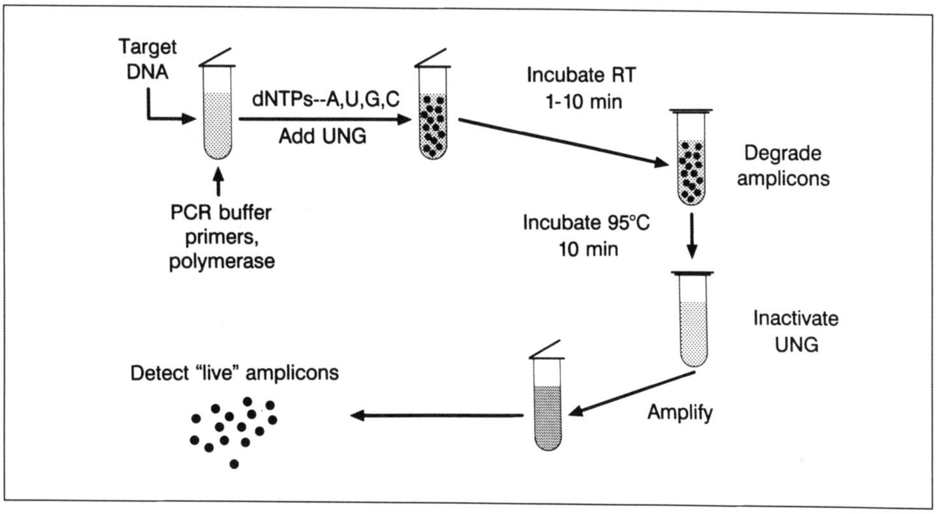


also reduce sensitivity because of competition between specific and nonspecific products for reaction components. To reduce the effects of low-stringency primer extension, the "hot-start technique" has been developed (2, 5); in this technique, an essential reaction component is withheld until elevated temperatures are attained (2, 28).

The UNG inactivation protocol takes advantage of the differences in temperature for optimal UNG and *Taq* polymerase activity to produce the enzymatic equivalent of a hot start. At ambient temperatures, nonspecific extension products are subject to degradation by UNG, because optimal UNG activity is expressed at around 25°C. UNG retains partial activity up to 55°C, which approaches the optimal annealing temperatures for most primers. Above 55°C, UNG loses activity, and primer extension by *Taq* polymerase becomes more efficient. Thus, prior to the first cycle, the contribution of nonspecific extension products to poor reaction specificity is markedly reduced. The practical impact of the enzymatic hot-start technique is discussed further in protocol 2.1.

## Optimization of Reactions for UNG Inactivation

As for all of the inactivation methods, amplification reactions incorporating UNG inactivation should be carefully optimized (18). Primers should be chosen such that amplification products are large enough to be inactivated efficiently, yet not so large as to reduce amplification efficiency (chapter 4). The efficiency of the UNG protocol also depends to some degree on the concentration of target residues in the product. G+C-rich products may be more difficult to inactivate because of the lower concentration of uracil residues.

Some investigators have observed a loss of sensitivity after incorporation of the UNG protocol (31). However, this is not likely to be a universal experience, considering the advantages of the enzymatic hot start. Losses of sensitivity and specificity can usually be avoided if several factors are considered. (i) The optimal dUTP concentration that will allow efficient amplification must be determined empirically. Many protocols will tolerate complete substitution of dUTP, but some will require either an increase in dUTP concentration relative to the ot'

**Figure 8.** Inactivation of previously amplified DNA by primer hydrolysis. Primers containing single or multiple ribose residues at their 3′ end were used to amplify the target nucleic acid. After amplification, the ribose linkage is cleaved by alkaline hydrolysis. This results in cleavage products that can be amplified only linearly, which will not result in false positivity due to geometric amplification of contaminating sequences. See the text for additional details.

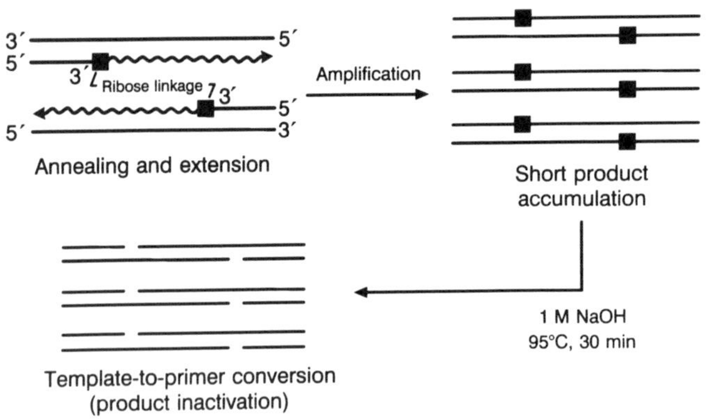


dNTPs or a mixture of TTP and dUTP. This may depend to some degree on the G+C content of the target sequence. (ii) UNG will regain activity, even after multiple cycles of amplification, if temperatures within the reaction mixture drop below 55°C. Residual UNG activity may affect the reaction sensitivity if the products are held at room temperature prior to gel loading or slot blotting (34, 47). Thus, reaction components should be held up to overnight at 72°C prior to analysis or chemically treated with NaOH before slot blotting. This will prevent postamplification degradation and the resultant loss of signal. In addition, primers and protocols should be designed so that primer annealing temperatures greater than 55°C are used in thermal-cycling profiles. (iii) dUTP substitution may reduce the affinity of internal oligonucleotide probes that are used to confirm the presence of specific PCR products. In general, the reduced affinity can be overcome by reducing the stringency of probe hybridization or washing.

By carefully optimizing each reaction for the incorporation of dUTP and the addition of UNG, the amplification efficiency of both new and existing protocols can meet that of unmodified protocols. By virtue of the enzymatic hot-start effect, the efficiency of amplification from complex mixtures (i.e., those containing large amounts of human genomic DNA) may actually be improved. Future improvements to this technology will probably include the use of more thermolabile forms of UNG or the use of phage-derived UNG inhibitors (48–50).

## Chemical Inactivation Protocols

### Primer Hydrolysis

Recently, an amplification product inactivation method based on selective base hydrolysis of ribose linkages was described (19). In this protocol, special primers that contain one or more ribose linkages at their 3′ end are synthesized. Despite the presence of a functional RNA-DNA duplex formed between the 3′ end of the primer and the template, *Taq* polymerase is capable of carrying out primer extension. When both primers contain a 3′ ribose linkage, each strand of the

**Figure 9.** Schematic diagram of inactivation of amplification products by primer hydrolysis. See the text for additional information.

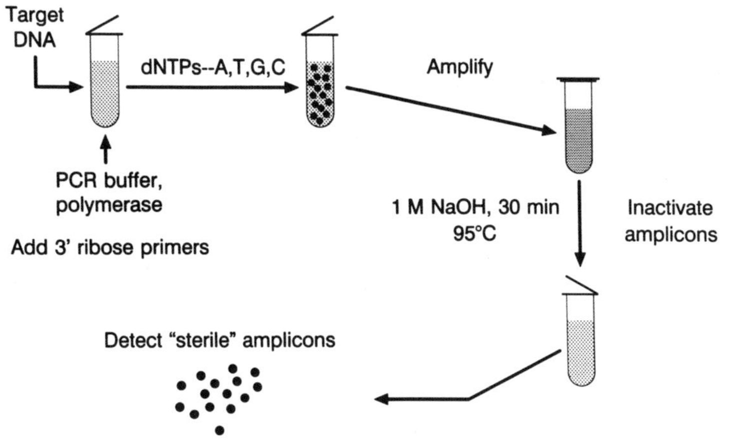


amplification product contains a ribose residue(s) (Fig. 8). Sugar-phosphate backbones composed of ribose residues are, in contrast to their deoxyribose counterparts, susceptible to cleavage by alkaline hydrolysis. Thus, if NaOH is added to reaction mixtures after amplification, the backbone of each strand is cleaved to separate the primer from its extension product. The primer hydrolysis protocol is shown schematically in Fig. 9.

The truncated template does not serve as an efficient template for amplification because primer extension reactions initiated on the cleaved strand do not result in the synthesis of a binding site for the opposing primer; the latter is the key component of geometric accumulation of amplification product. Instead, a linear increase in the primer extension product will result, which does not contribute significantly to the signal. Although the cleavage products could contribute to synthesis of a full-length product via discontinuous priming, the concentration of cleaved strands is too low to efficiently form such a product in most protocols. Thus, the primer hydrolysis method serves to effectively control the buildup of reamplifiable product.

Our laboratory (DHP) recently evaluated the primer hydrolysis method on a 156-bp *B. burgdorferi*-specific amplification product amplified with primers containing 3′ ribose residues (34). Amplification with ribose-containing primers was as efficient as with conventional deoxyribose-containing primers. After cleavage of reaction products with 1 M NaOH for 30 min at 95°C, reaction products were diluted in a 10-fold dilution series and reamplified to determine the efficiency of inactivation. Primer hydrolysis consistently eliminated $10^8$ to $10^9$ copies of amplification product. However, the efficiency of the method may be highly primer dependent; some primers may produce products that are much more refractory to inactivation. Nevertheless, with careful selection of primers and optimization of hydrolysis conditions, primer hydrolysis may prove to be a valuable method for preventing false-positive results due to amplicon carryover.

## Hydroxylamine Treatment

Another postamplification inactivation method relies on the ability of hydroxylamine to produce covalent modifications of DNA (1, 4, 15). Hydroxylamine preferentially reacts with the oxygen atoms in cytosine residues, creating covalent adducts that prevent base pairing with guanine residues (4). The presumed mode of action of hydroxylamine for inactivation of PCR products is to modify primer-binding sites and intervening template sequences so that annealing and reamplification cannot occur.

Aslanzadeh (1) used concentrations of hydroxylamine ranging from 100 to 250 mM to inactivate a 156-bp PCR amplification product of the *ospA* gene of *B. burgdorferi* and a 92-bp amplification product of herpes simplex virus. Final hydroxylamine levels of 250 mM resulted in a reduction in the amount of reamplifiable product by several orders of magnitude. A dose-dependent increase in the apparent molecular weight of the products was observed, as was a reduced affinity for internal oligonucleotide probes that could be overcome by lowering hybridization stringency. Interestingly, the selective modification of cytosine residues may make this method useful for inactivation of short G+C products; such products represent the most difficult to inactivate by other methods.

Inactivation of amplification products by hydroxylamine requires the post-amplification addition of an aqueous solution of the compound. In the initial description, a hydroxylamine solution was inserted with a tuberculin syringe needle through the top of reaction tubes, which minimized the risk of creating aerosols (1). However, both of the chemical inactivation methods (primer hydrolysis and hydroxylamine treatment) described in this section require addition of a reagent after amplification and thus involve exposure to the contents of amplification reactions. Future improvements may include the use of reaction tubes with injection ports that will help to avoid contact with the products of amplification.

## Topical Inactivation Methods

Although the ingenious approaches described in the preceding sections for the inactivation of amplified products are very effective, there remains a need for topical methods that can be used to clean up spills of reaction mixtures and for cleaning working surfaces as part of routine quality control programs (9, 38). The most useful and effective method to date is the application of dilute solutions (2 to 10%) of sodium hypochlorite (bleach) (17, 35). Bleach causes oxidative damage to DNA so that it cannot be amplified (17). The oxidative damage does not discriminate between previously amplified DNA and DNA extracted from a target organism; thus, the routine application of bleach to working surfaces is useful for cleaning up contamination arising from a variety of sources. Prince and Andrus (35) recently demonstrated the usefulness of bleach for inactivating amplified DNA and compared the effects of bleach and depurination by hydrochloric acid treatment. Bleach appeared to be several orders of magnitude more effective. However, bleach also attacks finishes and metal surfaces; it is important to rinse bleach-treated surfaces with autoclaved water or 70% ethanol after treatment. Regular use of 10% bleach is an essential component of PCR laboratory quality control (chapter 9).

UV lights placed over working surfaces may also be effective in controlling contamination. For maximum effectiveness, the UV source should be within a foot or two (30 to 60 cm) of the working surface to provide a dosage adequate to cross-link small DNA targets. Fairfax et al. (13) recently evaluated the effectiveness of UV lights in a biosafety cabinet for reducing contamination with amplified DNA. They concluded that UV light was only marginally effective when irradiation was limited to 1 or 2 h but that overnight irradiation was effective in eliminating amplification of a 200-bp herpes simplex virus target.

## Summary

This chapter has provided a review of most of the currently available methods for inactivating amplified DNA contamination, which is the most serious threat to the routine implementation of nucleic acid amplification protocols. Other methods for controlling nucleic acid contamination have recently been described (10, 11, 29, 45, 46, 51). However, no inactivation method should be a substitute for good laboratory technique. Strict separation of pre- and postamplification components should be maintained, and specific specimen-handling guidelines should be observed (chapters 8 and 9). In addition, laboratory personnel should be constantly aware of unexpected sources of contamination, such as specimen-to-specimen carryover, inadvertent contamination of samples during aliquoting and distribution (33), and the presence of cloned sequences in laboratories that have isolated and characterized their own target sequences. The best overall approach to identifying contamination problems is to place negative control reactions at every major step in the sample-processing scheme. Despite the best of intentions, however, false-positive results may still occur; it is important to realize that in this respect nucleic acid amplification methods are like any other diagnostic procedures and occasionally suffer from false-positive and false-negative results.

Since PCR is currently the most widely accepted amplification technique, it has also enjoyed the greatest development in the area of contamination control. However, all of the enzyme-catalyzed amplification methods discussed in chapter 3 are susceptible to false-positive results due to amplified substrate accumulation. The applicability of any of the procedures described herein to these amplification methods is unknown. Theoretically, the short amplification products of the ligase chain reaction and strand displacement amplification may be more difficult to inactivate than the products of the other methods. RNA product contamination may prove to be easier to control because of the inherent lability of RNA compared with DNA. Inactivation techniques, some of which will be based on principles discussed in this chapter, will most probably be necessary for all of the nucleic acid amplification methods before they can become clinically useful. The PCR paradigm has already taught us valuable lessons about the magnitude of the contamination problem and the means for its control; amplification product inactivation methods will probably be adapted to other nucleic acid amplification protocols over the next few years as these methods are refined further.

## References

1. **Aslanzadeh, J.** 1992. Application of hydroxylamine hydrochloride for post-PCR sterilization. *Ann. Clin. Lab. Sci.* **22:**280.
2. **Bloch, W.** 1992. Wax-mediated hot start PCR: AmpliWax™ PCR gems permit nonisotopic, unprobed detection of low-copy-number targets, p. 6–9. *In* L. S. Kelly (ed.), *Amplifications. A Forum for PCR Users*, vol. 8. Perkin Elmer Corp.
3. **Böttger, E.** 1990. Frequent contamination of Taq polymerase with DNA. *Clin. Chem.* **36:**1258–1259.
4. **Brown, D. M., and P. Schell.** 1961. The reaction of hydroxylamine with cytosine and related compounds. *J. Mol. Biol.* **3:**709–710.
5. **Chou, Q., M. Russell, D. E. Birch, J. Raymond, and W. Bloch.** 1992. Prevention of

pre-PCR mis-priming and primer dimerization improves low-copy-number amplifications. *Nucleic Acids Res.* **20:**1717–1723.
6. **Cimino, G. D., H. B. Gamper, S. T. Isaacs, and J. E. Hearst.** 1985. Psoralens as photoactive probes of nucleic acid structure and function: organic chemistry, photochemistry, and biochemistry. *Annu. Rev. Biochem.* **54:**151–1193.
7. **Cimino G. D., K. C. Metchette, S. T. Isaacs, and Y. S. Zhu.** 1990. More false-positive problems. *Nature* (London) **347:**340–341.
8. **Cimino G. D., K. C. Metchette, J. W. Tessman, J. E. Hearst, and S. T. Isaacs.** 1991. Post-PCR sterilization: a method to control carryover contamination for the polymerase chain reaction. *Nucleic Acids Res.* **19:**99–107.
9. **Cone, R., A. Hobson, M. L. Huang, and M. Fairfax.** 1990. Polymerase chain reaction decontamination: the wipe test. *Lancet* **316:**686–687.
10. **DeFilippes, F. M.** 1991. Decontaminating the polymerase chain reaction. *BioTechniques* **10:**26–30.
11. **Deragon, J. M., D. Sinnett, G. Mitchell, M. Potier, and D. Labuda.** 1990. Use of gamma irradiation to eliminate DNA contamination for PCR. *Nucleic Acids Res.* **18:**6149.
12. **Dwyer, D. E., and N. Saksena.** 1991. Failure of ultra-violet irradiation and autoclaving to eliminate PCR contamination. *Mol. Cell. Probes* **6:**87–88.
13. **Fairfax, M. R., M. A. Metcalf, and R. W. Cone.** 1991. Slow inactivation of dry PCR templates by UV light. *PCR Methods Applic.* **1:**142–143.
14. **Fox, J. C., M. Ait-Khalid, A. Webster, and V. C. Emery.** 1991. Eliminating PCR contamination: is UV irradiation the answer? *J. Virol. Methods* **33:**375–383.
15. **Freese, E., and H. B. Strack.** 1962. Induction of mutations in transforming DNA by hydroxylamine. *Proc. Natl. Acad. Sci. USA* **48:**1796–1803.
16. **Frothingham, R., R. B. Blitchington, D. H. Lee, R. C. Greene, and K. H. Wilson.** 1992. UV absorption complicates PCR decontamination. *BioTechniques* **13:**208–210.
17. **Hayatsu, H., S. K. Pan, and T. Ukita.** 1971. Reaction of sodium hypochlorite with nucleic acids and their constituents. *Chem. Pharm. Bull.* (Tokyo) **19:**2189–2192.
18. **Innis, M. A. and D. H. Gelfand.** 1990. Optimization of PCRs, p. 3–12. *In* M. A. Innis, D. H. Gelfand, J. J. Sninsky, and T. J. White (ed.), *PCR Protocols: A Guide to Methods and Applications*. Academic Press, Inc., San Diego, Calif.
19. **Integrated DNA Technology, Inc.** 1992. Product insert. Integrated DNA Technology, Inc., Coralville, Iowa.
20. **Isaacs, S. T., J. W. Tessman, K. C. Metchette, J. E. Hearst, and G. D. Cimino.** 1991. Post-PCR sterilization: development and application to an HIV-1 diagnostic assay. *Nucleic Acids Res.* **19:**109–116.
21. **Jinno, Y., K. Yoshiura, and N. Niikawa.** 1990. Use of psoralen as extinguisher of contaminated DNA in PCR. *Nucleic Acids Res.* **18:**6739.
22. **Kitchin, P. A., Z. Szotyori, C. Fromholc, and N. Almond.** 1990. Avoidance of false positives. *Nature* (London) **344:**201.
23. **Kwok, S., and R. Higuchi.** 1989. Avoiding false positives with PCR. *Nature* (London) **339:**237–238.
24. **Lindahl, T., S. Ljungquist, W. Siegert, B. Nyberg, and B. Sperens.** 1977. DNA N-glycosidases. Properties of uracil-DNA glycosidase from *Escherichia coli*. *J. Biol. Chem.* **252:**3286–3294.
25. **Lo, Y., W. Z. Mehal, and K. A. Fleming.** 1990. False positive results and the polymerase chain reaction. *Lancet* **i:**679.
26. **Longo, M. C., M. S. Berninger, and J. L. Hartley.** 1990. Use of uracil DNA glycosylase to control carry-over contamination in polymerase chain reactions. *Gene* **93:**125–128.
27. **Meier, A., D. H. Persing, and E. C. Böttger.** Elimination of contaminating DNA within PCR reagents—implications for a general approach to detect uncultured pathogens. *J. Clin. Microbiol.*, in press.
28. **Mullis, K.** 1991. The polymerase chain reaction in an anemic mode: how to avoid cold deoxyribonuclear fusion. *PCR Methods Applic.* **1:**1–4.

29. **Nerenberg, M. I., and T. Minor.** 1991. Detection of plasmid contamination in PCR samples. *BioTechniques* **11:**332.
30. **Ou, C. Y., J. J. Moore, and G. Schochetman.** 1991. Use of UV irradiation to reduce false positivity in the polymerase chain reaction. *BioTechniques* **10:**442–446.
31. **Pang, J., J. Modlin, and R. Yolken.** 1992. Use of modified nucleotides and uracil-DNA glycosylase (UNG) for the control of contamination in the PCR-based amplification of RNA. *Mol. Cell. Probes* **6:**251–256.
32. **Persing, D. H.** 1990. Polymerase chain reaction: trenches to benches. *J. Clin. Microbiol.* **29:**1281–1285.
33. **Persing, D. H., S. W. Barthold, and S. E. Malawista.** 1992. Molecular detection of *Borrelia burgdorferi*, p. 299–315. *In* S. Shutzer (ed.), *Lyme Disease: Molecular and Immunologic approaches*. Cold Spring Harbor Laboratory, Cold Spring Harbor, N.Y.
34. **Persing, D. H., and P. N. Rys.** Submitted for publication.
35. **Prince, A. M., and L. Andrus.** 1992. PCR: how to kill unwanted DNA. *BioTechniques* **12:**358–360.
36. **Rand, K. H., and H. Houck.** 1990. Taq polymerase contains bacterial DNA of unknown origin. *Mol. Cell. Probes* **4:**445–450.
37. **Saiki, R. K., D. H. Gelfand, S. Stoffel, S. J. Scharf, R. Higuchi, G. T. Horn, K. B. Mullis, and H. A. Erlich.** 1988. Primer-directed enzymatic amplification of DNA with a thermostable DNA polymerase. *Science* **239:**487–491.
38. **Saksena, N. K., D. Dwyer, and F. Barre-Sinoussi.** 1991. A "sentinel" technique for monitoring viral aerosol contamination. *J. Infect. Dis.* **164:**1021–1022.
39. **Sarkar, G., and S. S. Sommer.** 1990. Shedding light on PCR contamination. *Nature* (London) **343:**27.
40. **Sarkar, G., and S. S. Sommer.** 1990. More light on PCR contamination. *Nature* (London) **347:**340–341.
41. **Sarkar, G., and S. S. Sommer.** 1991. Parameters affecting susceptibility of PCR contamination to UV inactivation. *BioTechniques* **10:**590–594.
42. **Schmidt, T. M., B. Pace, and N. R. Pace.** 1991. Detection of DNA contamination in Taq polymerase. *BioTechniques* **11:**176–177.
43. **Shi, Y., and J. E. Hearst.** 1986. Thermostability of double-stranded deoxyribonucleic acids: effects of covalent additions of a psoralen. *Biochemistry* **25:**5895–5902.
44. **Shlomai, J., and A. Kornberg.** 1978. Deoxyuridine triphosphatase of *Escherichia coli*. Purification, properties, and use as a reagent to reduce uracil incorporation into DNA. *J. Biol. Chem.* **253:**3305–3312.
45. **Shuldiner, A. R., A. Nirula, and J. Roth.** 1990. RNA template-specific polymerase chain reaction (RS-PCR): a novel strategy to reduce dramatically false positives. *Gene* **91:**139–142.
46. **Shuldiner, A. R., K. Tanner, C. A. Moore, and J. Roth.** 1991. RNA template-specific PCR: an improved method that dramatically reduces false positives in RT-PCR. *BioTechniques* **11:**760–763.
47. **Thornton, C. G., J. L. Hartley, and A. Rashtchian.** 1992. Utilizing uracil DNA glycosylase to control carryover contamination in PCR: characterization of residual UDG activity following thermal cycling. *BioTechniques* **13:**180–182.
48. **Wang, Z., and D. W. Mosbaugh.** 1988. Uracil-DNA glycosylase inhibitor of bacteriophage PBS2 encodes a binding protein specific for uracil-DNA glycosylase. *J. Biol. Chem.* **264:**1163–1171.
49. **Wang, Z., and D. W. Mosbaugh.** 1988. Uracil-DNA glycosylase inhibitor of bacteriophage PBS2: cloning and effects of expression of the inhibitor gene in *Escherichia coli*. *J. Bacteriol.* **170:**1082–1091.
50. **Wang, Z., D. G. Smith, and D. W. Mosbaugh.** 1991. Overproduction and characterization of the uracil-DNA glycosylase inhibitor of bacteriophage PBS2. *Gene* **99:**31–37.
51. **Zhu, Y. S., S. T. Isaacs, G. D. Cimino, and J. E. Hearst.** 1991. The use of exonuclease III for polymerase chain reaction sterilization. *Nucleic Acids Res.* **19:**2511.

# Sample Preparation Methods

# 6

Larry Greenfield and Thomas J. White

Although there has been progress in simplifying the release and purification of bacterial or viral nucleic acids from clinical specimens, many research procedures are still unsuitable for the clinical laboratory and a universal automated method for use with any specimen has not yet been devised. In this chapter, we review some of the basic principles that have been learned to date which may guide and encourage the reader to develop further improvements that eliminate the requirements for hazardous solutions, centrifugations, and multiple steps. A variety of approaches which may be appropriate for certain specimens and pathogens but not for others are then described. Finally, each specimen type (e.g., whole blood, urine, sputum) is discussed with regard to specific protocols and pathogens.

## Basic Principles

The ideal sample preparation method represents a trade-off between the requirements for the optimal method, the clinical specimen, and the target (Table 1). Although many of these considerations are interrelated, selection of a few crucial items helps define many others. Once the target organism is selected, the clinical pathogenesis of the infection generally dictates the appropriate specimen and number of microorganisms likely to be present. Determination of the desired assay sensitivity and the number of tests to be performed on the processed sample then dictates the required volume of specimen to be processed.

### Sample Size Versus Target Copy Number

Microbiological culture as a "gold standard" has directed our selection of the appropriate specimen for many infectious diseases, e.g., blood or plasma for human immunodeficiency virus (HIV) and hepatitis C virus (HCV) and endocervical swabs for chlamydiae. For other pathogens, e.g., *Borrelia burgdorferi*, the optimal specimen for diagnosing each stage of infection has not yet been identified. For molecular diagnostic tests that are based on amplification, a single copy or molecule of the genetic target from the pathogen, if present in the reaction, can be detected in a fully optimized procedure (51, 70). To maximize the chance of diagnosing an infection, the largest convenient sample volume should be screened. However, since typical molecular diagnostic test reaction volumes are 100 μl or less, one is faced with a choice between complex target concentration steps (e.g., ethanol precipitation, nucleic acid target capture, and centrifugation) and lowered assay sensitivity. Hence, if no amplifiable target is detected

**Table 1.** Considerations necessary to select appropriate sample preparation method

| |
|---|
| **Considerations for Development of Sample Preparation Method** |
| Specimen considerations |
| Specimen type (blood, plasma, serum, sputum, CSF, urine, etc.) |
| Processing of specimen during or after collection (collection of blood in tubes containing EDTA, citrate, heparin, etc.; liquefaction of sputum, etc.) |
| Volume of sample to be processed |
| Level of extraneous nucleic acid (serous versus purulent fluid; whole blood versus plasma or serum) |
| Physical properties of specimen (swab, fluid, tissue, viscosity, etc.) |
| Presence of amplification inhibitors |
| Organism considerations |
| Bacterial load within specimen (the number of pathogenic organisms per volume of specimen) |
| Durability of cell wall against lysis (difficulty in lysing fungi and *Mycobacterium* species) |
| Tendency of organism to lyse spontaneously (*Mycoplasma* species) |
| Desire to sterilize the specimen during sample preparation (*Mycobacterium* species, HCV, HIV, etc.) |
| Nucleic target characteristics |
| RNA versus DNA |
| Double stranded versus single stranded |
| Requirement of isolation of RNA without contaminating DNA (i.e., in quantifying HIV without amplifying the provirus) |
| **Criteria for Optimum Sample Preparation Method** |
| Choice of the appropriate clinical specimen |
| Desired sensitivity of the assay |
| Required volume of sample to be processed |
| Efficiency of amplifiable target recovery |
| Number of different tests to be run on the processed clinical sample |
| Stabilization of the target during sample collection and processing (inactivation of nucleases, particularly in the case of RNA) |
| Avoidance of toxic solutions |
| Avoidance of centrifugations |
| Removal by dilution of amplification inhibitors |
| Minimal amount of sample handling |
| Robustness of protocol |
| Absence of requirement for specialized equipment (freezers, refrigerators, centrifuges, etc.) |

within 50 μl of a 400-ml unit of blood, we can only state that there are no more than 20,000 organisms per unit. With the appropriate sample preparation procedure, it may be possible to process and amplify the target from large sample volumes. However, other factors may come into play. For example, purulent sputum may contain up to 14 mg of DNA per ml (43, 72). Extraction of large sample volumes would result in higher levels of nonspecific priming as a result of the relative abundance of extraneous DNA and could therefore compromise the sensitivity of the assay.

Sample size becomes critical when the number of infectious organisms per test volume is very small (chapters 3 and 4). If an amplification method is sensitive

**Table 2.** Probability of a negative result for low target copy numbers

| Copy number ($C$) | % of dropouts | Frequency of occurrence of a dropout |
|---|---|---|
| 1 | 36.788 | 1 in 3 |
| 5 | 0.6738 | 1 in 150 |
| 10 | 0.0045 | 1 in 22,000 |
| 20 | $2.06 \times 10^{-7}$ | 1 in $5 \times 10^{8}$ |

enough to detect the product generated from one input target molecule, the probability of a dropout (absence of the specific amplification signal) can be estimated from the Poisson distribution of targets in solution:

$$P_N = C^N/N!e^c$$

where $C$ is the average number of copies of target molecules in solution, $N$ is the actual target number taken in the sample, and $P_N$ is the probability of picking up $N$ copies.

The problem is compounded if different laboratories use tests with different sensitivities, e.g., when one laboratory can detect a single copy but another detects no fewer than 20 to 50 copies. This can cause discordant results from different laboratories even when the test procedure is optimal, as shown in Table 2. As mentioned in chapter 3, this may pose problems to interpretation of laboratory quality control programs.

## Inhibitors

Sample preparation procedures can contribute in a variety of ways to the reliability of target detection within clinical specimens (Table 3). If the efficiency of target recovery by a sample preparation procedure is low, the statistics of small numbers may result in a greater number of false-negative amplifications. A variety of components within clinical specimens may inhibit detection systems by interacting with components of the enzymatic reaction (Table 3). On the other hand, if the target amounts within a specimen are sufficiently large, dilution of the specimen may eliminate the inhibitory effect while maintaining the nucleic acid at a level at which the frequency of sampling zero copies is low (i.e., 20 copies). In contrast, enzymatic or chemical destruction of the nucleic acid will reduce the effective number of amplifiable nucleic acid targets.

There has been little basic research on the biochemistry of the substances in biological fluids that may inhibit enzymes used in amplification reactions. Heme (at 0.8 μM) and its metabolic products are known inhibitors of DNA polymerases (42), and polymerases from different sources may be affected by different

**Table 3.** Role of sample preparation in target detection

| |
|---|
| Efficient target recovery |
| Maintenance of intact nucleic acid target |
| Removal of amplification inhibitors which act by interacting with nucleic acid |
| Removal of amplification inhibitors which act by interacting with polymerase |
| Removal of components which affect other enzymatic substrates (e.g., metals) |
| Sterilization of clinical samples |

concentrations (65). Cerebrospinal fluid (CSF), urine, and sputum may also contain inhibitors of the *Taq* DNA polymerase, but the inhibitory constituents have not been characterized. Acidic polysaccharides, which are a component of glycoproteins present in sputum (53), are known inhibitors of polymerases (27, 34). In addition, many of the reagents traditionally used by molecular biologists in the purification of nucleic acids, e.g., EDTA, detergents such as sodium dodecyl sulfate (SDS), and chaotropes such as guanidinium HCl, are inhibitory to amplification enzymes. Hence, the use of these compounds to solubilize cell membranes or inactivate nucleases requires additional steps to remove or neutralize them.

## DNA versus RNA

A further problem arises with procedures such as boiling, alkaline lysis, and drying that may produce partially single-stranded DNA. Denatured templates serve more readily as nonspecific substrates for annealing of primers under the low-stringency conditions occurring during reaction setup, thereby lowering the specificity of the system (8, 31, 63).

Pathogens that have both an RNA and a DNA stage in their life cycle, such as HIV, present an additional challenge if one desires to detect only viral RNA. Reverse transcription polymerase chain reaction (PCR) detects both proviral DNA and viral RNA unless steps are taken to destroy the DNA target (26). Amplification methods, such as self-sustaining sequence replication (3SR), that are theoretically capable of using solely RNA templates, should not amplify proviral DNA unless a denaturation step is included as part of the sample extraction technique. It will be important to determine whether single-stranded DNA actually interferes with this RNA-specific approach. Analogously, PCR would detect only proviral DNA unless a reverse transcription step precedes the PCR. Generally speaking, either selective extraction or destruction of DNA or RNA allows one to target the appropriate nucleic acid target. A largely unaddressed issue concerns the relative stability of RNA during sample storage prior to extractions. rRNA has been stabilized in chaotrope solutions for months at room temperature. Busch et al. (17) and Cuypers et al. (25a) have shown that serum and plasma storage conditions can seriously affect HCV RNA detection by PCR.

## Additional Considerations

When working with pathogens, it may be desirable for the sample preparation procedure to include sterilization of the specimen to protect laboratory personnel. Initial studies are frequently carried out in research laboratories in a carefully controlled environment to reduce the risk of exposure of laboratory personnel to pathogens. However, in clinical laboratories, such facilities may not always be available; it may therefore be preferable to include sterilization steps within the sample preparation protocols.

A final sample preparation issue concerns differentiation of viable from nonviable organisms. Amplification methods that target stable nucleic acids can detect dead, dormant, and replicating organisms. Detection of all forms of the microorganism may be diagnostically advantageous when viability is rapidly lost during transport or storage or when the organism cannot be cultured (i.e., a culturing method is not available or the organism is latent). Discrimination of dead from viable organisms would present a problem if nonviable organisms remain

**Table 4.** Fundamental goals of sample preparation protocols

| |
|---|
| Release of nucleic acid from bacteria, viruses, or fungi |
| Stabilization of nucleic acid against degradation |
| Removal of amplification inhibitors |
| Concentration of target into a small volume |
| Placement of target into an aqueous environment compatible with amplification |

once an active infection has cleared. A study of the course of bacteremia by culture of specimens following oral trauma showed that viable bacteria were rapidly cleared from the blood (40). In one study of chlamydia infections following treatment, results of PCR and culture showed perfect concordance (23). However, these may be best-case scenarios; other studies have documented late persistence of nucleic acid (64). In some cases it is beneficial to detect dead microorganisms, e.g., when an inaccessible reservoir of live organisms sheds bacteria that are rapidly cleared. Many more longitudinal studies of treatment to cure for various diseases will be necessary to determine the clinical significance of DNAemia.

## Overview of Approaches

Sample preparation methods can be divided into a number of generic steps (Table 4). The requirement for each step will depend on the organism and specimen. The release of nucleic acid may be easy for viruses and some bacteria (e.g., *Mycoplasma* species) but difficult for other bacteria (e.g., *Mycobacterium tuberculosis*) and fungi. RNA is more difficult to stabilize than DNA. More steps may be required to remove inhibitors from some specimens (e.g., sputum and blood) than from others (e.g., urine and CSF). Some specimens (e.g., sputum for *M. tuberculosis* and *Legionella pneumophila* and blood for sepsis) may require a greater degree of concentration than others (e.g., urethral swabs and urine for *Chlamydia* or *Gonococcus* species) to achieve the required sensitivity.

There are a variety of methods for the release of nucleic acid from microorganisms, including boiling in distilled water or PCR buffer (76), detergents with or without heat (76), sodium hydroxide with heat (13), freeze-thaw (15), SDS-proteinase K (51), perchloric acid (76), enzymes (30), sonication (15), and heat (55). Enzymatic digestion may be less desirable in that there may be components in the sample which prevent the action of the enzyme. For example, lysozyme has been used on liquified sputum (30), but sputum has a high content of mucopolysaccarides. Lysozyme is unstable following reduction (77) and forms complexes with dextrans and proteins (62). Many of the current and anticipated protocols require some method of separation for concentration of nucleic acid or removal of amplification inhibitors. Potential methods for separation include centrifugation, separation by magnetic particles, and separation by filtration.

### Crude Lysis

The simplest sample preparation method would entail only a crude lysis. However, such a method would require a high concentration of target in the specimen and/or small amounts of amplification inhibitors. If no other separation step is included, the volume of the specimen to be processed is limited by the volume of the amplification reaction. Simple lysis methods typically use detergents such as SDS or Triton X-100, chaotropes such as guanidinium isothiocyanate or sodium iodide, proteases such as proteinase K (which must be inactivated before the sample is added to the diagnostic reaction mixture) (42), substances such as saponin which lyse erythrocytes and leukocytes (e.g., the Wampole Isostat Microbial System), or heat (33). Such methods are generally suitable when the clinically significant number of infectious organisms per sample volume is large (e.g., *Chlamydia trachomatis* in endocervical swabs), so that the lysed specimen does not require significant further dilution (57). If the level of target is low, it is frequently necessary to remove amplification inhibitors by additional extraction (phenol-chloroform) steps or concentration of the target by alcohol precipitation. In addition, detergents are known to inhibit many enzymes, and high temperatures may result in degradation of nucleic acids (29).

### Target Capture

Target capture or cycling offers the possible advantages of automation, universality for all specimens, and concentration of target into a small volume. This approach has been investigated by Gillespie et al. (36), Hunsaker et al. (44), and Lanciotti et al. (54). However, to date there are no published studies that demonstrate efficient capture and detection of fewer than 100 target molecules, and automated instruments and reagents for this approach are not yet commercially available. Derivatized magnetic particles can be coupled to oligonucleotide capture probes and combined with manual washing steps to remove extraneous materials (2, 19). These approaches have their own problems, though, since manual washing causes aerosols that may result in sample-to-sample contamination.

Other matrices have been tested for general adsorption of nucleic acids. Glass matrices, Sephadex, and diatomaceous earth bind nucleic acids in chaotropic solutions (11, 16, 59, 79). Following binding of the nucleic acids to the solid-phase matrix, the impurities and amplification inhibitors are removed by centrifugation and washing and the nucleic acids are eluted in an amplification-compatible buffer. Such approaches are promising since they are relatively simple, can be automated, and do not require hazardous reagents.

Finally, filtration may become a useful approach for certain kinds of specimens if it can be automated and made rapid (7a). Cost will be a problem unless disposable devices can be manufactured cheaply, and the requirement for a vacuum or centrifuge could be a burden for many laboratories.

## Recommended Protocols for Various Specimens

### Whole Blood

Even after it is decided that the desired specimen for a given target is blood, there still remain a number of choices: plasma, serum, whole blood, leukocyte fractions, etc. Furthermore, there is a choice of anticoagulants if the specimen is plasma: EDTA, heparin, or citrate. The anticoagulant used for plasma collection and the method of storage may affect the ability of the assay to detect the presence of target sequences (17, 80). Heparin was found to inhibit the activity of both murine leukemia virus reverse transcriptase and *Taq* DNA polymerase (46). In addition, the inhibitory effect of heparin does not appear to be removed by extraction of RNA by a modification of the acid-phenol-guanidinium method. For EDTA-containing tubes, it is recommended that the final concentration of EDTA be 1 to 2 mg/ml of blood (final concentration, 6.8 mM). For heparin-containing

**Table 5.** Cellular components of blood

| Cell type | No./μl |
|---|---|
| Leukocytes | $(4.5–10.8) \times 10^3$ |
| Segmented neutrophils | $3.8 \times 10^3$ |
| Band neutrophils | $0.62 \times 10^3$ |
| Eosinophils | $0.2 \times 10^3$ |
| Basophils | $0.04 \times 10^3$ |
| Lymphocytes | $2.5 \times 10^3$ |
| Monocytes | $0.3 \times 10^3$ |
| Erythrocytes | $(4.2–6.2) \times 10^6$ |

tubes, the recommended final concentration of heparin is 0.1 to 0.2 mg/ml. Since the molecular weight of heparin ranges between 6,000 and 20,000, this translates to a final concentration of 5 to 33 μM. Finally, the recommended final concentration of trisodium citrate is 0.32 to 0.38% (12 to 15 mM).

Let us first examine the composition of blood and its components. Table 5 lists the cellular components of blood, and Table 6 outlines the concentrations of the predominant ions in plasma and erythrocytes. The amount of DNA in a diploid mammalian cell is approximately 7 pg. Therefore, the contribution of leukocyte DNA is 32 to 76 μg/ml of blood. RNA is also contributed by the leukocytes and erythrocytes. Thus, the amplification of blood without purification may significantly alter the components of an amplification reaction.

An important consideration in comparing techniques for the preparation of blood specimens is that many of them were designed for use in human genetic analyses. In these analyses, a relatively high copy number of targets ($10^3$ to $10^5$) was recovered, and so the applicability of these methods to samples containing few microorganisms (1 to 100) of different kinds is, in most cases, still to be defined. As mentioned in chapter 4, the requirement for a highly optimized amplification system becomes more critical when large amounts of nontarget nucleic acid are present.

As little as 1% (vol/vol) whole blood can inhibit a typical PCR (42). It has been suggested that the inhibition is due to the binding of heme and/or porphyrin to *Taq* DNA polymerase (42). Polymerases from other sources may be less

**Table 6.** Concentrations of ions in plasma and erythrocytes

| Component | Concn (mM) in: | |
|---|---|---|
| | Plasma | Erythrocytes |
| Sodium | 151 | 6.2 |
| Potassium | 4.3 | 102 |
| Calcium | 2.7 | 0.0089 |
| Magnesium | 1.6 | 3.06 |
| Manganese | | 0.0034 |
| Chloride | 109.7 | 78 |
| Bicarbonate | 28.7 | 8 |
| Phosphate | 0.7 | 13.2 |
| Sulfate | 0.55 | |
| Organic acid | 5.3 | |
| DNA | | Trace |
| RNA | | 1.55 μg/μl |

susceptible to such inhibition, and the *Thermus thermophilus* DNA polymerase is reported to tolerate up to 4% (vol/vol) whole blood or 8% if the single-stranded DNA-binding protein (phage T4 gene 32 protein) is added to the reaction (65). In that study, $10^3$ copies of target (*Campylobacter fetus*) bacteria were detected. Other investigators have successfully amplified the cystic fibrosis gene locus from 1 μl of whole blood following pretreatment with a thermophilic protease (61).

A number of methods have been developed in which DNA is isolated after nuclei are recovered following lysis of erythrocytes. In one method, a simple procedure for amplifying DNA from whole blood without extraction was successfully applied to cystic fibrosis and Duchenne muscular dystrophy targets (5). Whole blood was treated with a lysis buffer (0.32 M sucrose, 10 mM Tris-HCl [pH 7.5], 5 mM $MgCl_2$, 1% Triton X-100) and then subjected to centrifugation. The above buffer conditions have been reported to lyse all cells, leaving nuclei intact (22). The pellet was washed twice in lysis buffer and resuspended in PCR buffer with nonionic detergents plus proteinase K. Prior to amplification, the sample was boiled. In a second method, cells were lysed by the addition of Nonidet P-40 (52). Persing et al. (66; see also protocol 4.5) used hypotonic lysis to concentrate *Babesia microti* merozoites from whole blood, followed by protease digestion, boiling, and PCR amplification.

A number of more complicated methods have been used to obtain amplifiable nucleic acid from whole blood. DNA for the human leukocyte antigen DQA1 (HLA-DQA1) gene has been successfully extracted from bloodstains by a modification of a Chelex or Centricon extraction method (48). For the Chelex method, a portion of the bloodstain in 200 μl of 5% (wt/vol) Chelex was incubated at 56°C for 30 min to overnight. The extract was then vortexed and boiled. The Centricon extraction method involved extraction with phenol-chloroform-isoamyl alcohol followed by *n*-butanol. Finally, the buffer was changed by washing in a Centricon unit. A similar method was applied to the detection of *Ehrlichia chaffeensis* target from whole blood (3). *Treponema pallidum* has been detected from whole blood by lysis with SDS-proteinase K followed by adsorption to a silica matrix (81). Amplifiable *Plasmodium vivax* target was isolated from whole blood by processing blood spotted onto Whatman 3M chromatography paper (49). Gustincich (38) has used a lysis-denaturation extraction with cetyltrimethylammonium bromide and chloroform followed by DNA precipitation and resuspension. This method takes about 1 h and yields high-quality DNA but is still rather involved for a clinical test. RNA has been extracted from whole blood for PCR amplification by a modification of the acid-phenol-guanidinium method (21, 46, 69).

Recently, a novel approach to inactivating inhibitors in whole-blood samples was described (56). This procedure uses transferrin to complex heme compounds and is compatible with up to 20 μl of whole blood. The procedure has been applied to the detection of HIV in clinical specimens and, despite the relatively small volume of blood samples, showed excellent concordance with procedures that involve isolation and lysis of peripheral blood mononuclear cells (51). Other methods for whole blood are described in protocol 1.3.

## Purified Leukocytes

The earliest methods required isolation of lymphocytes on Ficoll-Hypaque gradients followed by lysis with detergent and digestion with proteinase K (1, 9, 51) or detergent alone (24). Alternative procedures use agents to selectively lyse erythrocytes; leukocytes are then collected by centrifugation, washed, and lysed in detergent (18). By using this procedure, 96 samples can be prepared for HIV analysis in about 3 h. Amplifiable DNA has also been extracted from isolated peripheral blood mononuclear cells by lysis with 1% SDS followed by adsorption onto glass powder (82). Protocols 2.1, 2.2, and 2.7 provide detailed methods for isolating leukocytes.

## Plasma or Serum

Plasma or serum is preferred over whole blood or purified leukocytes when the target microorganism is predominantly extracellular, e.g., hepatitis B virus (HBV) or HCV. HBV DNA, HCV RNA, and HIV RNA have been prepared from serum or plasma specimens (4, 45).

By heating sera prior to amplification to inactivate inhibitors of PCR, Frickhofen and Young (33) were able to detect the presence of $10^4$ copies of viral targets. Such an approach may not be applicable to samples containing lower target copy numbers or RNA targets. Similarly, small volumes of serum (3 to 12 μl) have been denatured and used directly in a reverse transcriptase PCR by Ravaggi et al. (68) to detect HCV. However, patients with low titers of circulating virus would be missed by this method of preparing the sample.

Many of the current protocols incorporate extraction of plasma sample with organic reagents. Amplifiable dengue virus RNA has been prepared from serum by using a modification (54) of the acid-phenol protocol described by Chomczynski and Sacchi (21). Extraction with phenol, phenol-chloroform-isoamyl alcohol, and/or chloroform-isoamyl alcohol followed by ethanol precipitation has been applied to the isolation of HCV target RNA from serum and plasma (17, 80). In general, standard guanidinium isothiocyanate or cesium chloride gradient procedures may be too difficult or hazardous for routine use (20).

Binding of nucleic acid to solid-phase extractants can circumvent removal of proteins and inhibitors with organic solvents. HIV RNA, hepatitis E virus RNA, and HBV DNA have been extracted from plasma by using commercially available glass powder or silica suspensions (10, 60, 82).

There is great need to improve on RNA target extraction techniques and to develop ways to stabilize the viral capsids or the RNA itself in the specimen during collection and transport. Busch et al. (17) and Wang et al. (80) have demonstrated that room temperature storage of serum results in a substantial decrease in HCV RNA detection and that repeated freezing and thawing cause a moderate reduction. The amount of contaminating DNA from lysed cells may also change depending on the method of storage and extraction.

## Urine

Purification of amplifiable nucleic acid is complicated by the formation of insoluble amorphous salts which precipitate during urine sample storage. Unless these are resolubilized, they will pellet with bacteria or cells during centrifugation. Some bacteria may lyse during storage of urine, so that nucleic acids will be lost in the supernatant in any procedure that depends on an initial centrifugation step to concentrate the microorganisms. Studies with normal urine samples that have been spiked with bacteria or extracted by various methods and spiked with purified DNA indicate that urinary inhibitors vary widely among individuals and may vary from day to day in a single individual. These cautions should be considered when evaluating negative molecular detection results relative to other diagnostic techniques. Until better methods are available, it may be useful to retest spiked negatives or include an internal control template to confirm that amplification could have occurred had the target been present.

Although urine specimens have been extracted by a variety of methods that have yielded amplifiable nucleic acid, a method that routinely eliminates inhibitory substances has yet to be published. PCR has been performed directly on urine for cytomegalovirus (CMV), but the presence of inhibitors usually requires clarification of the urine and pelleting of the virus by centrifugation. However, these steps often fail to remove inhibitors, which can even partition with nucleic acid during phenol-chloroform extraction.

Amplifiable *Leptospira interrogans* serovar hardjo DNA was prepared from bovine urine by repeated centrifugations and washing (35). *Borrelia burgdorferi* target was isolated from urine by centrifugation and boiling in Chelex-100 resin (55). CMV target has been prepared from urine by using glass beads (16).

Recently, a urine extraction technique for chlamydia detection in male urine specimens has become available (7). This procedure uses an equal volume of ethanol as a fixative to prevent degradation during transport and storage. A filtration or centrifugation step collects cellular material, which is then solubilized with detergents. This method takes about 60 min and involves a small number of steps.

### Cerebrospinal Fluid

CSF has been used as a specimen for preparing templates for the detection of herpes simplex virus, enteroviruses, *B. burgdorferi*, CMV, *T. pallidum*, and bacteria that cause meningitis (50, 64, 73). Small volumes of CSF (10 μl or less) can be used directly in some amplification systems. Extraction procedures have included standard phenol-chloroform extraction (73), ethanol precipitation (55), use of diatom or glass matrices (64), and pelleting and cell lysis (55). The latter method often yields extracts that are refractory to amplification. Protocols 1.4, 1.7, 1.21, 2.6, and 2.16 describe additional methods for the extraction of CSF specimens.

### Endocervical Swabs

One of the most rapid and simple extraction procedures for a complex specimen is that developed for the detection of *C. trachomatis* from endocervical swabs (57). A swab is taken and briefly put into a collection tube containing a detergent lysis solution. In the clinical laboratory, a second detergent solution is added and the solutions are mixed and allowed to stand for 5 min. The specimen is then ready for PCR; 96 samples can be prepared for amplification in 30 to 45 min. Blood and mucus generally do not interfere with this extraction method, although some samples remain inhibitory. More complex techniques for detecting human papillomavirus infection in genital swabs or saline cervicovaginal lavage samples have been described by Bauer et al. (6).

### Sputum

Sputum samples have been widely tested for a suitable extraction procedure that is compatible with both culture and DNA analysis for *Mycobacterium tuberculosis* (14, 39, 41, 67, 75; see also protocols 1.1, 1.2, and 6.1). This specimen presents a complex challenge including the need to prevent aerosols, solubilize viscous mucus, and lyse the mycobacterial cell wall. Furthermore, preliminary steps to decontaminate sputum samples with compounds such as *N*-acetyl-L-cysteine and alkali are different in Europe and North America, so that laboratory extraction methods may not always start with equivalent specimens. Generally, phenol-chloroform techniques followed by ethanol precipitation seem to be necessary to remove inhibitors and concentrate the specimen (13, 25). Powdered glass offers a method for removal of amplification inhibitors without the use of organic solvents (30). In some cases, repeated washing of pellets with detergents appears to remove amplification inhibitors (76).

With mycobacteria, many culture-positive sputum samples yield only a few colonies, so it is critical to maximize the equivalent volume of sputum that goes into the amplification reaction. Although we might expect that one bacterium will generally yield a single colony, we know little about the fraction of bacteria in a specimen that is capable of propagation in the laboratory. Perhaps we will find a situation like that with viruses, in which the number of viral particles may exceed the number that can be cultivated.

## Feces

Fecal specimens remain the most difficult and least studied with respect to the development of extraction methods that would allow DNA diagnosis to replace stool culture. Although some workers have used phenol-chloroform techniques, inhibitory substances are often coextracted with DNA and only dilution seems to help (12, 32). Saulnier and Andremont (71) describe a relatively simple extraction method with a sensitivity of about 10,000 bacterial cells per g of feces. Immunomagnetic enrichment has also been tested by Shirai et al. (74) for detecting the cholera enterotoxin gene. Immunologic separation of the target organism (58, 60) and glass powder have also been used to retrieve target sequences (60). A complicated method involving extraction with 1,1,2-trichloro-1,2,2-trifluoroethane, precipitation with polyethylene glycol 6000, digestion with proteinase K in the presence of detergent, precipitation with cetyltrimethylammonium bromide, extraction with phenol-chloroform and chloroform, and alcohol precipitation has been described (47). Another challenge will be to separate or enrich the pathogen DNA relative to that normally present from shed mucosal cells. Protocols 1.17, 1.19, 2.13, and 4.6 describe extraction methods for fecal specimens.

## Other Specimens

In addition to the more common types of clinical specimens, amplification techniques can be used on a multitude of unusual tissues and secretions, such as synovial fluid, amniotic fluid, bone, saliva, semen, nasal washes (28), and buccal swabs. Even archival specimens such as formalin-fixed paraffin-embedded tissues can be extracted and analyzed retrospectively (37, 78).

In conclusion, most of the sample preparation techniques described in later chapters of this book are still more suitable for relatively highly skilled laboratories and technical staff. As diagnostic test kits that have been approved by the Food and Drug Administration become available, it is likely that more simple and semiautomated extraction methods will become the standard for use on all pathogens found in a particular type of specimen.

## References


1. **Albert, J., and E. M. Fenyo.** 1990. Simple, sensitive, and specific detection of human immunodeficiency virus type 1 in clinical specimens by polymerase chain reaction with nested primers. *J. Clin. Microbiol.* **28:**1560–1564.
2. **Albert, J., J. Wahlberg, J. Lundeberg, S. Cox, E. Sandstrom, B. Wahren, and M. Uhlen.** 1992. Persistence of azidothymidine-resistant human immunodeficiency virus type 1 RNA genotypes in posttreatment sera. *J. Virol.* **66:**5627–5630.
3. **Anderson, B. E., J. W. Sumner, J. E. Dawson, T. Tzianabos, C. R. Greene, J. G. Olson, D. B. Fishbein, M. Olsen-Rasmussen, B. P. Holloway, E. H. George, and A. F. Azad.** 1992. Detection of the etiologic agent of human ehrlichiosis by polymerase chain reaction. *J. Clin. Microbiol.* **30:**775–780.
4. **Baginski, U., A. Ferrie, R. Watson, and D. Mack.** 1990. Detection of hepatitis B virus, p. 348–355. *In* M. A. Innis, D. H. Gelfand, J. J. Sninsky, and T. J. White (ed.), *PCR Protocols: A Guide to Methods and Applications*. Academic Press, Inc., San Diego, Calif.
5. **Balnaves, M. E., S. Nasioulas, H.-H. M. Dahl, and S. Forrest.** 1991. Direct PCR from CVS and blood lysates for detection of cystic fibrosis and Duchenne muscular dystrophy deletions. *Nucleic Acids Res.* **19:**1155.
6. **Bauer, H. M., C. E. Greer, and M. M. Manos.** 1992. Determination of genital human papillomavirus infection by consensus polymerase chain reaction amplification, p. 131–152. *In* C. S. Herrington and J. O. McGee (ed.), *Diagnostic Molecular Pathology: A Practical Approach*, vol. II. Oxford University Press, New York.
7. **Bauwens, J. E., A. M. Clark, M. J. Loeffelholz, S. A. Herman, and W. E. Stamm.** Diagnosis of *Chlamydia trachomatis* (CT) urethritis in men by PCR assay of first-void urine. Submitted for publication.

7a. **Bej, A. K., M. H. Mahbubani, J. L. Dicesare, and R. M. Atlas.** 1991. Polymerase chain reaction-gene probe detection of microorganisms by using filter-concentrated samples. *Appl. Environ. Microbiol.* **57:**3529–3534.

8. **Bloch, W.** 1991. A biochemical perspective of the polymerase chain reaction. *Biochemistry* **30:**2735–2747.
9. **Boland, G. J., R. A. de Weger, M. G. J. Tilanus, C. Ververs, K. Bosboom-Kalsbbek, and G. C. de Gast.** 1992. Detection of cytomegalovirus (CMV) in granulocytes by polymerase chain reaction compared with the CMV antigen test. *J. Clin. Microbiol.* **30:**1763–1767.
10. **Boom, R., C. J. A. Sol, R. Heijtink, P. M. E. Wertheim-van Dillen, and J. van der Noordaa.** 1991. Rapid purification of hepatitis B virus DNA from serum. *J. Clin. Microbiol.* **29:**1804–1811.
11. **Boom, R., C. J. A. Sol, M. M. M. Salimans, C. L. Jansen, and P. M. E. Wertheim-van Dillen.** 1990. Rapid and simple method for purification of nucleic acids. *J. Clin. Microbiol.* **28:**495–503.
12. **Brian, M. J., M. Frosolono, B. E. Murray, A. Miranda, E. L. Lopez, H. F. Gomez, and T. G. Cleary.** 1992. Polymerase chain reaction for diagnosis of enterohemorrhagic *Escherichia coli* infection and hemolytic-uremic syndrome. *J. Clin. Microbiol.* **30:**1801–1806.
13. **Brisson-Nöel, A., C. Aznar, C. Chureau, S. Nguyen, C. Pierre, M. Bartoli, R. Bonete, G. Pealoux, B. Gicquel, and G. Garrigue.** 1991. Diagnosis of tuberculosis by DNA amplification in clinical practice evaluation. *Lancet* **338:**364–366.
14. **Brisson-Nöel, A., D. Lecossier, X. Nassif, B. Gicquel, V. Lévy-Frébault, and A. J. Hance.** 1989. Rapid diagnosis of tuberculosis by amplification of mycobacterial DNA in clinical samples. *Lancet* **i:**1069–1071.
15. **Buck, G. E., L. C. O'Hara, and J. T. Summersgill.** 1992. Rapid, simple method for treating clinical specimens containing *Mycobacterium tuberculosis* to remove DNA for polymerase chain reaction. *J. Clin. Microbiol.* **30:**1331–1334.
16. **Buffone, G. J., G. J. Demmler, C. M. Schimbor, and J. Greer.** 1991. Improved

amplification of cytomegalovirus DNA from urine after purification of DNA with glass beads. *Clin. Chem.* **37:**1945–1949.

17. **Busch, M. P., J. C. Wilber, P. Johnson, L. Tobler, and C. S. Evans.** 1992. Impact of specimen handling and storage on detection of hepatitis C virus RNA. *Transfusion* **32:**420–425.
18. **Butcher, A., and J. Spadoro.** 1992. Using PCR for detection of HIV-1 infection. *Clin. Immunol. Newsl.* **12:**73–76.
19. **Chiodi, F., B. Keys, J. Albert, L. Hagberg, J. Lundeberg, M. Uhlen, E. M. Fenyo, and G. Norkrans.** 1992. Human immunodeficiency virus type 1 is present in the cerebrospinal fluid of a majority of infected individuals *J. Clin. Microbiol.* **30:**1768–1771.
20. **Chirgwin, J. M., A. E. Przybyla, R. J. MacDonald, and W. J. Rutter.** 1979. Isolation of biologically active ribonucleic acid from sources enriched in ribonuclease. *Biochemistry* **18:**5294–5299.
21. **Chomczynski, P., and N. Sacchi.** 1987. Single-step method of RNA isolation by acid guanidinium thiocyanate-phenol-chloroform extraction *Anal. Biochem.* **162:**156–159.
22. **Ciulla, T. A., R. M. Sklar, and S. L. Hauser.** 1988. A simple method for DNA purification from peripheral blood. *Anal. Biochem.* **174:**485–488.
23. **Claas, H. C. J., J. H. T. Wagenvoort, H. G. M. Niesters, T. T. Tio, J. H. Van Rijsoort-Vos, and W. G. V. Quint.** 1991. Diagnostic value of the polymerase chain reaction for *Chlamydia* detection as determined in a follow-up study. *J. Clin. Microbiol.* **29:**42–45.
24. **Conway, B., L. J. Bechtel, K. A. Adler, R. T. D'Aquila, J. C. Kaplan, and M. S. Hirsch.** 1992. Comparison of spot-blot and microtiter plate methods for the detection of HIV-1 PCR products *Mol. Cell. Probes* **6:**245–249.
25. **Cousins, D. V., S. D. Wilton, B. R. Francis, and B. L. Gow.** 1992. Use of polymerase chain reaction for rapid diagnosis of tuberculosis. *J. Clin. Microbiol.* **30:**255–258.

25a. **Cuypers, H. T. M., D. Bresters, I. N. Winkel, H. W. Reesink, A. J. Weiner, M. Houghton, C. L. van der Poel, and P. N. Lelie.** 1992. Storage conditions of blood samples and primer selection affect the yield of cDNA polymerase chain reaction products of hepatitis C virus. *J. Clin Microbiol.* **30:**3220–3224.

26. **Dilworth, D. D., and J. R. McCarrey.** 1992. Single-step elimination of contaminating DNA prior to reverse transcriptase-PCR. *PCR Methods Applic.* **1:**279–282.
27. **Do, N., and R. P. Adams.** 1991. A simple technique for removing plant polysaccharide contaminants from DNA. *BioTechniques* **10:**162–166.
28. **Donofrio, J. C., J. D. Coonrod, J. N. Davidson, and R. F. Betts.** 1992. Detection of influenza A and B in respiratory secretions with the polymerase chain reaction. *PCR Methods Applic.* **1:**263–268.
29. **Eigner, J., H. Boedtker, and G. Michaels.** 1961. The thermal degradation of nucleic acids. *Biochim. Biophys. Acta* **51:**165–168.
30. **Eisenach, K. D., M. D. Sifford, M. D. Cave, J. H. Bates, and J. T. Crawford.** 1991. Detection of *Mycobacterium tuberculosis* in sputum samples using a polymerase chain reaction. *Am. Rev. Respir. Dis.* **144:**1160–1163.
31. **Faloona, F., S. Weiss, F. Ferre, and K. Mullis.** 1990. Direct detection of HIV sequences in blood: high-gain polymerase chain reaction. *Abstr. 6th Int. Conf. AIDS*. **2:**318.
32. **Frankel, G., L. Riley, J. A. Giron, J. Valmassoi, A. Freidmann, N. Stokbine, S. Falkow, and G. K. Schoolnik.** 1990. Detection of *Shigella* in feces using DNA amplification. *J. Infect. Dis.* **161:**1252–1256.
33. **Frickhofen, N., and N. S. Young.** 1991. A rapid method of sample preparation for detection of DNA viruses in human serum by polymerase chain reaction. *J. Virol. Methods* **35:**65–72.
34. **Furukawa, K., and V. P. Bhavanandan.** 1983. Influences of anionic polysaccharides on DNA synthesis in isolated nuclei and by DNA polymerase α: correlation of observed effects with properties of the polysaccharides. *Biochim. Biophys. Acta* **740:**466–475.
35. **Gerritsen, M. J., T. Olyhoek, M. A. Smits, and B. A. Bokhout.** 1991. Sample preparation method for polymerase chain reaction-based semiquantitative detection of

*Leptospira interrogans* serovar Hardjo subtype hardjobovis in bovine urine. *J. Clin. Microbiol.* **29:**2805–2808.

36. **Gillespie, D., J. Thompson, and R. Solomon.** 1989. Probes for quantitating subpicogram amounts of HIV-1 RNA by molecular hybridization. *Mol. Cell Probes* **3:**73–86.
37. **Greer, C. E., S. L. Peterson, N. B. Kiviat, and M. M. Manos.** 1991. PCR amplification from paraffin-embedded tissues: effects of fixative and fixation time. *Am. J. Clin. Pathol.* **95:**117–124.
38. **Gustincich, S.** 1991. A fast method for high-quality genomic DNA extraction from whole human blood. *BioTechniques* **11:**298–301.
39. **Hance, A. J., B. Grandchamp, V. Lévy-Frébault, D. Lecossier, J. Rauzier, D. Bocart, and B. Gicquel.** 1989. Detection and identification of mycobacteria by amplification of mycobacterial DNA. *Mol. Microbiol.* **3:**843–849.
40. **Heimdahl, A., G. Hall, M. Hedberg, H. Sandberg, P. Söder, K. Tunér, and C. E. Nord.** 1990. Detection and quantitation by lysis-filtration of bacteremia after different oral surgical procedures. *J. Clin. Microbiol.* **28:**2205–2209.
41. **Hermans, P. W. M., R. A. Hartskeerl, J. E. Thole, and P. R. Klatser.** 1990. Development of diagnostic tests for leprosy and tuberculosis. *Trop. Med. Parasitol.* **41:**301–303.
42. **Higuchi, R.** 1989. Simple and rapid preparation of samples for PCR, p. 31–38. *In* H. A. Erlich (ed.), *PCR Technology: Principles and Applications for DNA Amplification.* Stockton Press, New York.
43. **Hubbard, R. C., N. G. McElvaney, P. Birrer, S. Shak, W. W. Robinson, C. Jolley, M. Wu, M. S. Chernick, and R. G. Crystal.** 1992. A preliminary study of aerosolized recombinant human deoxyribonuclease I in the treatment of cystic fibrosis. *N. Engl. J. Med.* **326:**812–815.
44. **Hunsaker, W. R., H. Badri, M. Lombardo, and M. L. Collins.** 1989. Nucleic acid hybridization assays employing dA-tailed capture probes. *Anal. Biochem.* **181:**360–370.
45. **Ishizawa, M. Y., T. Kobayashi, and S. Matsuuva.** 1991. Simple procedure of DNA isolation from human serum. *Nucleic Acids Res.* **19:**5792.
46. **Izraeli, S., C. Pfleiderer, and T. Lion.** 1991. Detection of gene expression by PCR amplification of RNA derived from frozen heparinized whole blood. *Nucleic Acids Res.* **19:**6051.
47. **Jiang, X., J. Wang, D. Y. Graham, and M. K. Estes.** 1992. Detection of Norwalk virus in stool by polymerase chain reaction. *J. Clin. Microbiol.* **30:**2529–2534.
48. **Jung, J. M., C. T. Comey, D. B. Baer, and B. Budowle.** 1991. Extraction strategy for obtaining DNA from blood stains for PCR amplification and typing of the HLA-DQα gene. *Int. J. Legal Med.* **10:**145–148.
49. **Kain, K. C., A. E. Brown, H. K. Webster, R. A. Wirtz, J. S. Keystone, M. H. Rodriguez, J. Kinahan, M. Rowland, and D. E. Lanar.** 1992. Circumsporozoite genotyping of global isolates of *Plasmodium vivax* from dried blood specimens. *J. Clin. Microbiol.* **30:**1863–1866.
50. **Kaneko, K., O. Onodera, M. Miyatake, and S. Tsuji.** 1990. Rapid diagnosis of tuberculosis meningitis by polymerase chain reaction. *Neurology* **40:**1617–1618.
51. **Kwok, S., D. H. Mack, K. B. Mullis, B. Poiesz, G. Ehrlich, D. Blair, A. Friedman-Kien, and J. J. Sninsky.** 1987. Identification of human immunodeficiency virus sequences by using in vitro enzymatic amplification and oligomer cleavage detection. *J. Virol.* **61:**1690–1694.
52. **Lahiri, D. K., and J. I. Nurnberger, Jr.** 1991. A rapid non-enzymatic method for the preparation of HMW DNA from blood for RFLP studies. *Nucleic Acids Res.* **19:**5444.
53. **Lamblin, G., H. Rahmoune, J.-M. Wieruszeski, M. Lhermitte, G. Strecker, and P. Roussel.** 1991. Structure of two sulphated oligosaccharides from respiratory mucins of a patient suffering from cystic fibrosis: a fast-atom-bombardment m.s. and $^{1}$H-n.m.r. spectroscopic study. *Biochem. J.* **275:**199–206.
54. **Lanciotti, R. S., C. H. Calisher, D. J. Gubler, G.-J. Chang, and A. V. Vornadam.** 1992.

Rapid detection and typing of Dengue viruses from clinical samples by using reverse transcription-polymerase chain reaction. *J. Clin. Microbiol.* **30:**545–551.

55. **Lebech, A.-M., and K. Hansen.** 1992. Detection of *Borrelia burgdorferi* DNA in urine samples and cerebrospinal fluid samples from patients with early and late Lyme neuroborreliosis by polymerase chain reaction. *J. Clin. Microbiol.* **30:**1646–1653.
56. **Lin, L., Y. Gong, G. D. Cimino, J. E. Hearst, and S. T. Isaacs.** 1990. Two novel, rapid, high yield sample preparation methods for the PCR. *Abstr. 5th Conf. Nucleic Acids.*
57. **Loeffelholz, M. J., C. A. Lewinski, S. R. Silver, A. P. Purohit, S. A. Herman, D. A. Buonagurio, and E. A. Dragon.** 1992. Detection of *Chlamydia trachomatis* in endocervical specimens by polymerase chain reaction. *J. Clin. Microbiol.* **30:**2847–2851.
58. **Lund, A., Y. Wasteson, and Ø. Olsvik.** 1991. Immunomagnetic separation and DNA hybridization for detection of enterotoxigenic *Escherichia coli* in a piglet model. *J. Clin. Microbiol.* **29:**2259–2262.
59. **Marko, M. A., R. Chipperfield, and H. C. Birnboim.** 1982. A procedure for the large-scale isolation of highly purified plasmid DNA using alkaline extraction and binding to glass powder *Anal. Biochem.* **121:**382–387.
60. **McCaustland, K. A., S. Bi, M. A. Purdy, and D. W. Bradley.** 1991. Application of two RNA extraction methods prior to amplification of hepatitis E virus nucleic acid by the polymerase chain reaction. *J. Virol. Methods* **35:**331–342.
61. **McHale, R. H., P. M. Stapleton, and P. L. Bergquist.** 1991. Rapid preparation of blood and tissue samples for polymerase chain reaction. *BioTechniques* **10:**20–22.
62. **McKenzie, H. A., and F. H. White, Jr.** 1991. Lysozyme and α-lactalbumin: structure, function, and interrelationships. *Adv. Protein Chem.* **41:**173–315.
63. **Mullis, K. B.** 1991. The polymerase chain reaction in an anemic mode: how to avoid oligodeoxyribonuclear fusion. *PCR Methods Applic.* **1:**1–4.
64. **Noordhoek, G. T., E. C. Wolters, M. E. J. de Jonge, and J. D. A. van Embden.** 1991. Detection by polymerase chain reaction of *Treponema pallidum* DNA in cerebrospinal fluid from neurosyphilis patients before and after antibiotic treatment. *J. Clin. Microbiol.* **29:**1976–1984.
65. **Panaccio, M., and A. Lew.** 1991. PCR based diagnosis in the presence of 8% (v/v) blood. *Nucleic Acids Res.* **19:**1151.
66. **Persing, D. H., D. Mathiesen, W. F. Marshall, S. R. Telford, A. Spielman, J. W. Thomford, and P. A. Conrad.** 1992. Detection of *Babesia microti* by polymerase chain reaction. *J. Clin Microbiol.* **30:**2097–2103.
67. **Pierre, C., D. Lecossier, Y. Bousougant, D. Bocart, V. Joly, P. Yeni, and A. J. Hance.** 1991. Use of a reamplification protocol improves sensitivity of detection of *Mycobacterium tuberculosis* in clinical samples by amplification of DNA. *J. Clin. Microbiol.* **29:**712–717.
68. **Ravaggi, A., D. Primi, and E. Cariani.** 1992. Direct PCR amplification of HCV RNA from human serum. *PCR Methods Applic.* **1:**291–292.
69. **Reddy, G. R., and J. B. Dame.** 1992. rRNA-based method for sensitive detection of *Babesia bigemina* in bovine blood. *J. Clin. Microbiol.* **30:**1811–1814.
70. **Saiki, R. K., D. H. Gelfand, S. Stoffel, S. J. Scharf, R. Higuchi, G. T. Horn, K. B. Mullis, and H. A. Erlich.** 1988. Primer-directed enzymatic amplification of DNA with a thermostable DNA polymerase. *Science* **239:**487–491.
71. **Saulnier, P., and A. Andremont.** 1992. Detection of genes in feces by booster polymerase chain reaction. *J. Clin. Microbiol.* **30:**2080–2083.
72. **Shak, S., D. J. Capon, R. Hellmiss, S. A. Marsters, and C. L. Baker.** 1990. Recombinant human DNase I reduces the viscosity of cystic fibrosis sputum. *Proc. Natl. Acad. Sci. USA* **87:**9188–9192.
73. **Shankar, P., N. Manjunath, K. K. Mohan, K. Prasad, M. Behari, Shriniwas, and G. K. Ahuja.** 1991. Rapid diagnosis of tuberculous meningitis by polymerase chain reaction. *Lancet* **337:**5–7.
74. **Shirai, H., M. Nishibuchi, T. Ramamurthy, S. K. Bhattacharya, S. C. Pal, and Y.**

**Takeda.** 1991. Polymerase chain reaction for detection of the cholera enterotoxin operon of *Vibrio cholerae. J. Clin. Microbiol.* **29:**2517–2521.

75. **Sjöbring, J., M. Mecklenburg, A. B. Andersen, and H. Miörner.** 1990. Polymerase chain reaction for detection of *Mycobacterium tuberculosis. J. Clin. Microbiol.* **28:**2200–2204.
76. **Sritharan, V., and R. H. Barker, Jr.** 1991. A simple method for diagnosis of *M. tuberculosis* infection in clinical samples using PCR. *Mol. Cell. Probes* **5:**385–395.
77. **Swaisgood, H. E., and G. L. Catignani.** 1987. Use of immobilized proteinases and peptidases to study structural changes in proteins. *Methods Enzymol.* **135:**596–604.
78. **Ting, Y., and M. Manos.** 1990. Detection and typing of genital human papillomaviruses, p. 356–367. *In* M. A. Innis, D. H. Gelfand, J. J. Sninsky, and T. J. White (ed.), *PCR Protocols: A Guide to Methods and Applications*. Academic Press, Inc., San Diego, Calif.
79. **Vogelstein, B., and D. Gillespie.** 1979. Preparative and analytical purification of DNA from agarose. *Proc. Natl. Acad. Sci. USA* **76:**615–619.
80. **Wang, J.-T., T.-H. Wang, J.-C. Sheu, S.-M. Lin, J.-T. Lin, and D.-S. Chen.** 1992. Effects of anticoagulants and storage of blood samples on efficacy of the polymerase chain reaction assay for hepatitis C virus. *J. Clin. Microbiol.* **30:**750–753.
81. **Wicher, K., G. T. Noordhoek, F. Abbruscato, and V. Wicher.** 1992. Detection of *Treponema pallidum* in early syphilis by DNA amplification. *J. Clin. Microbiol.* **30:**497–500.
82. **Yamada, O., T. Matsumoto, M. Nakashima, S. Hagari, T. Kamahora, H. Ueyama, Y. Kishi, H. Uemura, and T. Kurimaura.** 1990. A new method for extraction of DNA or RNA for polymerase chain reaction. *J. Virol. Methods* **27:**203–210.

# Amplification Product Detection Methods

# 7

Thomas J. White

The choice of the appropriate detection method for a particular molecular diagnostic technique depends on the level of specificity in the amplification and whether (i) target or probe amplification is used, (ii) the nucleic acid being detected is RNA or DNA, (iii) the sequence being detected is heterogeneous, (iv) rare sequence variants within the amplified products have to be detected, and (v) new sequence information beyond that which is inherent in the primers or probes is generated (chapter 4).

Target amplification systems not only provide for dramatic increases in the sequence to be detected but also, and equally importantly, reduce the complexity of the nucleic acid, thereby permitting much less stringent hybridization. Less stringent probe hybridization may prove critical to the detection of targets that are expected to contain multiple sequence variants. The RNA viruses in general, and human immunodeficiency virus in particular, vary in sequence within an infected individual and among infected individuals. In addition, searches for new pathogens require more specific amplification coupled with low-stringency hybridization so as not to miss a potentially interesting microorganism. The isothermal amplification reactions (e.g., self-sustaining sequence replication [3SR] and strand displacement amplification [SDA]) presently are carried out at about 37°C (as for the polymerase chain reaction [PCR] when the thermolabile DNA polymerase from *Escherichia coli* was used) and amplify not only the desired sequence but also a large number of additional sequences. As an example, the analysis of the PCR-amplified products in the early studies indicated that perhaps only 1% of the products were the targeted region (30). In brief, although the reduction in complexity for PCR is in the same order of magnitude as the level of amplification, the isothermal amplification strategies may reduce complexity 100-fold less than the amplification factor. It will be important to follow the progress made in the isothermal approaches for their incorporation of thermoactive and stable enzymes to allow more specific amplification to be performed.

Prior to PCR and the other amplification strategies, hybridization had to be driven by using vast excesses of the tagged or labeled probe. The requirement for such large amounts of labeled probe invariably led to background noise, which compromised the analytical sensitivity of the assay. Target amplification circumvents the need to use the labeled probe at high levels, and, in fact, the levels of amplification permit detection to be driven by the amplified products. This has allowed the use of novel detection formats that previously could not have been considered. An example is the use of a reverse dot blot detection format (Fig. 1) (29). This format allows the interrogation of the sample with multiple probes

**Figure 1.** Reverse dot blot hybridization. Sequence-specific capture probes are immobilized on a solid support such as nylon strips. Attachment as shown in the schematic is via 3′-terminal poly(T) tails that can be covalently linked to a matrix by exposure to UV light. Other modes of attachment have also been used. If biotinylated primers are used in amplification, the products hybridize to the capture probes, and following washing steps to remove the unbound amplified product and primers, the duplex binds an avidin-horseradish peroxidase conjugate. This enzyme then converts tetramethylbenzidine or other colorless substrate to a colored precipitate that can be clearly visualized as a blue dot. The example shown is for the typing of specific HLA-DQα alleles in samples of DNA from two different individuals.

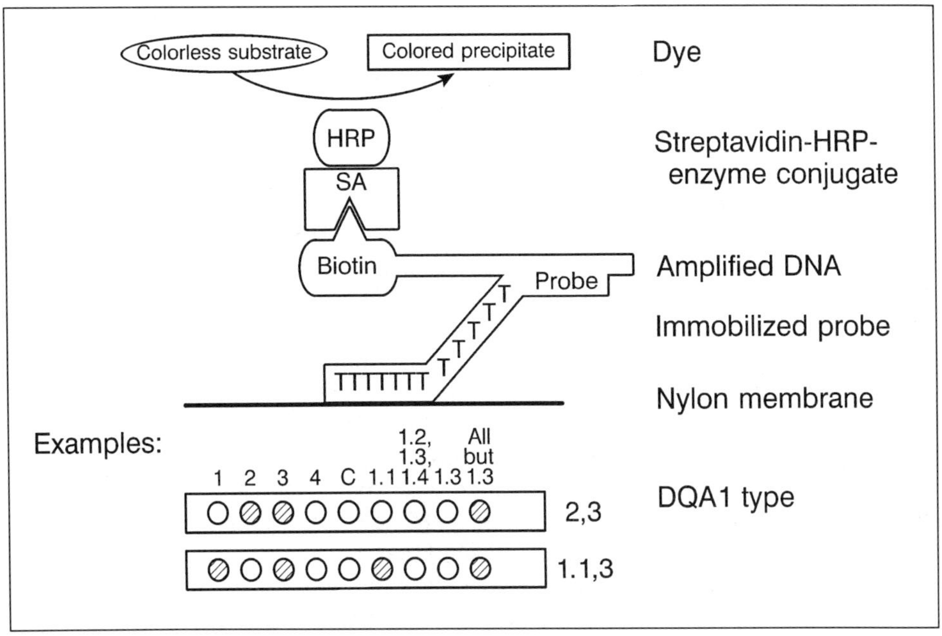


(protocol 1.21), which simultaneously avoids the need to use all of the amplification reaction for multiple probings and the possible mixup of the multiply spotted amplified products.

Although PCR, SDA, and ligase chain reaction (LCR) yield amplified DNA, 3SR generates RNA products. There are multiple implications to the detection of RNA or DNA. First, classic studies of duplex stability indicate that RNA-DNA hybrids are generally more stable than DNA-DNA hybrids. Second, although electrophoretic and high-pressure liquid chromatography (HPLC) systems that accurately size DNA products have been developed, to date it has proven difficult to size RNA products, presumably because of the various secondary-structure conformations that form. As a result, although amplified DNA products can be conveniently visualized as a band following electrophoresis on acrylamide or agarose, RNA products from 3SR fractionated electrophoretically smear on the gel. Thus, detection approaches that involve sizing are not recommended for 3SR. Third, the secondary-structure conformations readily formed with RNA may interfere with the availability of various regions of the amplified material to hybridize to a probe.

Although conventional bases have almost exclusively been used in amplification and detection, atypical bases are increasingly being used. The amplification strategies that involve a DNA polymerase and yield DNA (e.g., PCR and SDA) are likely to tolerate base substitutions more readily because of the malleable nature of this enzyme. On the other hand, 3SR may prove less malleable since RNA polymerases are less accommodating of modified bases. Extensive biophysical studies over the years investigated the effect of various base substituents on helix stability. These studies range from infrared solution studies of bases in apolar solvents to the use of the thermal stability of different homoduplexes to the use of bacteriophages that contain unconventional bases. More recently, the availability of defined sequence oligonucleotides permitted the examination of specific base pair alterations and the contribution of sequence context effects (8). In brief, these studies demonstrated that methyl or bromine substituents on the C-5 position of pyrimidines (e.g., dU and dC) increase the stability of the resulting duplex. Wong et al. (38) and Wong and McClelland (39) used 5-methyl-dCTP in PCR and demonstrated the increased thermal stability of the amplified products. In fact, these laboratories showed that for larger PCR products, the incorporation of 5-methyl-dC could prevent efficient amplification since the temperatures typically used for denaturation were not sufficient to accomplish strand separation, an event integral to amplification. Hoheisel et al. (10) recently suggested the use of 5-methyl-dC- and 5-bromo-dC-containing oligonucleotides as primers to enhance thermal stability. The recent use of dUTP to substitute for TTP, coupled with pretreatment of samples with the enzyme uracil *N*-glycosylase (UNG) to specifically degrade errant PCR products from previous amplifications to avoid false-positive results (19, 31), prompts the question of differential hybridization of dU- versus T-containing products. The differential PCR efficiency sometimes observed if dUTP is substituted for TTP confounds simple interpretation of the hybridization results. In general, the experience to date indicates that substitution of dU into PCR products has only a modest impact on overall duplex stability. Regardless, investigators should bear in mind that unconventional bases may alter the hybridization properties in their detection system. As is always the case, sequence context effects may lead to unexpected results. Other bases which have been examined either for their increased or decreased base-pairing discrimination include dI, diaminopurine (DAP), and deaza-G.

Beyond the use of modified bases to alter the hybridization properties of the amplified product or probes, altered bases can be used to carry in specific tags or labels. The most successful tagged bases to be used in PCR are digoxigenin-dUTP and biotin-dUTP. Once incorporated, amplification products can be captured or detected by using avidin or streptavidin. Another base that can be substituted for TTP is 5-bromo-dUTP. Since antibodies that recognize this base analog are available, this modified base-antibody pair may be recruited for detection. Efforts to efficiently incorporate fluorescently labeled bases have been less successful but merit further study.

Another concern that arises in considering the detection of amplified products is the fidelity of the amplification process. Fidelity is a measure of the accuracy of replication of the original template targets; it is a complex process with multiple elements. The enzyme is only one of the elements. Other elements critical to faithful copying include (i) the buffer conditions, (ii) the thermal cycling parameters, (iii) the number of cycles, (iv) the efficiency of amplification, and (v) the sequence of the DNA being copied. Several excellent reviews that summarize the relative importance of each parameter have appeared (1, 6). The high fidelity of *Taq* DNA polymerase led Eckert and Kunkel (7) to query whether a cryptic proofreading activity was present. In summary, it is highly unlikely that replication errors will affect detection under standard PCR conditions. RNA polymerases and reverse transcriptases are known to catalyze less faithful copying than

DNA polymerases. Whether this compromised fidelity will affect the utility of 3SR is unclear.

Some amplification procedures do not provide sequence information beyond that which is inherent in the primers and probes. In LCR and the oligomer ligation assay (OLA), for example, the ligation product of two labeled probes can be detected by a variety of methods which measure only the fusion product of both probes. A similar approach can be taken in PCR by using two labeled primers as long as the specificity of the amplification is high. These approaches are not suitable for 3SR since the single-stranded RNA product does not contain labels from both primers. On the other hand, PCR, 3SR, and SDA can all incorporate an additional level of certainty that the amplified fragment is the intended one by using an oligonucleotide probe to detect sequence information located between the two primers. The notion that detection is simultaneous with amplification is not accurate for most methods described to date, since they still require additional steps involving ligand binding, washing, and enzymatic conversion of colorimetric substrates.

The simplest method of detecting amplified DNA measures the increase in the amount of double-stranded DNA by monitoring the increase in fluorescence of bound ethidium bromide (9, 22). PCR can occur without significant inhibition in the presence of the dye, and the fluorescence can be detected without opening the reaction tube. The primers can be designed and optimized to provide a high degree of specificity for the intended target and to minimize artifacts such as primer multimers.

Higuchi et al. illustrated this approach with "allele-specific" primers for β-globin A or S in PCR and detected the fluorescence with a fiber-optic device. Other dyes could also be used (32), and a thermocycler could be fitted with optical fibers (or other excitation and detection devices) so that the reactions can be continuously monitored over the course of an amplification. This method may be suitable for specimens, such as serum or plasma, that do not initially contain large amounts of DNA, since such specimens would cause a decrease in the background fluorescence over which any additional fluorescence produced by amplification must be detected.

The next level of complexity in detection techniques involves primers or probes which are derivatized with ligands or signal moieties. Chehab and Kan labeled each primer with a different fluorescent dye; after separation of the unincorporated primers by selective filtration, they detected the amplified DNA by using gel electrophoresis and UV light (Fig. 2) (5).

OLA can be used to detect DNA amplified by either PCR or LCR (Fig. 3) (25, 36a). This format involves two oligonucleotide probes complementary to adjacent regions of a template; the upstream and downstream oligonucleotides are labeled at the 5′ and 3′ termini, respectively, with a ligand or a reporter molecule. If the amplified target is present, the probes are ligated. In subsequent steps the detection probes are captured by a receptor for the ligand on a solid support and the reporter molecule is detected directly (if fluorescent) or via an antibody-enzyme conjugate which uses a colorimetric substrate. OLA can be thought of as the last step of an LCR, and, as a result, detection approaches used for OLA can be used with LCR. If used to detect a PCR product, OLA probes will add a measure of additional confirmation of specificity to the analysis. With LCR, or PCR if the primers are labeled, OLA is inherent in the amplification. Attributes of the OLA approach include the potential for multiplex analysis of different alleles or species-specific sequences contained within the amplified region.

**Figure 2.** Strategy for the color complementation assay detection of point mutations. A cytidine-to-thymidine mutation is illustrated in the example. Two allele-specific primers corresponding to this region are labeled with red dye (corresponding to the wild-type allele) or green dye (corresponding to the mutant allele). The primer amplifying the opposite strand is unlabeled. After PCR and removal of unincorporated primers, the amplified products for normal, heterozygous, and homozygous DNA are red, yellow, and green, respectively. Reprinted from reference 5 with permission.

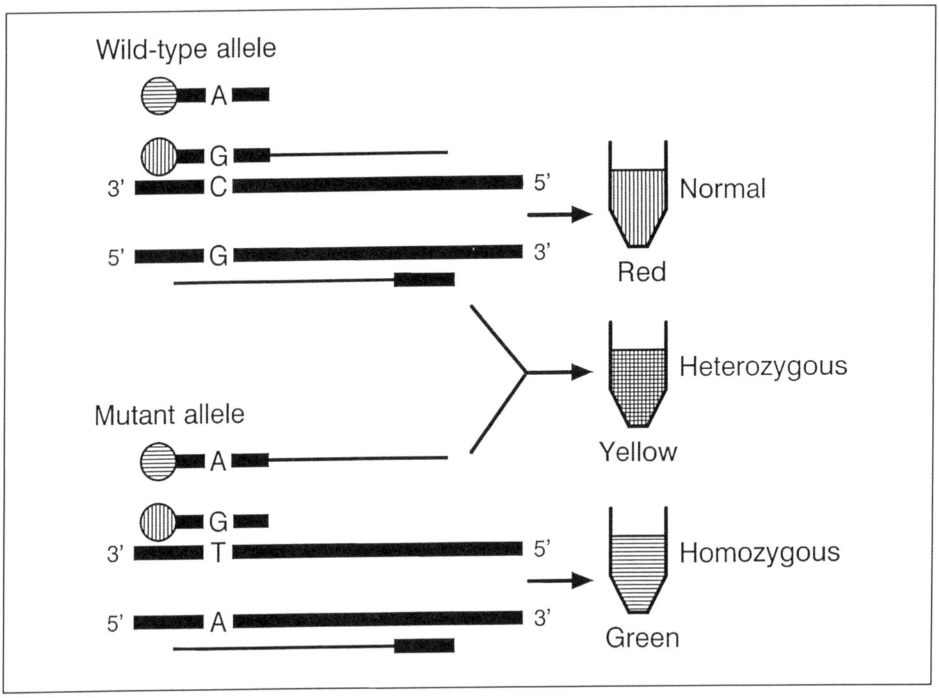


Other variations on labeling the primers include the incorporation of sequences that are recognized by DNA-binding proteins. Kemp and coworkers have used either the *GCN4* gene from *Saccharomyces cerevisiae* or the TyrR-binding site to label one primer and biotin to label the other (14, 34). Amplified DNA is captured by the DNA-binding protein on the wall of a microtiter plate. The biotin moiety is then detected with standard avidin-peroxidase reagents. Wahlberg et al. (35) used a similar approach with the *lac* operator sequence and a *lacI-lacZ* fusion protein for detection. A final version that does not use an internal probe involves the incorporation of digoxigenin-11-dUTP or biotin-dUTP during the PCR (2). Following gel electrophoresis, the gel is blotted onto a nylon membrane and the labeled DNA is detected with an anti-digoxigenin antibody: alkaline phosphatase conjugate and a chemiluminescent substrate. The gel-based methods still are relatively cumbersome for the clinical laboratory and offer mainly an increased sensitivity over the original method of simply detecting a band of the expected size on an agarose gel by staining it with ethidium bromide and detecting the fluorescence under UV light (23).

Probe-based detection methods run the gamut from Southern blots, oligomer restriction (28), oligomer hybridization (17), enzyme immunoassay (21), nonisotopic dot blot (3), reverse dot blot (29), acridinium ester derivatives (Fig. 4) (27), probe shift assay (15), and microwell plates (36). Some of these approaches are more suited for batch analysis of multiple samples in which a single pathogenic agent is being detected, whereas others are better suited for multiplex differenti-

**Figure 3.** Schematic diagram of the steps in the automated PCR/OLA procedure performed with a robotic workstation. The assay has three steps: (i) DNA target amplification; (ii) analysis of target nucleotide sequences with biotin (B)-labeled and digoxigenin (D)-labeled oligonucleotide probes and T4 DNA ligase (L); and (iii) capture of the biotin (B)-labeled probes on streptavidin (SA)-coated microtiter wells and analysis for covalently linked digoxigenin (D) by using an enzyme-linked immunosorbent assay (ELISA) procedure with alkaline phosphatase (AP)-conjugated anti-digoxigenin (αD) antibodies and a substrate (S). Reprinted from reference 25 with permission.

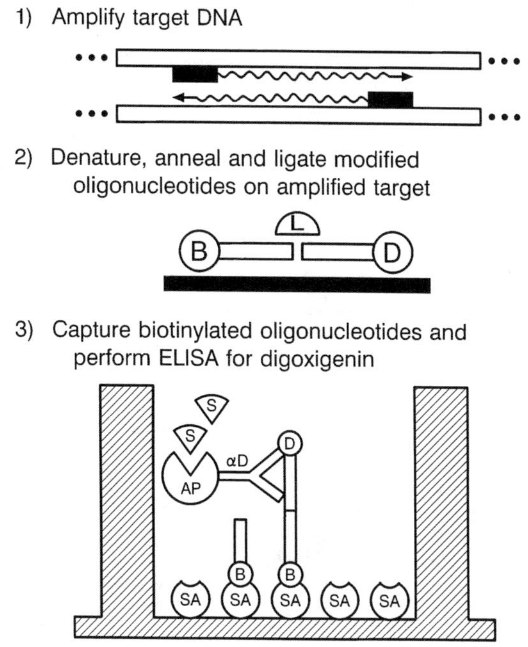


ation of multiple species that can be amplified with genus-specific primers, e.g., mycobacteria. As with amplification conditions, the specific hybridization parameters must be optimized for each pathogen target sequence. In general, isothermal amplification methods such as 3SR and SDA require probe-based detection methods because the heterogeneous single-stranded RNA or double-stranded DNA products, respectively, are not resolved by size-based methods. HPLC and capillary zone electrophoresis may have future potential as size-based detection methods if automation can increase their sample throughput to a practical level (1).

A novel detection method that makes use of an exonuclease activity of some thermostable DNA polymerases was recently described (11). Holland et al. used the 5′-to-3′ exonuclease activity of the *Taq* DNA polymerase to generate a sequence-specific probe signal during a PCR (11). During each cycle of amplification, the enzyme degrades the probe and releases a smaller fluorescent fragment which can be separated from uncleaved probe and measured through the wall of the reaction tube (Fig. 5). More recently, Holland et al. (12) demonstrated that this novel detection strategy can discern templates that vary by a single base. This method has potential for giving a very broad dynamic range for quantitative assays and requires little or no handling of the sample after amplification. Such approaches may reduce the chance of assay contamination and, by combining

**Figure 4.** Acridinium ester-labeled DNA probe method. Following amplification, a DNA probe with an attached acridinium is hybridized with the resulting amplified products. The acridinium is resistant to alkaline hydrolysis if it is in a DNA duplex (step 1) and therefore can chemiluminesce after oxidation by the addition of hydrogen peroxide (step 2). If the labeled probe does not hybridize, hydrolysis does occur (step 1).

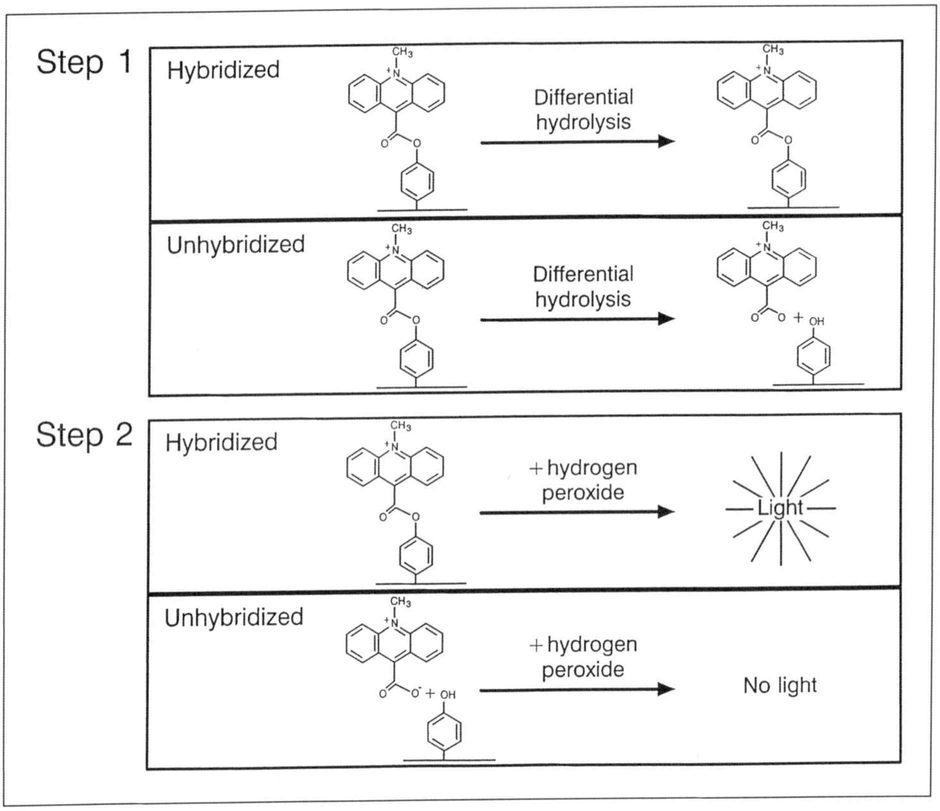


several differentially labeled probes in a single reaction, permit multiplex analysis of several agents.

Detection of rare sequence variants within amplified products presents additional challenges to the detection format. If the sequence variants differ by a single base pair, discrimination becomes difficult. Table 1 summarizes the approaches and the relative fraction of the amplified products that must contain the sequence variant to ensure detection. Clearly, the most sensitive method for single nucleotide variants is "allele-specific" PCR. This appellation is something of a misnomer since this method can also be used to detect somatic mutations. The penchant of investigators to coin terms for their own modifications of PCR has led to this single-base discrimination assay being referred to as allele-specific PCR (40), amplification refractory mutation system (24), and mismatch amplification (4). Sequencing, dot blot, single-strand conformation polymorphism (SSCP) (20), and heteroduplex analysis (13) require that a significant faction of the amplified products contain the variant sequence. OLA, LCR, and the single-nucleotide extension procedure (16, 33) allow as little as 1 to 2% of the products to be detected, even though they vary by a single base pair. Although improvements in the cost, simplicity, and speed of SSCP and DNA-sequencing instruments may permit these analytical methods to eventually find use in the clinical

**Figure 5.** Recruitment of the 5′-to-3′ exonuclease activity of *Taq* DNA polymerase for detection.

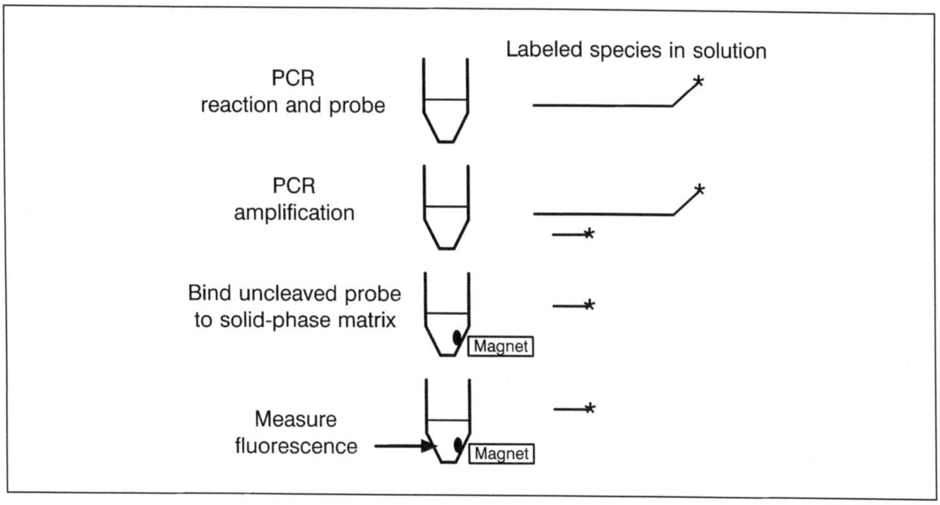


laboratory, they are currently limited to targets, such as rifampin-resistant tuberculosis bacilli, that require the analysis of multiple potential mutations throughout a particular locus (protocol 6.1). Recent studies in which LCR was used for the detection of single-base-pair changes indicate that a 100-fold excess of the unaltered gene will lead to nonspecific amplification (37). A further challenge is the need to determine not only whether various single-base-pair changes are present in the population but also whether they reside on the same molecule. A representative example of this is the detection of mutant human immunodeficiency virus reverse transcriptase genes. It may be important to detect an extremely small fraction of the population of proviral DNA that harbors multiple mutations critical to making the virus resistant to nucleotide analogs such as azidothymidine, ddC, and ddI (protocol 6.2). Another example of the need to detect a rare sequence variant within a population is the detection in peripheral blood of metastatic cancer cells shed from tissues or organs. The lower temperatures currently used for the isothermal amplification strategies (e.g., 3SR and SDA) preclude their use for single-base mutation-specific amplification until thermostable enzymes become available. None of the detection approaches, except for PCR, allow one to distinguish whether the mutations detected are on different molecules or on the same molecule. If the primers are positioned so that

**Table 1.** Detection of sequence variants in the DNA population to be amplified

| Method | % of variant | Quantitation | Linkage information | Detection |
|---|---|---|---|---|
| Sequencing | ≥33 | Yes | No | Fluorescent |
| SSCP | ≥15 | Possibly | No | Fluorescent |
| Heteroduplex | ≥15 | Possibly | No | Fluorescent |
| Reverse dot blot | ≥15 | Yes | No | Colorimetric |
| Hybridization | ≥15 | Yes | No | Radioactive |
| Nucleotide extension | >2 | Yes | No | Radioactive |
| OLA | ≥2 | Possibly | No | Colorimetric |
| Sequence-specific PCR | 0.0001–1 | No | Yes | Fluorescent |

**Table 2.** Detection strategies for amplified products

| Format | Hybridization | Detection | Signal | Elapsed time (from amplification to detection) |
|---|---|---|---|---|
| Electrophoretic | Post | Post | Radioactive | 1 day |
| Strip | Post | Post | Colorimetric | 5 h |
| Microtiter | Post | Post | Colorimetric | 2 h |
| Taqman | During | Post | Nonradioactive | <2 h |
| Simultaneous | | During | Nonradioactive | 1 h |

both 3′ termini are complementary to the nucleotide, successful amplification indicates that both mutations are on the same molecule (18). It is possible to design a nested system allowing the detection of four mutations on the same molecule.

The final detection system that merits comment is in situ PCR. The experience of several laboratories and a mounting publication list (for a review, see reference 26), principally by Nuovo and coworkers, suggests that this procedure will gain more widespread use in the near future. In situ PCR has been carried out both with tagged deoxynucleoside triphosphates and with subsequent probe hybridization to visualize the amplified material. An inherent limitation, given the estimated volume of cells, is that amplification of sequences is only approximately 100-fold; however, this level of amplification more than suffices for ready detection. A key advantage to this approach is the opportunity to superimpose cell morphology, epitope reactivity, and genetic sequences. Another advantage is that since amplification occurs within the cell, carryover of amplified products from previous reactions is extremely unlikely to affect detection.

Amplification strategies have permitted previous formats to be reexamined and new ones to be designed. Table 2 simply highlights the progress made to date in the exploration of various formats. We can expect to see continued improvements, simplification, and convenient quantification in detection formats. The challenge will be to use the resulting information to provide further understanding of pathogenic processes, differential treatment of patients, and improved individual and public health care.

## References


1. **Bloch, W.** 1991. A biochemical perspective of the polymerase chain reaction. *Biochemistry* **30:**2735–2747.
2. **Boehringer Mannheim Biochemicals.** 1991. *Biochemica* **8:**6–7 package insert. Sensitive, non-radioactive detection of PCR-amplified DNA. Boehringer Mannheim Biochemicals, Indianapolis, Ind.
3. **Bugawan, D., R. K. Saiki, C. H. Levenson, R. M. Watson, and H. A. Erlich.** 1988. The use of non-radioactive oligonucleotide probes to analyze enzymatically amplified DNA for prenatal diagnosis and forensic HLA typing. *Bio/Technology* **6:**943–947.
4. **Cha, R. S., H. Zarbl, P. Keohavong, and W. G. Thilly.** 1992. Mismatch amplification mutation assay (MAMA): application to the c-H-*ras* gene. *PCR Methods Applic.* **2:**14–10.
5. **Chehab, F. F., and Y. W. Kan.** 1989. Detection of specific DNA sequences by fluorescence amplification: a color complementation assay. *Proc. Natl. Acad. Sci. USA* **86:**9178–9182.

6. **Eckert, K. A., and T. A. Kunkel.** 1991. DNA polymerase fidelity and the polymerase chain reaction. *PCR Methods Applic.* **1:**17–24.
7. **Eckert, K. A., and T. A. Kunkel.** 1990. High fidelity DNA synthesis by the *Thermus aquaticus* DNA polymerase. *Nucleic Acids Res.* **18:**3739–3744.
8. **Fareed, A., D. Koh, and I. Tinoco, Jr.** 1985. Base-base mismatches. Thermodynamics of double helix formation for $dCA_3XA_3G + dCT_3YT_3G$ (X,Y = A,C,G,T). *Nucleic Acids Res.* **13:**4811–4824.
9. **Higuchi, R., G. Dollinger, P. S. Walsh, and R. Griffith.** 1992. Simultaneous amplification and detection of specific DNA sequences. *Bio/Technology* **10:**413–417.
10. **Hoheisel, J. D., A. G. Craig, and H. Lehrach.** 1990. Effect of 5-bromo- and 5-methyldeoxycytosine on duplex stability and discrimination of the *Not*I octadeoxynucleotide. *J. Biol. Chem.* **265:**16656–16660.
11. **Holland, P. M., R. D. Abramson, R. Watson, and D. H. Gelfand.** 1991. Detection of specific polymerase chain reaction product by utilizing the 5′ → 3′ exonuclease activity of *Thermus aquaticus* DNA polymerase. *Proc. Natl. Acad. Sci. USA* **88:**7276–7280.
12. **Holland, P. M., R. D. Abramson, R. Watson, S. Will, R. K. Saiki, and D. H. Gelfand.** 1992. Detection of specific polymerase chain reaction product by utilizing the 5′ → 3′ exonuclease activity of *Thermus aquaticus* DNA polymerase. *Clin. Chem.* **38:**462–463.
13. **Keen, J., D. Lester, C. Inglehearn, A. Curtis, and S. Bhattacharya.** 1991. Rapid detection of single-base mismatches as heteroduplexes on hydrolink gels. *Trends Genet.* **7:**5.
14. **Kemp, D. J., D. B. Smith, S. J. Foote, N. Samaras, and M. G. Peterson.** 1989. Colorimetric detection of specific DNA segments amplified by polymerase chain reactions. *Proc. Natl. Acad. Sci. USA* **86:**2423–2427.
15. **Kumar, R.** 1989. The technique of polymerase chain reaction. *Technique* **1:**133–152.
16. **Kuppuswamy, M. N., J. W. Hoffmann, C. K. Kasper, S. G. Spitzer, S. L. Groce, and S. P. Bajaj.** 1991. Single nucleotide primer extension to detect genetic diseases: experimental application to hemophilia B (factor IX) and cystic fibrosis genes. *Proc. Natl. Acad. Sci. USA* **88:**1143–1147.
17. **Kwok, S., G. Ehrlich, B. Poiesz, R. Kalish, and J. J. Sninsky.** 1988. Enzymatic amplification of HTLV-I viral sequences from peripheral blood mononuclear cells and infected tissues. *Blood* **72:**1117–1123.
18. **Lo, Y.-M. D., P. Patel, C. R. Newton, A. F. Markham, K. A. Fleming, and J. S. Wainscoat.** 1991. Direct haplotype determination by double ARMS: specificity, sensitivity and genetic applications. *Nucleic Acids Res.* **19:**3561–3567.
19. **Longo, M. C., M. S. Berninger, and J. L. Hartley.** 1990. Use of uracil DNA glycosylase to control carry-over contamination in polymerase chain reactions. *Gene* **93:**125–128.
20. **Makino, R., H. Yazyu, Y. Kishimoto, T. Sekiya, and K. Hayashi.** 1992. F-SSCP: fluorescence-based polymerase chain reaction-single-strand conformation polymorphism (PCR-SSCP) analysis. *PCR Methods Applic.* **2:**10–13.
21. **Mantero, G., A. Zonaro, A. Albertini, P. Bertolo, and D. Primi.** 1991. DNA enzyme immunoassay: general method for detecting products of polymerase chain reaction. *Clin. Chem.* **37:**422–429.
22. **Mitoma, Y.** 1992. Method for detecting or quantifying target nucleic acid. European patent application 91310062.4.
23. **Mullis, K. B., and F. A. Faloona.** 1987. Specific synthesis of DNA *in vitro* via a polymerase catalysed chain reaction. *Methods Enzymol.* **155:**335–350.
24. **Newton, C. R., A. Graham, L. E. Heptinstall, S. J. Powell, C. Summers, N. Kalsheker, J. C. Smith, and A. F. Markham.** 1989. Analysis of any point mutation in DNA. The amplification refractory mutation system. *Nucleic Acids Res.* **17:**2503–2516.
25. **Nickerson, D. A., R. Kaiser, S. Lappin, J. Stewart, L. Hood, and U. Landegren.** 1990. Automated DNA diagnostics using an ELISA-based oligonucleotide ligation assay. *Proc. Natl. Acad. Sci. USA* **87:**8923–8929.
26. **Nuovo, G. J. (ed.).** 1992. *PCR In Situ Hybridization: Protocols and Applications*. Raven Press, New York.

27. **Ou, C.-Y., S. H. McDonough, D. Cabanas, T. B. Ryder, M. Harper, J. Moore, and G. Schochetman.** 1990. Rapid and quantitative detection of enzymatically amplified HIV-1 DNA using chemiluminescent oligonucleotide probes. *AIDS Res. Hum. Retroviruses* **6:**1323–1329.
28. **Saiki, R. K., S. J. Scharf, F. Faloona, K. B. Mullis, G. T. Horn, H. A. Erlich, and N. Arnheim.** 1985. Enzymatic amplification of beta-globin genomic sequences and restriction site analysis for diagnosis of sickle cell anemia. *Science* **230:**1350–1354.
29. **Saiki, R. K., D. S. Walsh, and H. A. Erlich.** 1989. Genetic analysis of amplified DNA with immobilized sequence-specific oligonucleotide probes. *Proc. Natl. Acad. Sci. USA* **86:**6230–6234.
30. **Scharf, S. J., G. T. Horn, and H. A. Erlich.** 1986. Direct cloning and sequence analysis of enzymatically amplified genome sequences. *Science* **233:**1076–1078.
31. **Schochetman, G., and J. J. Sninsky.** 1992. Direct detection of human immunodeficiency virus infection using the polymerase chain reaction, p. 90–110. *In* G. Schochetman and J. R. George (ed.), *AIDS Testing: Methodology and Management Issues*. Springer-Verlag, New York.
32. **Searle, M. S., and K. J. Embrey.** 1990. Sequence-specific interaction of Hoescht 33258 with the minor groove of an adenine-tract DNA duplex studied in solution by $^{1}$H NMR spectroscopy. *Nucleic Acids Res.* **18:**3753–3762.
33. **Syvänen, A.-C., K. Aalto-Setälä, L. Harju, K. Kontula, and H. Söderlund.** 1990. A primer-guided nucleotide incorporation assay in the genotyping of apolipoprotein E. *Genomics* **8:**684–692.
34. **Triglia, T., V. P. Argyropoulos, B. E. Davidson, and D. J. Kemp.** 1990. Colourimetric detection of PCR products using the DNA-binding protein TyrR. *Nucleic Acids Res.* **18:**1080.
35. **Wahlberg, J., J. Lundeberg, T. Hultman, and M. Uhlén.** 1990. General colorimetric method for DNA diagnostics allowing direct solid-phase genomic sequencing of the positive samples. *Proc. Natl. Acad. Sci. USA* **87:**6569–6573.
36. **White, T. J., R. Madej, and D. H. Persing.** 1992. The polymerase chain action: clinical applications. *Adv. Clin. Chem.* **29:**161–196.
36a. **Whitely, N., M. W. Hunkapillar, and G. Glazer.** 1989. *Detection of Specific Sequences in Nucleic Acids.* U.S. patent 4,883,750.
37. **Wiedmann, M., J. Czajka, F. Barany, and C. A. Batt.** 1992. Discrimination of *Listeria monocytogenes* from other *Listeria* species by ligase chain reaction. *Appl. Environ. Microbiol.* **58:**3443–3447.
38. **Wong, D. M., T. H. Weinstock, and J. G. Wetmur.** 1991. Branch capture reactions: displacers derived from asymmetric PCR. *Nucleic Acids Res.* **19:**2251–2259.
39. **Wong, K. K., and M. McClelland.** 1991. PCR with 5-methyl-dCTP replacing dCTP. *Nucleic Acids Res.* **19:**1081–1085.
40. **Wu, D. Y., L. Ugozzoli, B. K. Pal, and R. B. Wallace.** 1989. Allele-specific enzymatic amplification of β-globin genomic DNA for diagnosis of sickle cell anemia. *Proc. Natl. Acad. Sci. USA* **86:**2757–2760.

**Acknowledgment.** I thank John Sninsky for helpful contributions to this chapter.

# Laboratory Design and Work Flow

# 8

Bruce J. McCreedy and Thomas H. Callaway

Clinical diagnostic assays that use molecular probes are now commonly performed in most clinical laboratories. With the development of nucleic acid amplification technologies, the potential applications for DNA probe-based diagnostic testing are expanding into areas previously thought to be unsuitable for such methods (2, 12).

The potential advantages offered by nucleic acid amplification and detection systems for clinical testing include high analytical sensitivity coupled with species-specific identification capability. Although amplification of a nucleic acid target increases assay sensitivity, clinical laboratories that use nucleic acid amplification technologies such as polymerase chain reaction (PCR) (see chapter 4 for a technology review) must take precautions to prevent false-positive results that may arise from contamination of specimens by amplification products of previously analyzed specimens (8). Careful consideration should be given to facility design and operation for clinical laboratories in which nucleic acid amplification-based assays are performed.

An optimal facility design is described in this chapter. However, recent improvements in amplification technology and the future availability of licensed diagnostic kits may require less stringent conditions, especially for low-volume users and laboratories not involved in routine performance or development of non-kit-based amplification assays.

## Specimen Receipt and Accessioning

Specimens should be received and accessioned in an area of the facility that is isolated from any testing areas. Whenever possible, only original specimen containers should be accepted. Pouring off of blood specimens or other splitting of original specimens for use in other clinical testing should be avoided. Because of the high sensitivity of amplification-based testing, minute quantities of contaminants introduced into a specimen before receipt in the testing laboratory could produce a positive result from a clinically negative individual.

Gloves and laboratory coats should be worn by accessioning personnel when they are sorting specimens. Specimens should be assigned unique identifiers and placed in appropriate containers (i.e., foam blocks or tube racks) that will prevent leakage of liquid specimens. Frozen specimens should be kept frozen at −20 to −70°C. Freeze-thawing of clinical specimens can have deleterious effects on the ability to recover amplifiable nucleic acid from some organisms; therefore it should be avoided. Accessioning personnel should not open any original contain-

**Figure 1.** Floor plan for the ideal clinical PCR laboratory.

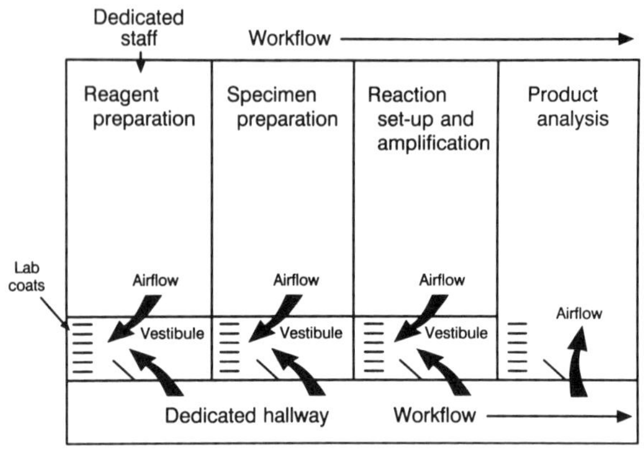


ers or split specimens for multiple testing; this precaution is taken to prevent cross-contamination of original specimens and to avoid any accumulation of target nucleic acids originating from specimens in the accessioning area. When accessioning is completed, personnel from the laboratory testing area should be notified to pick up the specimens. Personnel who work in the accessioning department should not deliver samples to testing areas; this minimizes the potential for transfer of amplification products from previously analyzed specimens into the accessioning area. Instead, accessioned specimens ready to be picked up for testing should be placed in a designated area for retrieval by the appropriate personnel.

## General Design Considerations

Because of the large number of copies of amplification products that can be generated from specimens harboring the nucleic acid target in question ($10^{10}$ to $10^{12}$ copies), it is essential that physical separation of preamplification and postamplification areas be included in the design of the testing facility (7). A unidirectional work flow concept, which includes separate testing areas, should be followed. Each phase of the testing process, including the movement of specimens and the movement of personnel, should proceed from areas free of amplification products to areas where amplification products are analyzed throughout the work day. Any movement of personnel in the reverse direction (i.e., from product-containing to product-free areas) should be kept to a minimum and eliminated if possible. A separate laboratory should be set up for preparation, dispensing, and storage of reagents to be used in amplification assays. Personnel working in product-containing areas should not be allowed access to the reagent preparation laboratory. Conversely, personnel from the reagent preparation laboratory should not work in product-containing areas and the reagent preparation laboratory concomitantly.

Figure 1 illustrates an ideal facility design which incorporates separate laboratories dedicated to each of the processes involved in clinical testing by PCR amplification: reagent preparation, specimen preparation, amplification, and product analysis. The unidirectional work flow concept is illustrated by the succession of the laboratory areas, beginning with the reagent preparation laboratory and ending with the PCR product analysis laboratory. Each of the four laboratories is designed to isolate potentially contaminating aerosols by using a combination of controlled airflow, two-door vestibule-type entrances with magnetic interlocks, and UV light fixtures installed in the overhead lighting.

Each laboratory, with the exception of the postamplification analysis area, consists of a positive- or neutral-pressure inner working laboratory connected to a positive- or neutral-pressure hallway by a vestibule-type anteroom which is maintained under negative pressure. Magnetic interlocks control the opening of the two doors connecting the vestibule to the inner laboratory and the outer hallway. The interlocks permit only one door to the vestibule to be opened at a time. The combination of negative pressure (0.05 to 0.1 in. [0.125 to 0.25 cm], water gauge) in the vestibule and positive or neutral pressure in the inner laboratory and hallway creates a net air flow into the vestibule from either area, depending on which door is opened. Therefore, each time a door to the vestibule is opened, air moves from the positive- or neutral-pressure area into the vestibule and is exhausted to the outside air. Depending on which door is opened, this pressure differential prevents aerosols within the inner laboratory from escaping into the hallway and prevents potentially contaminating aerosols from the hallway from entering the inner laboratory. Outside air which is drawn into the facility is HEPA filtered to minimize the potential for recirculation of contaminants exhausted from the laboratories through the negative-pressure vestibules.

Although the inclusion of an entrance vestibule under negative pressure relative to the working laboratory and common hallway is not essential, such a chamber also provides a convenient area for storage of supplies and can be used to hang coats and umbrellas to avoid bringing these items into the inner laboratory. Positive to neutral pressure is the normal state of heated and air-conditioned structures, and negative-pressure areas are easily established by venting combinations and speed settings for fan motors to achieve the desired movement of air. Although the primary goal of protecting against product carryover contamination can be achieved by physical separation of the preamplification and postamplification areas, the added protection against transfer of aerosol contaminants afforded by controlled airflow is well worth the investment when possible, especially for large-volume laboratories.

Another desirable design feature is the inclusion of UV light fixtures connected to timer switches in the overhead lighting for the inner laboratories. This permits the control of surface contamination by exposing the laboratory surfaces to 15 to 30 min of UV irradiation when the area is vacant. Nightly UV treatment of work surfaces can be incorporated into a quality control program for the facility. Shortwave UV irradiation, such as the wavelength of light emitted by UV lamps in overhead lighting fixtures, can significantly decrease the amount of amplifiable DNA deposited on laboratory surfaces without causing significant deterioration of common laboratory countertop construction material (15, 16). Frequent changing of absorbent laboratory soak coverings and wiping of benchtops with 10% bleach (0.5% hypochlorite) followed by rinsing with ethanol or distilled water are also effective for controlling surface contamination.

Laboratory coats should be dedicated for use in each testing area and should remain in those areas except to be cleaned. Laboratory coats designated for use in each laboratory should be worn at all times within the laboratory and should be removed before leaving the laboratory. Amplification products can be introduced into an assay by a shirtsleeve previously dragged across a contaminated benchtop.

Wearing laboratory coats and removing them when leaving individual areas of the facility helps to reduce the risk of contact contamination and transfer of amplification products to product-free areas.

## Reagent Preparation Laboratory

The reagent preparation laboratory is the heart of any facility in which nucleic acid amplification assays are performed. Each reagent to be used in amplification assays must be carefully quality controlled to ensure maximum assay sensitivity. In addition, multiple aliquots should be tested to ensure that low levels of contaminating amplification products have not been introduced during the preparation process. A dedicated reagent preparation laboratory is not necessary if amplification-based testing is performed with commercially produced kits, in which case production and quality control of assay reagents are performed and documented by the manufacturer.

Equipment should be dedicated for use in the reagent preparation laboratory and should be cleaned and calibrated frequently (pipettors, balances, centrifuges) as part of a quality assurance program. Whenever possible, disposable supplies such as plastic vials, pipettes, and centrifuge tubes should be used. Molecular biology grade chemicals should be used for preparation of solutions to protect against introduction of nucleases into reagents for testing.

Master mixes for each amplification-based assay should be produced containing as many of the individual reagent components as is feasible for performance and stability requirements. Providing testing components in the form of master mixes minimizes the number of pipetting and mixing steps required in later processes (sample preparation, amplification, analysis), thereby decreasing the potential for introduction of contaminants and mistakes due to the omission of a necessary reagent component. In addition, the use of master mixes decreases the variability of performance observed for amplification assays. Solutions should be aseptically dispensed in small-volume, single-use aliquots to minimize the number of times a particular solution is opened. Some reagents such as wash buffers need not be dispensed in single-use aliquots, but appropriate controls must be included in the amplification assay to detect contamination of these reagents.

Quality control of chemical components purchased from commercial vendors has usually been performed by the manufacturer. Certificates of analysis can be obtained from the manufacturers for individual lots as proof of quality control. Individual components synthesized by the reagent preparation laboratory, such as oligonucleotide primers, should have appropriate documentation of purity. For many primer sequences, high-performance liquid chromatography (HPLC) with anion-exchange columns can be used to separate out full-length primer sequences from shorter failure sequences. Only primer batches collected from the full-length primer sequence peak following separation by HPLC should be used for preparation of master mixes. In some instances, such as highly degenerate (mixed-base) primers, this is not possible or feasible.

Quality control of reagents prepared by the reagent preparation laboratory should be performed by testing personnel on multiple portions of each reagent batch provided by the reagent preparation laboratory. Personnel working in testing areas can perform the necessary quality control tests of reagents and report results to the reagent laboratory director and the testing laboratory director. All master mixes and buffers used in amplification testing should be tested for low-level product contamination by multiple amplification reactions (at least 10) without addition of any target nucleic acid sequence for each aliquot to be tested. Master mixes should also be checked for optimal performance by using positive controls of low copy number.

Positive controls should consist of a low copy number of the desired nucleic acid target. It is prudent to include only low copy numbers of positive controls, both to demonstrate appropriate sensitivity when validating the performance of master mixes and to keep the quantity of potential contaminating target nucleic acid to a minimum. Positive-control samples may consist of dilutions of plasmids (5) or dilutions of cell lines (1, 3) containing the target DNA sequence. Stocks of positive-control samples should never be prepared or stored in the reagent preparation laboratory. Positive-control stocks and aliquots of low-copy-number working dilutions should be prepared in a separate amplification-product-free area of the facility. Small-volume aliquots of diluted controls can be stored in the specimen preparation laboratory. Reagents that have successfully passed quality control checks should be clearly labeled, including expiration dates previously determined by stability studies, and stored under appropriate conditions in the reagent preparation laboratory.

## Specimen Preparation Laboratory

The specimen preparation laboratory is the starting point for clinical-testing personnel. Only personnel involved in testing of specimens should have access to the specimen preparation laboratory, and, as described above, laboratory coats and equipment should be dedicated for use in only this laboratory to prevent transfer of amplification products into this area. In keeping with the unidirectional work flow concept, personnel should begin the work day in this area before moving into amplification and analysis areas. In laboratories that do a large volume of tests, such as commercial reference laboratories, staff may be rotated so that personnel work in one particular area (i.e., specimen preparation, amplification, analysis) on a given day or week. Rotating personnel among laboratories decreases traffic between preamplification and postamplification areas and helps to prevent burnout associated with routine performance of the same tasks.

Testing personnel should pick up specimens from a designated area after accessioning and bring them to the specimen preparation laboratory. While specimens are being recorded for preparation, refrigerated reagents should be removed from storage and allowed to come to room temperature. Specimens should be prepared within biologic-safety enclosures to prevent and contain the spread of aerosols. Whole blood, urine, and other biologic fluids should be handled in laminar-flow hoods. Disposable gloves should be worn during all phases of sample preparation and should be changed frequently. It is a good idea to wear two pairs of gloves when working with potentially infectious specimens and to dispose of the outer pair if it has come in contact with the specimen. Disposable sleeve protectors also provide additional protection to the technician and lessen the likelihood of contaminating the sleeves of the laboratory coat while the technician is working inside laminar-flow hoods. Centrifuges should be equipped with aerosol containment rotors to protect personnel from aerosolized pathogens and to prevent cross-contamination of specimens. Alternatively, centrifuge tubes with compressible O-ring-type seals may be used. Supernatant fluids should be aspirated, not decanted, into receptacles containing bleach. Specimens such as whole blood in Vacutainer tubes should be uncapped within the laminar-flow hood with gauze or disposable stopper removers to prevent contamination of gloves and exposure of technicians to potential pathogens if glass containers are broken when stoppers and caps are removed.

Whenever possible, pipetting should be done with disposable plastic pipettes. Small-volume pipetting should be done with positive-displacement pipettes with disposable pistons and sleeves. Alternatively, micropipettors and pipette tips containing aerosol-resistant barriers can be used. Several manufacturers now provide this type of micropipette tip, but caution should be exercised when selecting tips. Only tips containing hydrophobic membranes bonded to the walls of the pipette tip have proved to be effective in preventing aerosol contamination of pipettors. Pipette tips containing filter plugs that are inserted (not bonded) into the tip do not provide an airtight seal and have proved unreliable in our experience.

The accuracy of the results obtained by DNA probe analysis of amplified nucleic acids depends largely on the quality of the specimen used for the test reaction. Positive controls, as described above, and negative controls, usually consisting of mock specimens prepared along with patient samples, should be processed with each batch of specimens. Controls should be constructed to ensure that sufficient quantities of nucleic acid free of amplification inhibitors are being generated by the reagents and the procedure used.

Incubation of specimens at increased temperatures for extraction and enzymatic digestion protocols can be carried out in either heating blocks or water baths. When heating blocks are used, more uniform temperatures can be obtained by filling the wells with silica sand and placing specimen tubes into the sand. Floating-type tube racks should be used to incubate sample tubes in water baths. Temperature settings for heating blocks and water baths should be verified with a calibrated thermometer each time the equipment is used. It is important to tighten the caps of specimen tubes securely before incubation at elevated temperatures. After incubation, samples should be allowed to cool to room temperature and then be centrifuged so that sample liquids that are in the caps as a result of refluxing during incubation at elevated temperatures are forced to the bottom of the tubes. Centrifugation after the sample cools helps to minimize the potential for contamination when the caps are removed for amplification of the prepared specimens. Specimens can now be removed to an area where setup for amplification is performed. Biologic-safety cabinets, centrifuges, and other equipment used during the preparation of specimens should be decontaminated with 10% bleach and then rinsed with distilled water or 70% ethanol to remove residual bleach, which will oxidize metal surfaces (13).

## Amplification Laboratory

Amplification reactions should be set up in a product-free area of the facility. Depending on available space, a separate laboratory for amplification setup or a small area dedicated for this purpose, such as a dead-air culture enclosure equipped with a UV lamp on a benchtop in the specimen preparation laboratory, can serve the purpose. In the example shown in Fig. 1, after sample preparation, technicians transfer the samples to the amplification laboratory (unidirectional movement), where reactions are set up in benchtop enclosures. The laboratory depicted in the figure is designed with the same features (e.g., negative-pressure vestibule, UV lights) as the reagent and specimen preparation laboratories, although this extra level of protection is not required.

Amplification reaction reagents and prepared specimens should be pipetted by using positive-displacement pipettes or micropipettors with aerosol-resistant tips. Master mixes complete with the required enzymes should be pipetted into all reaction tubes first, and then all the tubes should be capped. Specimens can then be pipetted individually into the appropriate reaction tubes. When opening and closing specimen tubes, care must be taken to minimize the potential for the production of aerosols and contamination of the fingertips of gloves. One method to help avoid contamination of gloves is to use a piece of sterile gauze (2.5 by 2.5 cm) to cover the cap of each specimen tube when removing and replacing caps. To decrease the potential for carryover contamination before amplification, some laboratories may elect to expose reaction tubes and trays to irradiation in a UV cross-linking apparatus before adding reaction components and specimens (11). Each reaction tube should be capped before the next specimen is pipetted. Positive controls should be pipetted after specimens, and then specimen preparation negative controls and amplification negative controls (master mix with diluent as specimen) should be pipetted, to maximize the ability to detect carryover contamination.

Sterilization protocols such as uracil *N*-glycosylase restriction (10) and incorporation of isopsoralens (6) should be followed whenever possible for a given amplification assay (see chapter 9 for a discussion of sterilization protocols). To demonstrate the integrity of nucleic acid from the prepared specimens and to protect against false-negative results due to inhibitors of the amplification reaction present in the specimens, a control amplification reaction can be performed for each specimen. Control reactions consist of an amplification from the specimen of a resident target gene sequence such as β-globin or human leukocyte antigen DQα (HLA-DQα) (14) for specimens known to contain human genomic DNA. In some instances, an additional aliquot of a specimen can be spiked with a positive control of low copy number for the target in question to demonstrate the ability to amplify the target from the specimen as a control for the presence of inhibitors. Control amplification reactions may be performed before or at the same time as reactions intended to amplify the desired target from the prepared specimens. Amplification reaction tubes containing added specimens can now be removed from the enclosures and placed into amplification equipment to begin the reaction process. All tube caps must be tightly closed.

Amplification factors, such as cycling parameters for DNA thermocyclers, should be optimized for each target nucleic acid, as previously described (chapter 5). Equipment such as DNA thermocyclers must be carefully maintained and checked for performance as part of a quality assurance program. Thermocyclers typically have high amperage consumption and are sensitive to voltage fluctuations in supply lines. For this reason, it is a good idea to provide dedicated circuits for these machines so that cycling on and off of heating units from other machines does not adversely affect the performance of machines running simultaneously because of decreases in supply voltage. Thermocycling devices should be checked periodically to ensure that actual temperatures obtained in the sample blocks correspond to the digital readout of the instrument display panel (4, 9). In areas equipped with UV ceiling fixtures, thermocycler blocks can be decontaminated by exposure to UV light during routine UV treatment of rooms as part of quality control, as described above. Swabbing with cotton-tipped applicators soaked in 10% bleach followed by swabbing with water-soaked applicator tips is also effective for decontaminating thermocycler blocks.

When removing reaction tubes from thermocyclers, it is necessary to ensure that all caps on reaction tubes remain tightly closed. During thermocycling, sample refluxing within reaction tubes may cause some caps to pop up slightly, especially if the tubes used are not molded with snug-fitting caps. It is a worthwhile investment to purchase reaction tubes that fit tightly into the wells of thermocycler blocks and have securely fitting caps. The manufacturers of temperature-cycling devices usually provide recommendations of which specific type of tube gives optimum performance in their machines. After removal from thermocyclers, reaction tubes should be transferred immediately to an amplification product analysis area in a carrier rack that can be soaked in bleach and rinsed before being returned to the amplification setup area. Technicians transferring amplified specimens to product-containing analysis areas should plan their day such that analyzing amplified samples is done last to minimize or eliminate movement from product-containing to product-free areas.

## Analysis Laboratory

Following amplification of samples, the amplification reaction tubes are transferred to the analysis laboratory to determine whether the nucleic acid target in question was present in the original specimen. As shown in Fig. 1, the analysis laboratory is the final laboratory in the succession of testing areas and is the termination point of the unidirectional movement of personnel. Within the testing facility, the area designated for analysis of amplification products should be the only area where postamplification reaction tubes are opened.

Unlike the above-mentioned reagent preparation and testing laboratories, the analysis area does not contain a vestibule-type entrance chamber under negative pressure. There are two reasons why this area of the facility lacks the vestibule entrance design. First, several different groups may work in an analysis area and perform various clinical assays, thereby increasing traffic to and from this area and making a two-door vestibule-type entrance impractical. Second, and most important, the entire analysis area should be negative pressure (i.e., air should circulate into this area) relative to the remainder of the facility to contain any aerosolized amplification products within this area. The air-handling unit for this area is adjusted such that air is actually drawn into this area from the surrounding areas of the facility and is exhausted from the building without being recirculated. This simple adjustment in the air-handling system for the facility allows the door to the analysis area to be opened constantly without allowing potentially contaminating aerosols generated from within the analysis laboratory to escape into product-free areas.

Because the air from within the sample preparation and amplification laboratories exits through the vestibule (not into the outer hallway), personnel working in the common analysis area are not exposed to potential airborne contaminants from the specimen preparation areas. Again, it must be emphasized that such a facility design is optimal, especially for large-volume clinical laboratories, but it is neither required nor practical for many users.

The various methods used to detect nucleic acid amplification products were described earlier (chapter 7) and will not be discussed here. It is important to remember that any piece of equipment used in the analysis of amplification products can be a potential source of product carryover contamination. It is essential to provide controls within each assay, including the analysis steps, so that carryover can be detected and to clean and decontaminate all equipment thoroughly. All equipment used in product detection formats should be wiped or

**Figure 2.** Essential elements of PCR laboratory in which diagnostic kits are used.

Work Flow →

| Preamplification Areas | | Postamplification Area |
|---|---|---|
| Area 1 | Area 2 | Area 3 |
| 1. Reagent Preparation | 2. Specimen Preparation<br>3. Amplification Setup | 4. Amplification<br>5. Detection/Product Analysis |

soaked with 10% bleach (a 30-min soak is sufficient) and rinsed thoroughly with distilled water. Alternatively, bleach-sensitive components (e.g., blotting equipment) can be placed into a UV cross-linking device and exposed to irradiation to render contaminating nucleic acids nonamplifiable. Incorporation of amplification product inactivation protocols into assays largely negates the need to decontaminate equipment thoroughly. Likewise, the use of prepackaged kits for PCR will reduce the need for separate laboratories and stringent reagent preparation procedures because reagents will be prepared off site (Fig. 2). Once a result has been determined for a sample, the amplification products should be disposed of and not stored. Lysates from original samples can be reamplified to resolve discrepant results obtained after detection of amplification products.

All of the procedures done in the different areas of the facility and the location and movement of personnel during and after performance of tasks should be clearly outlined in a standard operating procedure document for each clinical assay. White et al. (17) described some important considerations and provided helpful suggestions concerning standard operating procedures and reporting algorithms for clinical applications of PCR amplification. Laboratory personnel should be thoroughly familiar with procedures, including equipment preventive maintenance, quality control, and quality assurance procedures. Records containing written documentation of performance of all maintenance and quality control procedures should be kept and reviewed on a regular basis.

An algorithm for testing of samples and reporting of results should be included for each assay. An example of such a testing and reporting algorithm is shown in Fig. 3. Testing personnel should verify that accessioned samples are of the correct type and of sufficient quantity for the sample preparation procedure. During sample preparation, lot numbers for positive and negative controls and all reagents should be recorded with specimen data. Before or at the same time as specific target amplifications, control amplifications should be performed to ensure competency of specimens and absence of inhibitors. Specimens that generate the expected product following control amplifications can be amplified and analyzed for the presence of the specific target nucleic acid. Specimens that fail to give positive signals for control amplifications may be diluted (1:2 to 1:5, depending on the analytical sensitivity of the assay) and reamplified to determine whether the effects of any inhibitory substances can be overcome. Some specimens may require organic extraction and precipitation of nucleic acids in order to remove inhibitors. Specimens with concordant results after duplicate amplifica-

**Figure 3.** Algorithm for specimen processing and reporting of results.

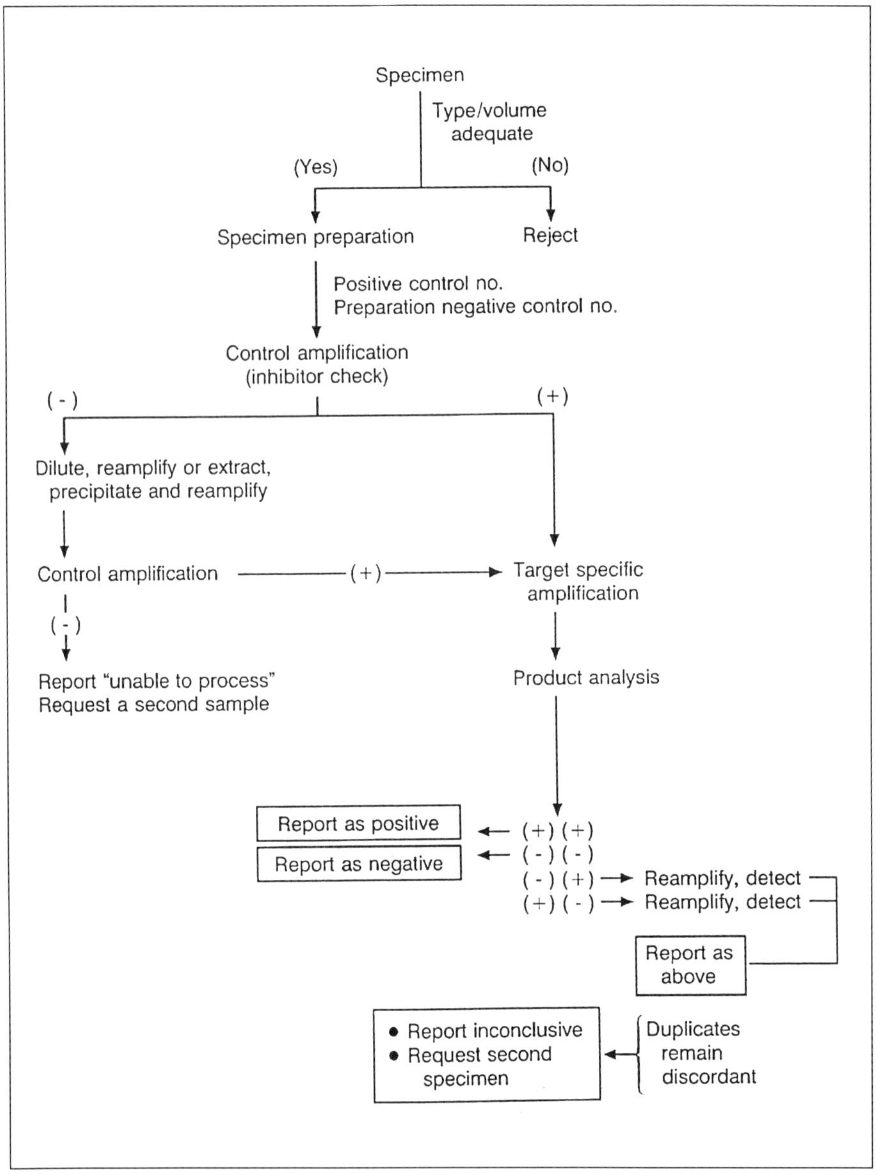


tion for the specific target can be reported as DNA detected or DNA not detected for the specific target in question. Specimens with discordant results after duplicate amplification should be reamplified in duplicate. If the results remain discordant, the specimen is reported as inconclusive for the target in question and a second sample should be requested for testing.

Advances in amplification technology and DNA probe detection formats are being reported frequently. The use of nucleic acid amplification-based testing in the clinical laboratory continues to expand and may become a dominant technology in the near future, with the introduction of user-friendly commercially produced kits. At present, with careful planning and consideration of design and operation of clinical laboratories performing nucleic acid amplification-based testing, highly accurate and reliable results can be obtained for organisms that previously were difficult to detect by more traditional testing methods.

## References

1. **Clouse, K. A., D. Powell, I. Washington, G. Poli, K. Strebel, W. Farrar, P. Barstad, J. Kovacs, A. S. Fauci, and T. M. Folks.** 1989. Monokine regulation of human immunodeficiency virus-1 expression in a chronically infected human T cell clone. *J. Immunol.* **142:**431–438.
2. **Erlich, H. A., D. Gelfand, and J. J. Sninsky.** 1991. Recent advances in the polymerase chain reaction. *Science* **252:**1643–1651.
3. **Folks, T. M., D. Powell, M. Lightfoote, S. Koenig, A. S. Fauci, S. Benn, A. Rabson, D. Daugherty, H. E. Gendelman, M. D. Hoggan, S. Venkatesan, and M. A. Martin.** 1986. Biological and biochemical characterization of a cloned Leu-3$^-$ cell surviving infection with the acquired immune deficiency syndrome retrovirus. *J. Exp. Med.* **164:**280–290.
4. **Haff, L., J. G. Atwood, J. DiCesare, E. Katz, E. Picozza, J. F. Williams, and T. Woudenberg.** 1991. A high-performance system for automation of the polymerase chain reaction. *BioTechniques* **10:**102–112.
5. **Hart, C., S.-Y. Chang, S. Kwok, J. Sninsky, C.-Y. Ou, and G. Schochetman.** 1990. A replication-deficient HIV-1 DNA used for quantitation of the polymerase chain reaction (PCR). *Nucleic Acids Res.* **18:**4029–4030.
6. **Isaacs, S. T., J. W. Tessman, K. C. Metchette, J. E. Hearst, and G. D. Cimino.** 1991. Post-PCR sterilization: development and application to an HIV-1 diagnostic assay. *Nucleic Acids Res.* **19:**109–116.
7. **Kwok, S.** 1990. Procedures to minimize PCR-product carry-over, p. 142–145. *In* M. A. Innis, D. H. Gelfand, J. J. Sninsky, and T. J. White (ed.), *PCR Protocols: A Guide to Methods and Applications*. Academic Press, Inc., San Diego, Calif.
8. **Kwok, S., and R. Higuchi.** 1989. Avoiding false positives with PCR. *Nature* (London) **339:**237–238.
9. **Linz, U.** 1990. Thermocycler temperature variation invalidates PCR results. *BioTechniques* **9:**286–293.
10. **Longo, M. C., M. S. Berninger, and J. L. Hartley.** 1990. Use of uracil DNA glycosylase to control carry-over contamination in polymerase chain reactions. *Gene* **93:**125–128.
11. **Ou, C.-Y., J. L. Moore, and G. Schochetman.** 1991. Use of UV irradiation to reduce false positivity in polymerase chain reaction. *BioTechniques* **10:**442–446.
12. **Persing, D. H., and M. L. Landry.** 1989. *In vitro* amplification techniques for the detection of nucleic acids: new tools for the diagnostic laboratory. *Yale J. Biol. Med.* **62:**159–171.
13. **Prince, A. M., and L. Andrus.** 1992. PCR: how to kill unwanted DNA. *BioTechniques* **12:**358–360.
14. **Saiki, R. K., T. L. Bugawan, G. T. Horn, K. B. Mullis, and H. A. Erlich.** 1986. Analysis of enzymatically amplified β-globin and HLA DQα with allele-specific oligonucleotide probes. *Nature* (London) **324:**163–166.
15. **Sarkar, G., and S. Sommer.** 1990. More light on PCR contamination. *Nature* (London) **347:**340–341. (Letter to the editor.)
16. **Sarkar, G., and S. Sommer.** 1991. Parameters affecting susceptibility of PCR contamination to UV inactivation. *BioTechniques* **10:**590–594.
17. **White, T. J., R. Madej, and D. H. Persing.** 1992. The polymerase chain reaction: clinical applications. *Adv. Clin. Chem.* **29:**161–196.

# Quality Control of Polymerase Chain Reaction

# 9

Elizabeth A. Dragon, Joanne P. Spadoro, and Roberta Madej

The polymerase chain reaction (PCR) method was first described by scientists from Cetus Corporation in the mid-1980s (11, 12, 14). Since its introduction, PCR has been widely recognized and accepted as a powerful molecular-biology method. PCR has been adapted for use in such diverse fields as human genetics, forensic science, clinical diagnostics, and environmental science. The use of nucleic acid amplification technology for clinical diagnostics has been limited primarily to research applications, because of the power and sensitivity inherent in the technique. The onus of false-positive reactions due to amplicon carryover has been a criticism of PCR for years (7). Recent developments (3–5, 9) addressed these potential shortcomings, enabling more practical use of PCR.

The focus of this chapter is on the implementation of adequate quality control measures in a PCR laboratory. The design of a PCR laboratory will be reviewed, with the goal of avoiding contamination of PCR reagents and samples for analysis. Quality control of reagents and equipment will be discussed to guarantee continued efficacy of PCR results. If these principles of proper PCR are followed, one should be able to perform accurate and sensitive PCR routinely.

## Design of a PCR Laboratory

Four separate functions must be accommodated in the design of a PCR laboratory: reagent preparation, sample preparation, template addition, and post-PCR detection. In an ideal laboratory, there would be a separate room for each of these four tasks; however, most laboratories have limited space and this separation can be accomplished by establishing individual workstations within one or two rooms. Each workstation should have dedicated equipment and supplies. For reagent preparation, sample preparation, and reaction setup, the use of benchtop hoods (with UV lights for decontamination) is highly recommended (Table 1).

### Pre-PCR

All reagents to be used for sample preparation and amplification should be made in the pre-PCR area with dedicated reagents and supplies. The use of aliquots of reagent stocks should be encouraged to decrease the possibility of contamination of stock reagents. PCR master mix should be dispensed into reaction tubes, and the tubes should be capped and removed to a separate area for the addition of sample. Amplicon and target DNA (clinical specimens, plasmid, or

**Table 1.** Guidelines for the operation of a PCR laboratory

| |
|---|
| Establish separate pre- and post-PCR work areas with dedicated supplies and reagents |
| Carefully plan experiments; do not enter the pre-PCR area after handling amplicons or target DNA |
| Use plugged tips or positive-displacement pipettes |
| Use aliquots of all reagents to limit handling |
| Incorporate enzymatic or chemical methods to control amplicon carryover |
| Always use a low-copy-number (10 to 50 copies per PCR) for positive controls, a large number of negative controls, and reagent controls with every amplification |

phage DNA containing target sequences) must be excluded from the pre-PCR area at all times. The flow of work in the laboratory should be organized in such a manner that individuals never enter the pre-PCR area after handling amplicon or target DNA, and supplies or reagents from sample preparation, template addition, or post-PCR workstations are not brought into the pre-PCR area. A useful and easy method to enforce separation of pre- and post-PCR functions is the color coding of supplies, reagents, and laboratory coats (chapter 8). Gloves and laboratory coats should always be worn in the pre-PCR laboratory; the pre-PCR laboratory coat should be removed from the laboratory only when being laundered.

## Specimen Preparation and Template Addition

All specimen preparation should be done in a dedicated area with appropriate biosafety precautions. The primary goal in specimen preparation is to prepare the target specimen in the most efficient manner possible without introducing, maintaining, or enhancing inhibitors or contaminants. All supplies and reagents should be dedicated for sample preparation and should never be placed in the pre-PCR workstation. To avoid amplicon contamination of the specimens, one should avoid preparing specimens after working in the post-PCR workstation.

Universal precautions should be observed when handling any clinical specimen. Most laboratories prefer to prepare all clinical specimens in a biologic-safety cabinet. The air system in the cabinet must be turned on at least 30 min before any procedure for specimen preparation, and the working surface in the cabinet must be decontaminated with a 10% bleach solution (made fresh daily) followed by a 70% ethanol rinse before and after specimen preparation.

In the template addition area, processed specimens or control nucleic acids (<50 copies) or both are added to reaction tubes containing master mix. Positive-displacement tips or plugged or barrier tips should be used for template addition. To avoid potential cross-contamination among samples, reaction tubes should be capped once the sample has been added. When strip caps are used (as for the Gene Amp TC9600 system; Perkin Elmer Corp., Norwalk, Conn.), samples should be added to the reaction tubes one at a time and then the tubes should be capped before samples are added to the next row.

If laboratory space is limited, sample preparation and template addition can be performed in the same area. It is critical to use positive-displacement pipettes or plugged tips to avoid sample contamination.

## Post-PCR

PCR amplifications are analyzed in the post-PCR workstation. This area contains high levels of amplicon; once individuals have worked in the post-PCR area, they should not enter the pre-PCR, sample preparation, or template addition areas. Laboratory coats and gloves should always be worn and should be removed immediately after completion of the analysis. One should exercise care when opening PCR reaction tubes, to avoid aerosolization or inadvertent contamination. The reagents and equipment used in the post-PCR area should never be removed from this area. Laboratory workbenches and pipettors should be cleaned with 10 to 20% bleach on a regularly scheduled basis. If there is a spill of amplified product, the spill should be contained and carefully removed. Any contaminated area should be treated with a bleach solution as part of the cleanup protocol. The same care and protocols should be used when working with amplified products as when handling radioactive material.

## Laboratory Technique and Practice

As described above, the physical separation of pre- and post-PCR activities is one important step toward the successful implementation of PCR in the laboratory. Additionally, other practices should be encouraged (6, 7). The careful selection and use of positive controls is extremely important. Generally, one should amplify fewer than 50 copies of a well-characterized positive control; however, to avoid false-negative results due to sampling bias, at least 10 copies of the control should be amplified.

Controls serve multiple functions in a PCR assay. First, an intrinsic control can aid in the determination of the PCR competency of a specimen. For human clinical specimens, one can check the quality of DNA preparation by amplifying either the β-globin or the human leukocyte antigen DQα (HLA-DQα) gene. Amplification controls include the intrinsic DNA control described above. There are two additional amplification controls: the low-copy-number positive control described above and the spiked control. When a clinical specimen does not amplify and one suspects that PCR inhibitors are present in the specimen, a low copy number (e.g., 20 copies) of a control sequence can be spiked into the clinical specimen and rerun. If the spiked positive control does not amplify, one can assume that the specimen is inhibitory.

The consequences of sloppy or casual laboratory technique are frustration, invalid experiments, and lost time. Because of the extreme sensitivity of PCR, aerosol contamination of reagents or equipment can lead to false-positive results. Once a reagent(s) becomes contaminated, the only alternative is to discard it and begin again. Most often it is impossible to accurately determine which reagent is contaminated and when the contamination occurred. All laboratory personnel must understand that their individual work habits directly affect everyone else in the laboratory. Complete compliance is the only way to ensure that PCR will be successful in the laboratory.

## Carryover Control

Enzymatic (5, 9) and chemical (3, 4) methods have been developed to control amplicon carryover. The enzymatic method involves the use of the enzyme uracil *N*-glycosylase (UNG) and the preferential restriction of uracil-containing DNA. dUTP is substituted in the master mix for TTP, thereby generating U-containing DNA. UNG, at higher pH and in the presence of heat, cleaves the DNA at the uracil residues, rendering the DNA nonamplifiable. Pretreatment of all PCRs with UNG eliminates the amplification of U-containing DNA but does not affect the amplification of T-containing DNA (target DNA). UNG is not active during the amplification reaction, so newly synthesized molecules are unaffected. The use of this technology is essentially invisible to the user; no additional manipulations or equipment is necessary.

The chemical method for carryover control is the isopsoralen photochemical postamplification procedure for the inactivation of amplified DNA. In this procedure, isopsoralen derivatives are used. Amplification occurs in the presence of the isopsoralens. Following amplification, the reaction tube is irradiated with UV light and the isopsoralens form cyclobutane monoadducts with pyrimidine bases in the DNA. If this modified DNA is carried over into subsequent amplification reactions, it will not serve as a template for amplification.

Both methods perform well with most PCR applications. When these methods are used in conjunction with the laboratory practices previously described, one should be able to perform PCR successfully. However, the use of double PCR (nested primers) is not readily adaptable to either of these carryover control procedures. The use of nested primers requires the opening of the reaction tube and the reamplification of a portion of the reaction with a new set of primers. Double PCR is prone to false-positive results because of the additional manipulations, and it is not recommended for routine diagnostic applications unless (i) great care is taken to avoid contamination and (ii) negative controls are interspersed regularly throughout the reaction series to monitor for the increased likelihood of false-positive results.

## Single Versus Multiple Primer Pair Issues

In developing a PCR assay for the detection of an infectious organism, one usually initiates analysis by using several primer pair systems. This analysis allows one to develop a data base that should identify a candidate single primer pair for the specific target organism. The need for the confirmatory primer pair may be eliminated if the data from multiple clinical isolates demonstrate perfect concordance with the candidate primer pair and alternative primer pairs as well as other clinical documentation (from culture, serologic testing, and symptoms). When the wealth of published PCR data is carefully examined, it is seen that when sample discrepancies occurred, the issue of sequence variability or deletion of DNA target sequence was generally not substantiated (16, 18, 19). Schochetman and Sninsky (16) reviewed 18 published studies regarding detection of human immunodeficiency virus type 1 (HIV-1) and concluded that use of the *gag* primer systems (either SK38-39 primer/SK19 probe or SK145-101 primer/SK102 probe) demonstrated the best concordance. Most often, the disagreement between primer pairs is the result of use of systems lacking in equivalent analytic sensitivities, use of unoptimized systems, sample bias due to low copy numbers, or false-positive results due to carryover contamination (these studies were done before the introduction of carryover control systems). Studies have been specifically initiated to address the issues of interlaboratory performance and primer pair concordance for the detection of HIV-1 (2, 17).

## Quality Control Procedures

### Validation of Plugged Tips

It is important to check the quality and performance of plugged (aerosol barrier) pipette tips. Both the integrity and the effectiveness of the barrier must be assessed. To assess the integrity of the barrier, the following steps should be performed. A solution containing water, 1 to 2% glycerol, and a dye (food coloring) is prepared. For a tip with a maximum volume of 50 μl, the tip is placed on the pipette, the pipette is set to 55 to 60 μl, and the colored solution is pipetted. A good tip will not allow any of the solution above the plug. Low-quality tips will draw the solution around the plug (up above the plug). Obviously, there is no airtight barrier in tips that fail this test. The same thing can be done with any volume tip, with the pipette always set approximately 10% over the maximum.

To test the effectiveness of the barrier, the following experiment should be used. Several strong positive samples (>1,000 copies) and several negative controls are prepared. A plugged tip is placed on a pipette, the volume is set for the normal sample volume that would be added to the amplification reaction, and a strong positive sample is pipetted up and down in the pipette tip 10 times (this mimics the pipetting of 10 positive samples). The positive sample is pipetted into an amplification reaction. The tip is then discarded; a fresh tip is placed on the same pipette, and 10 negative controls are repipetted consecutively (with the same or fresh tips), each into an amplification reaction. This series should be repeated five to eight times. Each sample is then amplified and detected by the standard detection method used in the laboratory. All of the positive samples should give strong signals; low or no signal in the positive reaction means that the tips contain inhibitory substances for PCR. All of the negative controls should have negative results. Any positive signals in the negative samples may indicate that the tips are not effective in preventing aerosol contamination of the pipette.

## PCR Reagents

All amplification reaction components and sample preparation reagents must be rigorously tested to demonstrate the absence of amplicon or target DNA contamination. If one has incorporated the UNG restriction system into the amplification system, testing of the reaction mixtures with and without UNG will aid in determining whether the reagents are contaminated with amplicon or template DNA (plasmid or clinical specimen). The use of dedicated pre-PCR reagents and supplies for sample preparation and amplification components cannot be overemphasized.

The oligonucleotides that are used as primers and probes should be at least 90% pure. An $A_{260}/A_{280}$ ratio ranging between 1.5 and 1.9 should be obtained for all oligonucleotides. Although it is not necessary to use dedicated chromatography columns for the purification of each oligonucleotide, it is strongly recommended that separate columns be used for primers and probes. All oligonucleotides should be analyzed for base composition and, if possible, sequenced. Each new lot of primers and probes is tested to confirm concentration, purity, and function. To assess purity, oligonucleotides can be run on acrylamide gels (a single band of the appropriate size should appear) or assessed by high-pressure liquid chromatography (a high percentage of specific-length oligomers should be present) without an excessive amount of failed sequences.

All reagents must be tested. Sample preparation reagents must be tested for their ability to liberate amplifiable DNA. For samples that contain cells, the use of β-globin or HLA-DQα as an internal control is recommended. Every time a large batch of master mix is prepared, it should be tested with a panel of well-characterized positive and negative controls. As different lots of primers, enzyme, or other key components of the reaction change, crossover experiments must be

**Figure 1.** Schematic representation of a thermocouple monitoring of a well of the thermocycler. The probe is inserted into a tube in a well of the thermocycler; the tube contains a buffer solution (Tris-EDTA) overlaid with oil (to prevent evaporation). The probe is then hooked to a chart recorder to record the temperature changes. An example of a temperature chart recording is also shown. (Figure courtesy of K. Legassic.)

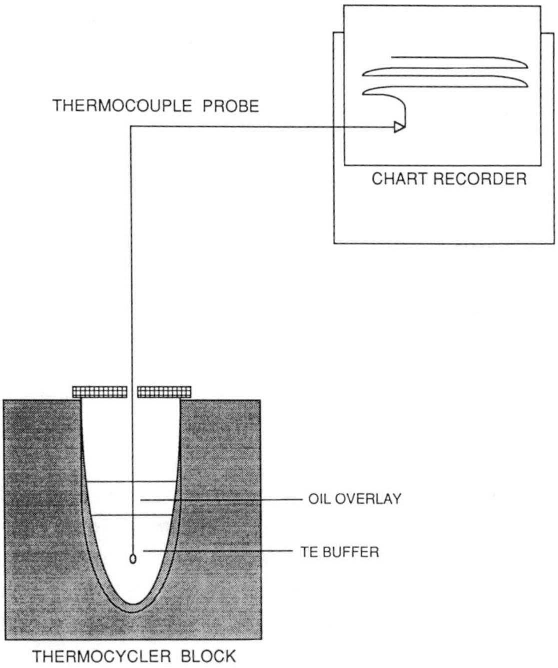


performed to ensure comparable performance. Inter- and intra-assay precision studies should be run with the positive and negative controls to ascertain the performance variation of the reaction.

## Thermocycler

A thermocycler that precisely regulates temperature and time cycles is needed to ensure the reproducibility and accuracy of the amplification reactions. There should be no more than a ±1°C variation between the set temperature and the actual temperature in all the wells of the instrument. The manufacturer's recommendations for regular servicing should be followed to ensure optimal performance.

One method of determining the reproducibility of the thermocycler is to monitor the internal temperature within the wells of thermocyclers by using a type K thermocouple probe, minivolt converter, and chart recorder (Fig. 1) (10). This setup can be used to detect large variations (>1°C) between wells. The thermocouple should be placed in a reaction tube and inserted into the well of the thermocycler; the instrument is then programmed for a typical run. The thermocouple should be moved from well to well; at least 6 distant wells in a 48-well instrument should be tested. Although the probe can be moved during the program, at least one cycle should be allowed for equilibration after the probe has been moved to a new well. The temperatures should be recorded for all

**Table 2.** Daily task checklist

1. Check and record temperatures in all water baths, heat blocks, and incubators
2. Make a fresh 10% bleach solution
3. Before beginning work, turn on the biologic-safety cabinet; let it run for at least 30 min before doing any work in the cabinet
4. Wipe all work surfaces with 10% bleach, and rinse them with 70% ethanol
5. Decontaminate all pipettors with 10% bleach, and rinse them with 70% ethanol
6. Follow unidirectional work flow procedures: always work from clean (pre-PCR) to dirty (post-PCR) sites

programmed steps. To determine functional specifications, the instrument manual should be consulted for the claims. Because temperature readings depend on the precision of the unit, a good thermocouple is necessary, and it must be calibrated against a temperature standard. Some instrument manufacturers (e.g., Perkin Elmer Corp.) sell temperature probes specifically designed to fit into the wells of their thermocyclers.

In addition to the temperature check, another quality assurance check is simply to print the programs periodically (or in some diagnostic assays, with each run). The printout can elucidate several things: it will prove that the correct program was used for the assay in question, it will record that the proper number of cycles was used, and it will record any breaks in the cycling; also, by examining the times recorded to obtain each of the temperature plateaus, one may be able to observe instrument malfunction before it becomes apparent elsewhere. For example, if an instrument usually ramps up to 94°C within 30 s but one printout shows that the time is becoming increasingly prolonged, this may signal that an instrument component is becoming worn.

An alternative method (a functional check) is to amplify routinely a known positive control in all wells of the instrument. One should obtain equivalent results from all of the replicates. PCR dropouts should not be observed; however, if dropouts are observed, one should determine whether specific wells repeatedly yield these false-negative reactions. If so, these specific wells should be avoided until the instrument can be repaired.

## Other Instruments and Equipment

Most of the other instruments and equipment in PCR-based assays are pipettors, incubators, heating blocks, water baths, and detection-based equipment. Because only small amounts of reagents and samples are used in PCR, pipettors should be calibrated at least twice a year. This can be done by the standard method: measurement of the weight of water delivered by the pipette. Pipette manufacturers have different claims and warranties, and these should be consulted.

Incubators, water baths, and heating blocks should be checked each time they are used; this is done by placing a thermometer (standardized against a National Bureau of Standards thermometer) in them. In laboratories in which assays are run continuously, temperatures are recorded daily. If hybridization or other reactions are performed at room temperature, the daily room temperature is recorded. This is consistent with current good laboratory procedures in clinical laboratories. Table 2 can be used as a checklist for various quality control and quality assurance procedures.

### Amplification Reaction

Every PCR run should include positive and negative controls. The negative control should always be replicated within a run and should yield only background signal. A general recommendation is that 5% of the total reactions in a run should be negative controls. The positive control, if possible, should generate a moderate signal. If the positive control yields a negative or unusually weak signal, the entire PCR run should be repeated. A false-positive result with the negative control necessitates that all reagents be checked for amplicon or target DNA contamination or both. When nested PCR is performed, every test sample should be alternated with a negative result to control for carryover during transfer of first-round PCR products into new tubes. Again, the entire PCR run must be repeated if the negative control gives a positive result.

### Detection Format

The earliest descriptions of PCR used radioactively labeled probes for the detection of the amplified products (11, 12). As amplification technology improved, it became possible to visualize the amplified DNA by using ethidium bromide-stained gels (14). The use of both techniques or a combination of these techniques (Southern blots, dot blots, oligomer hybridization analysis) is generally considered to be the usual research laboratory method for detection. The use of a probe is encouraged because it provides additional information and ensures sequence specificity. Nonisotopic methods for detection of PCR amplification reactions, such as dot blots (1), reverse dot blots (15), the colorimetric microtiter plate assay (8), and the hybridization protection assay (13), have all been applied to the detection of PCR amplification products.

## Conclusions

Over the past 8 years, PCR has proved to be a valuable tool for a wide variety of research and diagnostic applications. Setting up a PCR laboratory and initiating PCR experiments require special thought and effort to avoid potential problems. Dedicated pre- and post-PCR workstations and equipment, incorporation of carryover control methods, and modified laboratory techniques will significantly decrease the chance of false-positive results. Implementation and continued compliance with such measures will ensure the efficacy of all PCR results.

## References

1. **Bugawan, T. L., R. K. Saiki, C. H. Levenson, R. M. Watson, and H. A. Erlich.** 1988. The use of non-radioactive oligonucleotide probes to analyze enzymatically amplified DNA for prenatal diagnosis and forensic HLA typing. *Bio/Technology* **6:**943–947.
2. **Butcher, A., L. Salter, S. Kinard, K. Kung, B. McCreedy, and J. Spadoro.** 1992. Evaluation of a single primer pair for the detection of HIV-1 in clinical specimens by PCR amplification: comparison of a non-radioactive microtiter plate assay with solution hybridization, p. 417, abstr. T-5. *Abstr. 92nd Gen. Meet. Am. Soc. Microbiol. 1992.* American Society for Microbiology, Washington, D.C.
3. **Cimino, G. D., K. C. Metchette, J. W. Tessman, J. E. Hearst, and S. T. Isaacs.** 1991. Post-PCR sterilization: a method to control carryover contamination for the polymerase chain reaction. *Nucleic Acids Res.* **19:**99–107.
4. **Isaacs, S. T., J. W. Tessman, K. C. Metchette, J. E. Hearst, and G. D. Cimino.** 1991.

Post-PCR sterilization: development and application to an HIV-1 diagnostic assay. *Nucleic Acids Res.* **19:**109–116.

5. **Kellogg, D. E., C. Gates, B. Dragon, K. Kung, J. Wang, S. Kwok, and J. J. Sninsky.** 1991. Uracil-N-glycosylase (UNG) elimination of HIV PCR carryover, abstr. Th.A. 16, p. 59. *7th Int. Conf. AIDS.* Istituto Superiore di Sanita, Rome.
6. **Kwok, S.** 1990. Procedures to minimize PCR-product carry-over, p. 142–145. *In* M. A. Innis, D. H. Gelfand, J. J. Sninsky, and T. J. White (ed.), *PCR Protocols: A Guide to Methods and Applications.* Academic Press, Inc., San Diego, Calif.
7. **Kwok, S., and R. Higuchi.** 1989. Avoiding false positives with PCR. *Nature* (London) **339:**237–238.
8. **Loeffelholz, M. J., C. A. Lewinski, S. R. Silver, A. P. Purohit, S. A. Herman, D. A. Buonagurio, and E. A. Dragon.** 1992. Detection of *Chlamydia trachomatis* in endocervical specimens by polymerase chain reaction. *J. Clin. Microbiol.* **30:**2847–2851.
9. **Longo, M. C., M. S. Berninger, and J. L. Hartley.** 1990. Use of uracil DNA glycosylase to control carry-over contamination in polymerase chain reactions. *Gene* **93:**125–128.
10. **Louie, P., C. Dela Rosa, R. Madej, and G. Rodgers.** 1991. Evaluating the proficiency of PCR systems for clinical diagnostic testing. *Clin. Chem.* **37:**1059. (Abstract.)
11. **Mullis, K. B., and F. A. Faloona.** 1987. Specific synthesis of DNA *in vitro* via a polymerase-catalyzed chain reaction. *Methods Enzymol.* **155:**335–350.
12. **Mullis, K., F. Faloona, S. Scharf, R. Saiki, G. Horn, and H. Erlich.** 1986. Specific enzymatic amplification of DNA in vitro: the polymerase chain reaction. *Cold Spring Harbor Symp. Quant. Biol.* **51:**263–273.
13. **Ou, C.-Y., S. H. McDonough, D. Cabanas, T. B. Ryder, M. Harper, J. Moore, and G. Schochetman.** 1990. Rapid and quantitative detection of enzymatically amplified HIV-1 DNA using chemiluminescent oligonucleotide probes. *AIDS Res. Hum. Retroviruses* **6:**1323–1329.
14. **Saiki, R. K., S. Scharf, F. Faloona, K. B. Mullis, G. T. Horn, H. A. Erlich, and N. Arnheim.** 1985. Enzymatic amplification of β-globin genomic sequences and restriction site analysis for diagnosis of sickle cell anemia. *Science* **230:**1350–1354.
15. **Saiki, R. K., P. S. Walsh, C. H. Levenson, and H. A. Erlich.** 1989. Genetic analysis of amplified DNA with immobilized sequence-specific oligonucleotide probes. *Proc. Natl. Acad. Sci. USA* **86:**6230–6234.
16. **Schochetman, G., and J. J. Sninsky.** 1992. Direct detection of human-immunodeficiency-virus infection using the polymerase chain-reaction, p. 90–110. *In* G. Schochetman and J. R. George (ed.), *AIDS Testing: Methodology and Management Issues.* Springer-Verlag, New York.
17. **Sheppard, H. W., M. S. Ascher, M. P. Busch, P. R. Sohmer, M. Stanley, M. C. Luce, J. A. Chimera, R. Madej, G. C. Rodgers, C. Lynch, H. Khayam-Bashi, E. L. Murphy, B. Eble, W. Z. Bradford, R. A. Royce, and W. Winkelstein, Jr.** 1991. A multicenter proficiency trial of gene amplification (PCR) for the detection of HIV-1. *J. Acquired Immune Defic. Syndr.* **4:**277–283.
18. **Sninsky, J. J.** 1991. Application of the polymerase chain reaction to the detection of viruses, p. 799–805. *In* F. B. Hollinger, S. M. Lemon, and H. Margolis (ed.), *Viral Hepatitis and Liver Disease. Proceedings of the 1990 International Symposium on Viral Hepatitis and Liver Disease: Contemporary Issues and Future Prospects.* The Williams & Wilkins Co., Baltimore.
19. **White, T. J., R. Madej, and D. H. Persing.** 1992. The polymerase chain reaction: clinical applications. *Adv. Clin. Chem.* **29:**161–196.

PART II

# Applications of Diagnostic Molecular Microbiology

SECTION 1

# Bacterial Pathogens

# Genotypic Identification and Detection of Mycobacteria—Facing Novel and Uncultured Pathogens

# 1.1

PHILIP KIRSCHNER, ALBRECHT MEIER, AND ERIK C. BÖTTGER

## Background

The genus *Mycobacterium* consists of a diverse group of acid-fast bacilli which include a number of human and animal pathogens (20). For many years following the discovery of the human tubercle bacillus, only the three species known as *Mycobacterium tuberculosis*, *M. bovis*, and *M. leprae* were consistently recognized as human pathogens. During the last decades, however, it has become increasingly clear that the genus *Mycobacterium* also harbors a number of important opportunistic pathogens (e.g., *M. avium*, *M. intracellulare*, *M. kansasii*, and *M. simiae*), which can cause severe disease whenever the normal cellular defenses have been temporarily depressed. The most striking example is probably the prevalence of disseminated *M. avium* infection in patients with AIDS.

The definite laboratory diagnosis of mycobacterial infections rests on isolation by culture followed by identification. Owing to the long generation time of mycobacteria, culture-based isolation techniques require 3 to 8 weeks. In addition, standard protocols for definite identification include complex and time-consuming procedures—especially for nontuberculous mycobacteria—resulting in a further delay of several weeks. Identification by phenotyping contains an additional inherent problem: the phenotype of a species is not an absolute stable property but exhibits quite remarkable variability. Thus, what about isolates which cannot be assigned on the basis of their biochemical and phenotypic properties to an established taxon?

We have recently described a method for identification of isolated *Mycobacterium* species by direct sequencing of amplified rDNA (21). In contrast to standard identification protocols, direct 16S rDNA sequence determination is rapid (completion within 2 days) and results in a definite and unambiguous identification. Besides rapid identification of isolates, i.e., assignment to an established taxon, 16S rDNA sequencing offers unique advantages. In particular, this approach allows the prompt recognition and characterization of previously unrecognized species as analysis of 16S rRNA sequences permits us to study the phylogenetic relationships among bacteria (27). In contrast to phenotypic and biochemical features, the 16S rRNA sequence of a species is a stable property which is specific for microorganisms at a species level (27). Thus, in conjunction with the appropriate phenotypic data, the characterization of a novel 16S rRNA sequence within the background of a data base optimally including the sequences of all established species belonging to the genus of the isolate in question allows us to establish an isolate as a genetically distinct and unique taxon. This possibility will increase our understanding of microbial diversity and will result in the rapid and definitive characterization of isolates which are being kept in freezers in microbiology laboratories as nonidentified isolates by more conventional identification schemes.

For mycobacteria the signature sequences of all established slow-growing species have been determined (Tables 1 and 2). The finding of a novel sequence within the signature region hereby represents the first step. Microheterogeneity within the 16S rRNA molecule can easily be ruled out as a cause of sequence variation (13) by determination of phenotypic and biochemical properties of the isolate as well as by determination of the complete 16S rRNA sequence. Using this strategy, we have characterized a number of previously unrecognized species (Table 1; see also Fig. 2). In each case the finding of a unique 16S rRNA sequence was in good agreement with the failure to identify the isolate by phenotypic methods. Strain 1389/90 was isolated from sputum and was not associated with disease (14). Strain 1669/91 was isolated from a patient with acute exacerbation of chronic bronchitis whose sputum was repeatedly positive by the Ziehl-Neelsen stain (15). Strains 2554/91 and 0908/92 were associated with cervical lymphadenitis in children (10), the most common manifestation of infection with nontuberculous mycobacteria in childhood.

In addition, we and others have described the potential use of nucleic acid amplification techniques for detection of mycobacteria in clinical samples. A number of different target genes have been suggested, mainly for detection of *M. tuberculosis*, e.g., the 65-kDa heat shock protein (8), MBP 70 antigen, MBP 64 protein, IS*6110*, and a 38-kDa protein. In contrast to the above-mentioned targets, 16S rDNA offers the unique advantage of performing a single amplification reaction for the detection and identification of multiple mycobacterial species including all clinically relevant species (2).

Perhaps the most intriguing clinical application of 16S rRNA sequencing is the identification of novel pathogens that cannot be cultured (26). For example, in 1990 a patient infected with human immunodeficiency virus died of an overwhelming infection with an acid-fast microorganism that could not be propagated in vitro (11). Thus, biochemical analysis and characterization was impossible, and the nature of the microorganism remained elusive. It was quickly suggested that analysis of the rRNA sequence might help to identify the organism (6).

When primers are complementary to highly conserved regions of the 16S rRNA, rDNA gene fragments of unknown microorganisms can be amplified (9, 19). Amplification with such highly conserved regions as primer target sites thus theoretically permits amplification of any bacterial 16S rDNA. However, this approach is hampered by contamination and lack of specificity. For instance, *Taq* polymerase itself is frequently contaminated with bacterial DNA, which may result in amplification of DNA fragments without adding extraneous nucleic acids (5, 18).

Given the acid fastness of the uncultured microorganism, we decided to prime the PCR with a combination of oligonucleotides which will specifically amplify mycobacterial 16S rDNA (Fig. 1; Table 3). We were thus able to avoid problems with contaminating DNA in *Taq* polymerase as well as with contaminating members of the *Enterobacteriaceae* in autopsy samples. Complete sequence determination of the amplified 16S rRNA gene revealed that the microorganism in question represents a previously unknown mycobacterium, for which we suggested the name "*M. genavense*" (7) (Fig. 2). By definition of nucleic acid sequences which are highly specific for "*M. genavense*" (Table 1; Fig. 1), the molecular tools for sensitive detection of this microorganism were established. So far, we have been able to detect disseminated infection with "*M. genavense*" in more than 20 patients infected with human immunodeficiency virus, most of them suffering from a chronic syndrome characterized by fever, diarrhea, and loss of weight. The organism could be detected in blood, bone marrow, and multiple tissues (intestines, liver, spleen, and lymph nodes). In six of these patients no other life-threatening infection besides infection with "*M. genavense*" was found at the time of their death.

These and other results (19, 26) illustrate the fascinating perspectives but also the necessary precautions which have to be taken when using the power of gene amplification techniques for defining the etiologic agent of obscure infectious diseases and especially for characterizing otherwise elusive bacterial pathogens.

## Identification of Culture Isolates

### Materials and Reagents

Vortexer

Acid-washed glass beads with a diameter of 100 μm (Sigma, Munich, Germany)

Mickle Tissue Disintegrator (Mickle Laboratory, Gomshall, United Kingdom)

Dynabeads M-280 Streptavidin (Dynal, Germany); use 20 μl of beads (10 μg/μl) per PCR

PCR master mix: 50 mM KCl, 10 mM Tris-HCl (pH 8.3), 1.5 mM $MgCl_2$, 0.01% (wt/vol) gelatin, 200 μM each deoxynucleoside triphosphate, 1.25 U of *Taq* polymerase (Perkin-Elmer Corp.), 30 pmol of primer 264, 10 pmol of biotinylated primer 285 (per 50-μl reaction) (Fig. 1; Table 3). *Note: For a previously unrecognized species, the nearly complete 16S rRNA gene is amplified in two separate reactions by primers which direct the synthesis of two overlapping gene fragments, using 10 pmol of each primer. Primers 264 and 285 direct the synthesis of the 5' part of the gene, and primers 248 and 261 direct the synthesis of the 3' part of the gene. Typically, one of the primers is biotinylated to allow single-stranded DNA sequencing.*

TBE buffer: 89 mM Tris-borate, 89 mM boric acid, 8 mM EDTA

Orange G/Ficoll 20%: dissolve 5 g of Ficoll (type 400-DL; Sigma) in 25 ml of TE, and add Orange G (Sigma) until a good color is formed

Dynal MPC-E, magnetic separator for microcentrifuge tubes

Primers 244 and 259

Sequencing kit, e.g., Sequenase version 2.0 DNA sequencing kit (United States Biochemical Corp.)

0.1 M NaOH

[$\alpha$-$^{32}$P]dATP, 3,000 Ci/mmol (Amersham Buchler, Braunschweig, Germany)

Sequencing apparatus (e.g., IBI STS 45)

6% polyacrylamide–7 M urea solution

10% acetic acid–10% methanol

### Methods

#### *Sample Collection*

Use 0.5 ml of a liquid culture (12B BACTEC [Becton-Dickinson]; a growth index of ≥50 is sufficient) or a small colony from a slant (e.g., Löwenstein-Jensen)

**Table 1.** Alignment of 16S rDNA sequences including all described species of slow-growing mycobacteria[a]

| Type | Nucleotide sequence, 5′ to 3′[b] | Species |
|---|---|---|
| | 129 175 | |
| Slow growth | TGA TCT GCC CTG CAC TTC CGG ATA GG-ACCA CGG GAT GCA TG TCT-TGT GGT GGA AAG C | *M. tuberculosis* complex |
| | CA. ...A.. ... ... ... ... ... ..-..T .AA ..C ... .. ...-.C. ... ... ... . | *M. avium* |
| | CA. ... ... ... ... ... ... ... ..-..T .AA ..C ... .. ...-.C. ... ... ... . | *M. avium*, *M. paratuberculosis*, *M. lepraemurium* |
| | CA. ... ... ... ... ... ... ... ..-..T TTA .GC ... .. ...-.TA ... ... ... . | *M. intracellulare* |
| | CA. ... ... ... ... ... ... ... ..-..T TTA ..C ... .. ...-.T. ... ... ... . | *M. intracellulare* serovar 18 |
| | CA. ... ... ... ... ... ..A ... ..-..T TTA ..C ... .. ...-.T. ... ... ... . | *M. intracellulare* serovar 7 |
| | CA. ... ... ... ... ... ... ... ..-.... .TT .GC ... .. C..-... ... ... ... . | *M. scrofulaceum* |
| | CA. ... ... ... ... AC. ... ... ..-.... .TT .GC ... .. C..-... ... ... ... . | *M. gastri*, *M. kansasii* |
| | .A. ... ... ... ... ... ... ... ..-.... .TT .GC ... .. C..-... ... ... ... . | *M. simiae* |
| | .A. ... ... ... ... ... ..A ... ..-..C ..A .GC ... .. C..-..G ... ... ... . | *M. malmoense* |
| | .A. ... ... ... ... ... ... ... ..-..C ..A .GC ... .. C..-..G ... ... ... . | *M. szulgai* |
| | .A. ... ... ... ... ... ... ... ..-..T .AA .GC ... .. C..T... ... ... ... . | *M. haemophilum* |
| | .A. ... ... ... ... ... ... ... ..-.TT .AA .GC ... .. ...-... ... ... ... . | *M. leprae* |
| | ... ... ... ... ... ... ... ... ..-.... ... ... ... .. ..C-... ... ... ... . | *M. asiaticum* |
| | .A. ... ... ... ... A.. ..A ... ..-.... .A. ..C A.. .. ..C-... ... ... ... . | *M. gordonae* |
| | .A. ... ... ... ... A.. ..A ... ..-.... .A. A.C A.. .. ..C-... ... ... ... . | *M. gordonae* |
| | .A. ... ... ... ... A.. ... ... ..-.... .A. A.C A.. .. ..C-... ... ... ... . | *M. gordonae* |
| | C.. ... ... ... ... ... ... ... ..-.... ... ... T.. .. ..C-... ... ... ... . | *M. marinum*, *M. ulcerans* |
| | ... ... ... ... ... .CT ... ... ..-.... T.. ... ... .. .TC-... ... ... ... . | *M. terrae* |
| | ... ... ... ... ... ..T ..A ... ..-..G .AT .C. ... .. GTG-... ... ... ... . | *M. nonchromogenicum* |
| | ... ... ... ... ... ... ... ... ..-.... .AT .TC ... .. GTG-... ... ... ... . | *M. shimoidei*, *M. triviale* |
| | ... ... ... ... ... ..T ... ... ..-.T. ... AC. ... .. GTC-... ... ... ... . | *M. cookii* |
| | ...C.. ... ... ... ... ... ... ..-.... TTC TGC ... .. .GG-G.. ... ... ... . | *M. xenopi* |

| | | | | | | | | | | | | | | | | | |
|---|---|---|---|---|---|---|---|---|---|---|---|---|---|---|---|---|---|
| Rapid growth | ... | ... | ... | ... | ... | .T | .A | ... | TTCC .TN | NT. | .TC | .... .. | G.CTG.. | A.G | ... | .... | *M. flavescens* |
| | ... | ... | ... | ... | ... | .T | .A | ... | CACC .TG | .T. | .TC | .... .. | G.CTG.. | A.G | ... | .... | *M. smegmatis* |
| | ... | ... | ... | ... | ... | .T | .A | ... | CACC .TG | .T. | .TC | .... .. | G.CTG.. | ..G | ... | .... | *M. thermoresistibile* |
| | ... | ... | ... | ... | ... | .CT | ... | ... | CACC .TT | .T. | .T. | .... .. | G..-G.G | A.G | ... | .... | *M. phlei* |
| | ... | ... | ... | ... | ... | .T | .A | ... | CACC .TT | TT. | .TC | .... .. | G.C-GT. | A.G | ... | .... | *M. agri* |
| | ... | ... | ... | ... | ... | .T | .A | ... | T.-.T.. | T.C | TC. | T.. .. | GGG-... | ... | ... | .... | *M. sphagni* |
| | ... | ... | ... | ... | ... | .T | .A | ... | ..-.T. | ..C | ... | ... .. | ..G-... | ... | ... | .... | *M. aurum* |
| | ... | ... | ... | ... | ... | .T | .A | ... | T.-.T.. | TC. | .C. | T.. .. | GTC-G.. | ... | ... | .... | *M. neoaurum* |
| | ... | ... | ... | ... | ... | .CT | ... | ... | ..-.... | .AC | AC. | T.. .. | GTG-A.. | ... | .C. | .... | *M. chelonae* subsp. *chelonae* |
| | ... | ... | ... | ... | ... | .CT | ... | ... | ..-.... | .AC | AC. | T.. .. | GTG-A.. | ... | .C. | .... | *M. chelonae* subsp. *abscessus* |
| | ... | ... | ... | ... | ... | .T | ... | ... | ..-.... | ..C | .C. | T.. .. | GTG-... | ... | ... | .... | *M. senegalense* |
| | ... | ... | ... | ... | ... | .T | .A | ... | T.-.... | ..C | .C. | T.. .. | GTG-... | ... | ... | .... | *M. fortuitum* subsp. *fortuitum* |
| | ... | ... | ... | ... | ... | .T | .A | ... | T.-...G | ..C | AC. | T.C .. | GTG-... | ... | ... | .... | *M. fortuitum* subsp. *peregrinum* |
| | ... | ... | ... | ... | ... | ... | ... | ... | T.-.... | T.. | AC. | T.. C. | GT.-... | ... | ... | .... | *M. komossense* |
| | ... | ... | ... | ... | ... | .CT | ... | ... | T.-.T.. | T.C | .CC | T.. .. | GTG-... | ... | ... | .... | *M. fallax* |
| Novel species | ... | ... | ... | ... | ... | ..T | .A | ... | CACC .TT | .T. | ATC | .... .. | GTCTG.G | A.G | ... | .... | *Mycobacterium* strain 1389/90 |
| | A. | ... | ... | ... | ... | ACT | ... | ... | ..-..T | .TC | .GC | .... .. | C..AG.A | ... | ... | .... | *Mycobacterium* strain 1669/91 |
| | A. | ... | ... | ... | ... | A.. | .A | ... | ..-..T | ..A | .GC | .... .. | C..-... | ... | ... | .... | *Mycobacterium* strain 2554/91 |
| | A. | ... | .... | ... | ... | ... | ... | ... | ..-..T | ..A | .GC | .... .. | C..-... | ... | ... | .... | *Mycobacterium* strain 0908/92 |
| | A. | ... | ... | ... | ... | ... | ... | ... | T.-.... | ... | A.C | .... .. | .T.-... | ... | ... | .... | "*M. genavense*" |

[a] *M. tuberculosis* was used as the reference sequence. Nucleotides different from those of *M. tuberculosis* are indicated; bars indicate deletions (3, 7, 10, 13–15, 21, 22, 24, 25; unpublished sequences). The numbers indicate the respective *E. coli* 16S rRNA positions.

[b] For reasons of space limitation, only selected stretches of 16S rDNA sequences are shown. An alignment of the complete hypervariable region (corresponding to *E. coli* positions 129 to 266) is available from the authors upon request.

**Table 2.** Alignment of 16S rDNA sequences[a]

| Type | Nucleotide sequence, 5′ to 3′ | Species |
|---|---|---|
| | 439 | |
| Slow growth | TCA CCA TCG ACG AAG G-- TC CGG GTT CTC TCG GAT TGA CGG TAG GTG GAG AAG AAG CAC | *M. tuberculosis* |
| | ... ... ... ... ... .-- .. ... ... ... ... ... ... ... ... ... ... ... ... ... | *M. gastri*, *M. kansasii* |
| | ... ... ... ... ... .-- .. ... ... T.. ... ... ... ... ... ... ... ... ... ... | *M. avium*, *M. paratuberculosis*, *M. intracellulare*, *M. szulgai*, *M. malmoense*, *M. asiaticum* |
| | ... .T. ... ... ... .-- .. ... ... T.. ... ... ... ... ... ... ... ... ... ... | *M. lepraemurium* |
| | ... ... ... ... ... .-- .. T.. ... T.. ... ... ... ... ... ... ... ... ... ... | *M. leprae* |
| | ... ... ... ... ... .-- .T ... ... T.. ... ... ... ... ... ... ... ... ... ... | *M. marinum*, *M. haemophilum*, *M. ulcerans*, *M. intracellulare* |
| | ... ... ... ... ... .-- .. ... ... T.. ... .GC ... ... ... ... ... ... ... ... | *M. gordonae* |
| | ... G.C ... ... ... C-- .G ... ... T.. ... TGG ... ... ... .G. C.. ... ... ... | *M. xenopi* |
| | ... ... ... ... ... C-- .G ... ... T.. ... TGG ... ... ... ... ... ... ... ... | *M. shimoidei* |
| | ... ... ... ... ... .-- CT .-- -AC T.T GT. .G. ... ... ... ... ... ... ... ... | *M. scrofulaceum* |
| | ... GT. ... G.. ... CTC CA A.. T.. TCT .T. .GG ... ... ... ..A C.. ... ... ... | *M. nonchromogenicum* |
| | ... GT. ... G.. ... CTC CG T.. T.. TCT G.. .GG ... ... ... ..A C.. ... ... ... | *M. terrae* |
| | ... ... ... G.. ... CCG GG A.C T.. TGT .TC TGG ... ... ... ... ... ... ... ... | *M. cookii* |
| | ... G.. GG. ... ... C-- G. A-- --- --- --- -.G ... ... ..C C.. C.. ... ... ... | *M. simiae* |
| | ... GTC A.. ... ... C-- G. A-- --- --- --- -.G ... ... ... TGA C.. ... ... ... | *M. triviale* |

| | | |
|---|---|---|
| Rapid growth | ..G GT. GG. ... ... C-- G. A-- --- --- --- -.G ... ... ..C C.A C.. ... ... G.. | *M. flavescens* |
| | ... G.. CA. ... ... C-- G. A-- --- --- --- -.G ... ... ..T ... C.. ... ... G.. | *M. smegmatis*, *M. neoaurum* |
| | ... G.. GG. ... ... C-- G. A-- --- --- --- -.G ... ... ..C C.. T.. ... ... ... | *M. sphagni* |
| | ..G ... GG. ... ... C-- G. A-- --- --- --- -.G ... ... ..C C.. ... ... ... G.. | *M. aurum* |
| | ... AT. GG. ... ... C-- G. A-- --- --- --- -.G ... ... ..C C.A T.. ... ... G.. | *M. fortuitum* subsp. *fortuitum*, *M. fortuitum* subsp. *peregrinum*, *M. senegalense* |
| | ... GT. GG. ... ... C-- G. A-- --- --- --- -.G ... ... ..C C.A T.. ... ... G.. | *M. komossense* |
| | ... GT. C.. ... ... C-- GN A-- --- --- --- -.G ... ... ... ..A C.. ... ... ... | *M. fallax* |
| | ... GT. GG. ... ... C-- GA A-- --- --- --- -.G ... ... ..C C.A C.. ... ... G.. | *M. chelonae* subsp. *chelonae*, *M. chelonae* subsp. *abscessus* |
| | ..G .TC A.. ... ... .-- GT G-- --- --- --- -.C ... ... ... TGA ... ... ... G.. | *M. agri* |
| | ... GT. GG. ... ... C-- GT G-- --- --- --- -.G ... ... ..C C.A C.. ... ... ... | *M. phlei* |
| | ... GTG C.. ... ... C-- GG A-- --- --- --- -.G ... ... ... .CA C.. ... ... G.. | *M. thermoresistibile* |

[a] *M. tuberculosis* was used as the reference sequence. Nucleotides different from those of *M. tuberculosis* are indicated; bars indicate deletions (2, 21, 22, 24, 25; unpublished sequences). The number indicates the respective *E. coli* 16S rRNA position.

**Figure 1.** Schematic drawing of the target and the oligonucleotides used. The numbers in parentheses indicate the corresponding *E. coli* 16S rRNA positions. Oligonucleotides 264, 248, and 259 are specific for mycobacteria; oligonucleotides 285, 244, and 261 are broad-range eubacterial primers. The specificity of the amplification is defined by primers 264 and 248 when using primer combinations 285-264 and 248-261, respectively.

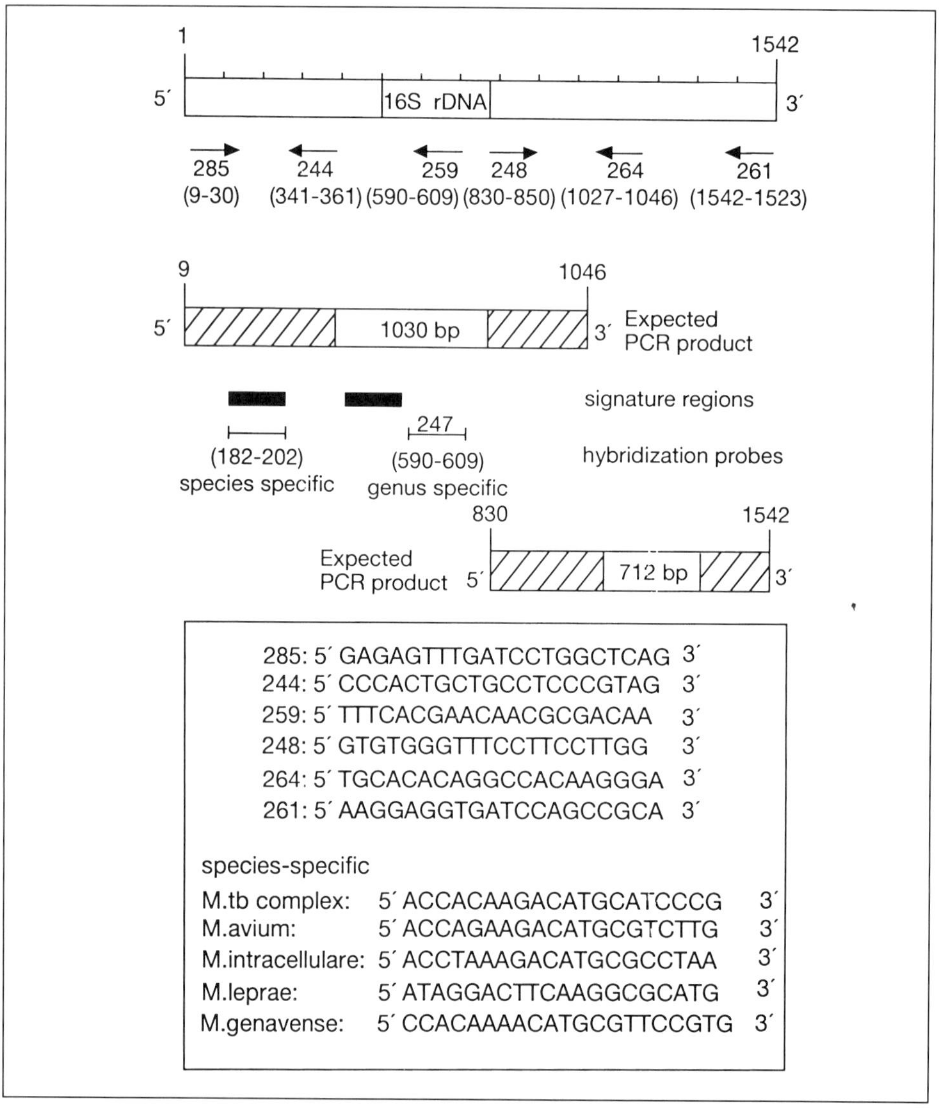


dissolved in 0.5 ml of TE (microscopically positive condensation fluid from a slant is also sufficient).

***Sample Preparation***

Prepare template DNA by the following procedure.

1. Transfer the sample into a 1.5-ml microcentrifuge tube. Incubate it for 10 min at 80°C to inactivate the mycobacteria.
2. Centrifuge the sample at maximum speed in a microcentrifuge for 10 min. Discard the supernatant. *Note: The supernatant should be treated as infectious material.*

**Table 3.** Oligonucleotides used in detection of mycobacteria

| Name or target | Nucleotide sequence, 5′ to 3′ |
|---|---|
| 285 | GAG AGT TTG ATC CTG GCT CAG |
| 244 | CCC ACT GCT GCC TCC CGT AG |
| 259 | TTT CAC GAA CAA CGC GAC AA |
| 248 | GTG TGG GTT TCC TTC CTT GG |
| 264 | TGC ACA CAG GCC ACA AGG GA |
| 261 | AAG GAG GTG ATC CAG CCG CA |
| Species specific | |
| *M. tuberculosis* complex | ACC ACA AGA CAT GCA TCC CG |
| *M. avium* | ACC AGA AGA CAT GCG TCT TG |
| *M. intracellulare* | ACC TAA AGA CAT GCG CCT AA |
| *M. leprae* | ATA GGA CTT CAA GGC GCA TG |
| *M. "genavense"* | CCA CAA AAC ATG CGT TCC GTG |

3. Add 100 μl of TE and a loopful of glass beads. Place tubes for 2 min into a Mickle apparatus at maximum speed. *Note: It may be sufficient to simply boil the samples for 5 min, although we have not yet tried this procedure.*
4. Centrifuge the sample at maximum speed in a microcentrifuge for 2 min.
5. Transfer the supernatant into a new microcentrifuge tube, and use a 5-μl aliquot in PCR.

***PCR Amplification***

1. Prepare PCR master mix as specified in the recipe. Place 45-μl aliquots of master mix in 0.5-ml microcentrifuge tubes, and overlay each with a drop of mineral oil. Heat the thermal cycler to 70°C, and add 5 μl of the sample to the preheated PCR mix. *Note: To add the sample to the bottom phase of the tube, which contains the PCR mix, simply place the pipette at the bottom of the tube and release the 5-μl volume.*
2. Amplify in a programmable thermal cycler by using a two-step profile: a 1-min denaturation step at 94°C followed by a 3-min annealing and extension step at 68°C, for a total of 39 cycles. *Note: Every thermal cycler is different. You may have to modify amplification conditions depending on your cycler.*
3. Prepare a 0.8% agarose gel in TBE buffer. Take a 5-μl aliquot of the PCR and add 5 μl of Orange G/Ficoll 20%. Electrophoretically separate the reaction products in TBE buffer; the Orange G marker should migrate approximately 6 to 10 cm. Visualize the 1-kb amplification product by UV. *Note: Agarose gel electrophoresis is only to control for successful amplification. To shorten the time necessary for this analysis, we routinely add ethidium bromide (final concentration, 0.1 μg/ml) to the gel as well as to the running buffer instead of staining with ethidium bromide after gel running.*

***Preparation of Biotinylated Amplification Product for Sequencing***

Prepare the Dynabead single-stranded DNA template essentially as described by the manufacturer (Dynal). Finally, resuspend the beads in 20 μl of sterile $H_2O$.

***DNA Sequencing***

Details on DNA sequencing are given in reference 23.

1. You can use common sequencing kits for manual and automated dideoxy sequencing procedures. Use 2 to 5 μl of the Dynabead–single-stranded DNA solution and 2 pmol of sequencing primer 244 per sequencing reaction. *Note:*

**Figure 2.** Phylogenetic tree displaying the relationships of *Mycobacterium* species. This tree was constructed as described previously (22) and rooted by using *Nocardia asteroides* as an outgroup. The bar indicates 10 nucleotide differences.

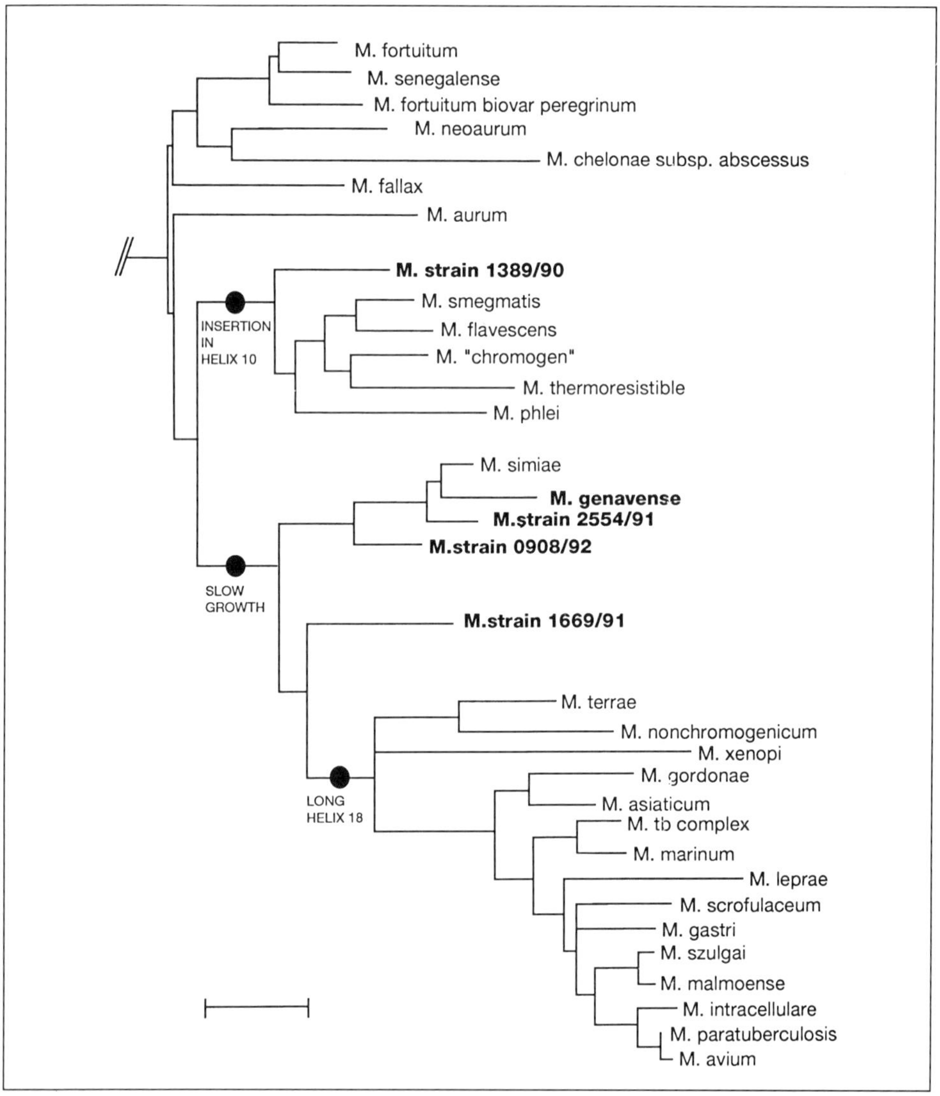


*We use Sequenase version 2.0 (United States Biochemical Corp.) and the conditions specified by the manufacturer.*

2. After completion of the termination reaction and addition of the stop solution, heat the sample at 75°C for 2 min to elute the newly synthesized DNA strands from the beads. Recover the supernatant by using the MPC-E.
3. Load 2.5 μl on a 6% polyacrylamide–7 M urea sequencing gel.
4. Expose overnight, and analyze the results by comparison with Table 1.

## Quality Control

### *Positive Control*

1 ng of *M. tuberculosis* H37 DNA (one reaction).

*Negative Control*

Use $H_2O$ instead of culture fluid, and carry it throughout the whole procedure (one reaction).

## Comments

1. Because of random misincorporation of nucleotides, inherent in the action of all DNA polymerases, every DNA copy made by the PCR may differ slightly from the original. Cloning selects one particular sequence among all those containing random misincorporations, and sequencing of amplified, cloned PCR products will reflect errors of *Taq* polymerase as well as real differences between the original DNAs. Thus, small sequence differences are sometimes difficult to interpret (19). In contrast, direct sequence determination of amplified PCR products is not hampered by such artifacts, since all random misincorporations are averaged out. In fact, this approach probably represents the most accurate sequencing method described (17), provided that enough template DNA is present (more than 100 molecules) when starting the PCR.
2. We perform manual sequencing and use [$^{32}$P]dATP instead of [$^{35}$S]dATP mainly for two reasons: (i) $^{32}$P can be easily monitored, and (ii) because of the higher-energy emission of $^{32}$P, less template DNA is needed for sequencing. In addition, the $^{32}$P signal can be enhanced by using amplifying screens. If you are lucky enough to have an automated DNA sequencing system in your laboratory, you may want to try appropriately labeled primer and *Taq* cycle sequencing technology.
3. Using this protocol, we obtain readable sequences of up to 300 nucleotides by doing a short and long polyacrylamide gel electrophoresis run. If primer 244 is used as the sequencing primer, only an intermediate run is necessary for determination of the signature sequence. *Note: In a case of previously unrecognized mycobacteria, extremely conserved regions, which are dispersed through the 16S rRNA sequence and allow rapid sequence determination of the complete 16S rRNA gene, have been described (9).*
4. For identification of cultured isolates, it is sufficient to use primer 244 for sequencing (Table 1). Sometimes it may be desirable to use primer 259 in addition (21), e.g., when differentiating *M. kansasii* from *M. scrofulaceum* and *M. simiae* or when differentiating *M. shimoidei* from *M. triviale* (Table 2).
5. It is possible to perform double-stranded DNA sequencing on PCR templates. This technique frequently requires gel purification of the amplified PCR product (21). We have found it more convenient to use the solid-phase sequencing technique developed by Uhlen and colleagues (12).
6. For samples from BACTEC 13A blood culture bottles, use a 100-μl aliquot. *Note: An optional step is hypotonic lysis of the erythrocytes. For this procedure add 900 μl of TE, hold samples for 5 min at room temperature to allow cells to lyse, centrifuge at maximum speed in a microcentrifuge for 10 min, and resuspend the pellet in 100 μl of TE.* To the 100-μl aliquot add 400 μl of GCN L6 and 2 μg of carrier DNA and follow steps 5 to 13 of sample preparation for detection of mycobacteria in clinical samples (see below).
7. Given the high accuracy of direct sequence determination, single nucleotide differences are statistically significant. In view of the possibility of microheterogeneity within the 16S rRNA sequence (13), exact 16S rRNA limits for defining a given taxon will have to be determined for each bacterial phylum.

For slow-growing mycobacteria, our experience would suggest that differences of more than five nucleotides within the complete 16S rRNA gene indicate a genetically unique and distinct taxon.

## Detection of Mycobacteria in Clinical Samples

### Materials and Reagents

Carrier DNA (200 μg/ml): human DNA or sheared herring sperm DNA (Boehringer GmbH, Mannheim, Germany)

GCN L6 (4): dissolve 120 g of guanidine isothiocyanate (GuSCN) (GIBCO-BRL, Eggenstein, Germany) in 100 ml of 0.1 M Tris-HCl (pH 6.4) (dissolution of GuSCN is facilitated by heating in a 60 to 65°C water bath with continuous shaking), and add 22 ml of 0.2 M EDTA adjusted to pH 8.0 with NaOH and 2.6 ml of Triton X-100

GCN L2 (4): dissolve 120 g of GuSCN in 100 ml of 0.1 M Tris-HCl (pH 6.4) (GCN L6 and GCN L2 are stable for at least 4 weeks at room temperature in the dark)

Diatom suspension (4): to 10 g of high-purity analytical grade Celite (Kieselgur; Merck, Darmstadt, Germany) add 50 ml of $H_2O$ and 500 μl of 32% (wt/vol) HCl; divide the suspension into aliquots in small glass bottles, close the bottles tightly, and autoclave

Glass bead suspension: to one part acid-washed glass beads with a diameter of 100 μm (Sigma), add one part GCN L6 and vortex before use, as the glass beads will quickly settle to the bottom

Phenol solution: phenol equilibrated with 0.1 M Tris-HCl (pH 7.5)–chloroform–isoamyl alcohol (25:24:1, vol/vol/vol)

Sterile TE buffer: 10 mM Tris-HCl (pH 7.4), 1 mM EDTA

Mickle Tissue Disintegrator (Mickle Laboratory)

Acetone

3 M sodium acetate (pH 5.2)

PCR master mix: 50 mM KCl, 10 mM Tris-HCl (pH 8.3), 1.5 mM $MgCl_2$, 0.01% (wt/vol) gelatin, 200 μM each deoxynucleoside triphosphate, 1.25 U of *Taq* polymerase (Perkin-Elmer Corp.), 50 pmol of each of primer 264 and primer 285 (per 50-μl reaction) (Fig. 1; Table 3)

0.5 M NaOH–25 mM EDTA

2 M ammonium acetate

Nitrocellulose sheets fitted for the slot blot manifold (e.g., BA 85; Schleicher & Schüll, Dassel, Germany)

1 M NaOH

1 M HCl

10× PNK buffer: 0.5 M Tris-HCl (pH 7.6), 0.1 M $MgCl_2$, 0.05 M dithiothreitol, 1.0 mM spermidine

Oligonucleotide for hybridization (259 and species specific [Fig. 1])

Nuctrap push columns (Stratagene, Heidelberg, Germany)

Scintillation counter

Labeling reaction mix: 1.5 μl of 10× PNK buffer, 1.0 μl of oligonucleotide (50 ng/μl), 3.0 μl of $H_2O$, 8.0 μl of [$\gamma$-$^{32}$P]ATP (10 μCi/μl), 1.5 μl of T4 polynucleotide kinase (10 U/μl)

Hybridization solution: 6× SSC (1× SSC is 0.15 M NaCl plus 0.015 M sodium citrate), 0.1% sodium dodecyl sulfate (SDS), 100 μg of tRNA (Boehringer) per ml, 0.1% Ficoll, 0.1% polyvinylpyrrolidone, 0.1% bovine serum albumin

## Methods

***Sample Preparation***

1. Decontaminate the sample (e.g., sputum, urine) by the method used in your laboratory, e.g., the *N*-acetyl-L-cysteine–sodium hydroxide method (20). Use half of the decontaminated sample for culture, and transfer the other half to a 1.5-ml screw-cap microcentrifuge tube for extraction of DNA. *Note: Nucleic acid amplification techniques for detection of mycobacteria in clinical samples still have to be validated by culture. Therefore, always prepare samples as for culture and split them after the appropriate pretreatment. For blood samples we use a lysis-centrifugation method (Isolator; Du Pont Co., Wilmington, Del.). Specimens collected aseptically (e.g., tissue, sterile body fluids) may be used directly without prior decontamination, provided that they are properly homogenized.*
2. Centrifuge the sample at maximum speed in a microcentrifuge for 20 min, and remove the supernatant. *Note: Blood samples are treated as samples from BACTEC 13A blood culture bottles (hypotonic lysis should be performed).*
3. Add 300 μl of GCN L6, resuspend the pellet by vortexing, and incubate the suspension for 10 min at room temperature.
4. Add carrier DNA: 10 μl of human DNA or herring sperm DNA (200 μg/ml). *Note: For quantitative binding of the sample's nucleic acids to the diatoms, it is important to add carrier DNA.*
5. Add 30 μl of glass-bead suspension.
6. Add 300 μl of phenol solution.
7. Place the sample for 2 min in a Mickle apparatus at maximum speed to disrupt cells.
8. Centrifuge the sample at maximum speed in a microcentrifuge for 5 min for phase separation. *Note: If the organic (bottom) and aqueous (top) phases are not well separated, centrifuge again for 15 min.*
9. Transfer the aqueous phase to a new 1.5-ml microcentrifuge tube.
10. Add 100 μl of GCN L6 to the organic phase for reextraction, vortex for 1 min, and centrifuge for phase separation.
11. Transfer the aqueous phase to the first aqueous phase from step 9.
12. Add 300 μl of phenol solution to the pooled aqueous phases, vortex for 1 min, and centrifuge for phase separation.
13. Transfer the aqueous phase to a new 1.5-ml microcentrifuge tube. Discard the interface and lower organic phase.

Nucleic acids may now be purified either by binding to diatoms or by precipitation with isopropanol.

***Binding of Nucleic Acids to Diatoms***

1. Add 40 μl of diatom suspension and incubate for 15 min at room temperature. Resuspend diatoms occasionally by brief vortexing. *Note: It is important that the diatoms be completely suspended.*

2. Centrifuge at maximum speed in a microcentrifuge for 30 s.
3. Remove the supernatant by suction.
4. Wash the diatom pellet twice with 300 μl of GCN L2 (add solution, vortex briefly for complete resuspension of the diatoms, and centrifuge at maximum speed for 30 s). Remove the supernatant by suction.
5. Wash the diatom pellet twice with 900 μl of 70% (vol/vol) ethanol.
6. Wash the pellet once with acetone.
7. Remove the supernatant by suction.
8. Place samples for 15 min with open lids in a 56°C heat block to dry the diatom pellet.
9. Add 40 μl of sterile $H_2O$, and resuspend the pellet by vortexing.
10. Place samples for 10 min in a 56°C heat block (caps closed), and resuspend the diatom pellet every 3 min by brief vortexing.
11. Centrifuge at maximum speed in a microcentrifuge for 2 min.
12. Collect 20 μl of supernatant carefully, and use a 5-μl portion for PCR. *Note: Avoid suction of diatoms, which may inhibit PCR.*

***Precipitation of Nucleic Acids with Isopropanol***

1. Add 20 μl of 3 M sodium acetate (pH 5.2) and 1 vol (400 μl) of cold isopropanol to the aqueous phase and mix well.
2. Precipitate nucleic acids at −20°C for at least 2 h.
3. Centrifuge at maximum speed in a microcentrifuge for 15 min at 4°C.
4. Remove the supernatant by suction. *Note: Be careful not to remove the pellet. As an aid in locating the pellets, which may be difficult to see, turn the neck of the microcentrifuge tube outward to serve as a marker for the expected location of the pellet.*
5. Wash the pellet twice with 400 μl of cold 70% ethanol (vol/vol).
6. Place samples for 10 min with caps open in a 56°C heat block to dry the pellet.
7. Add 40 μl of sterile $H_2O$, and resuspend the pellet by vortexing.
8. Use a 5-μl aliquot for PCR.

***PCR Amplification***

1. Prepare the PCR master mix as specified in the recipe. Place 45-μl aliquots of master mix into 0.5-ml microcentrifuge tubes, and overlay each with a drop of mineral oil. Heat the thermal cycler to 70°C, and add 5 μl of the sample to the preheated PCR mix (“hot start” [2]). *Note: To add the sample to the bottom phase of the tube, which contains the PCR mix, simply place the pipette at the bottom of the tube and release the 5-μl volume. Keep the plunger on the pipettor depressed as it is withdrawn from the tube.*
2. Amplify in a programmable thermal cycler by using a two-step profile: a 1-min denaturation step at 94°C followed by a 3-min annealing and extension step at 68°C, for a total of 39 cycles. *Note: Every thermal cycler is different. You may have to modify amplification conditions depending on your cycler.*
3. Transfer 10 μl of PCR product to a new microcentrifuge tube; add 100 μl of 0.5 M NaOH–25 mM EDTA for denaturation, and incubate the mixture for 30 min at room temperature.
4. Add 110 μl of 2 M ammonium acetate for neutralization.

5. Assemble the manifold as specified by the manufacturer, using a nitrocellulose membrane previously wetted in 10× SSC.
6. Load samples into wells of the manifold.
7. Rinse each well with 0.4 ml of 10× SSC.
8. Bake the nitrocellulose membrane for 2 h at 80°C. *Note: For that purpose, place the membrane between two sheets of 3MM paper and do not exceed the 2-h incubation time. It is not necessary to use a vacuum oven; a simple oven as used for drying of glassware is sufficient.*
9. The nitrocellulose membrane is now ready for hybridization and can be stored in a dry place.
10. Carefully clean the manifold with repeated and alternate washes of 1 M NaOH and 1 M HCl.

***Detection***

1. Label the oligonucleotide. Prepare the labeling reaction mix.
2. Mix and incubate for 30 min at 37°C.
3. Add an additional 0.5 μl of T4 polynucleotide kinase (10 U/μl).
4. Mix and incubate for 30 min at 37°C.
5. Add 1.0 μl of 0.5 M EDTA, and incubate for 10 min at 75°C to inactivate the kinase.
6. Separate the labeled oligonucleotide from residual [$\gamma$-$^{32}$P]ATP by using Nuctrap push columns as specified by the manufacturer. Count a 1-μl aliquot of a 1:10 dilution by using a scintillation counter. *Note: A good labeling reaction results in a total of $1 \times 10^7$ to $5 \times 10^7$ cpm, which corresponds to a specific activity of the labeled oligonucleotide of $2 \times 10^8$ to $1 \times 10^9$ cpm/μg.*
7. Prehybridize the nitrocellulose membrane for 1 h at 50°C in hybridization solution.
8. Add labeled oligonucleotide to the hybridization mix (final concentration, $1 \times 10^6$ to $5 \times 10^6$ cpm/ml), and incubate overnight at $T_d - 5$°C in a hybridization oven. *Note: A hybridization time of 3 to 4 h is sufficient because of the rapid hybridization kinetics of oligonucleotides.*
9. Wash the membranes three times for 20 min each at room temperature in 6× SSC–0.1% SDS, and then wash them three times for 30 min each in 6× SSC–0.1% SDS at the $T_d$ of the oligonucleotide used for hybridization. *Note: Be careful not to exceed the $T_d$. $T_d$ (the temperature at which half of the duplex is dissociated) is calculated by the formula $T_d$(°C) = 2°C (number of A+T residues) + 4°C (number of G+C residues)*
10. Expose plastic film-wrapped membranes at −70°C overnight to Kodak X-Omat film in cassettes containing an amplifying screen.

## Quality Control

***Positive Amplification Control***

1 ng and 10 pg of *M. tuberculosis* H37 DNA (one reaction each).

***Negative Amplification Control***

5 μl of distilled water added to PCR buffer (two reactions).

***Sensitivity Control***

$10^3$, $10^2$, and $10^1$ *M. tuberculosis* H37 cells, suspended in sputum from an uninfected person (one reaction each).

***Negative Control***

Sputum from an uninfected person (three times).

***Inhibitor Control***

For each sample from a patient, perform amplification of, e.g., a β-globin gene fragment in parallel to control for inhibitors (the respective primers can be obtained from Perkin-Elmer). *Note: As human DNA is present in vast excess over mycobacterial DNA in the samples, amplification of a* β*-globin fragment is only a crude indicator for inhibition. To control for amplification efficiency, it is preferable to spike an aliquot of the sample DNA with a low-copy-number target gene, e.g., 10 genomes of M. gordonae.*

## Comments

1. The main problem in using PCR-based techniques for sensitive detection of mycobacteria in clinical samples is the quantitative extraction of nucleic acids. Because of their unique lipid-rich cell wall, mycobacteria are rather difficult to lyse. The method described in this chapter is essentially a GuSCN lysis followed by purification of nucleic acids by binding to diatoms (4) or by precipitation with isopropanol. In our hands, precipitation with isopropanol is the more sensitive procedure (by a factor of 10). Alternative protocols, e.g., a lysozyme-proteinase K-SDS procedure (2) or a high-salt, alkaline lysis method (8), have been described. However, all protocols published so far have disadvantages; i.e., they are either laborious, insensitive, or complicated by the presence of inhibitors. Therefore, a simple method for quantitative extraction of nucleic acids from mycobacteria is still needed.
2. The assay is biased toward detection of mycobacterial DNA primarily because of the specificity of primer 264. As nonspecific amplification products may be seen when a vast excess of nontarget DNA (e.g., other bacterial DNA from related organisms) is present, it is necessary to hybridize to the internal probe.
3. Many different techniques of nonradioactive hybridization have been published as alternatives to $^{32}P$-labeled oligonucleotides (e.g., biotinylated probes, enzyme-linked immunosorbent assay-based solid-phase hybridization methods, and other protocols in this volume).
4. For screening for the presence of mycobacteria in clinical samples, we use hybridization to the genus-specific oligonucleotide 259. Positive samples are hybridized to the appropriate species-specific probes by using an aliquot (10 μl) of the amplification reaction.
5. Species-specific oligonucleotides for identification of virtually any described species of clinical importance can be deduced from published 16S rRNA sequences (2, 3, 7, 13, 14, 21, 22, 24, 25) (Tables 1 and 2).
6. The sensitivity of the procedure, as determined by dilution of mycobacteria with uninfected sputum as the diluent, is approximately 10 cells per sample. With rRNA as the target, followed by reverse transcription, the sensitivity can be theoretically enhanced to the suborganism level (2).
7. Until methods for elimination of contaminating DNA within the *Taq* polymerase and PCR reagents (16) are developed, the broad-range eubacterial primer

approach for amplification of the 16S rRNA gene of unknown microorganisms (9, 19, 26) will be limited. It has been estimated that the commercially available lots of *Taq* polymerase are contaminated with 100 to 1,000 genome equivalents of bacterial DNA per unit enzyme (18). To avoid amplification of the unwanted and contaminating DNA, as visualized by agarose gel electrophoresis, PCR has to be restricted to 25 to 30 cycles. Thus, one is limited to samples which show a large number of cells of the microorganism of interest. Samples which are positive by microscopy indicate the presence of at least $10^4$ to $10^5$ target organisms per ml. A 100-fold excess of input DNA over unwanted contaminating DNA will ensure that the input DNA will be preferentially amplified. The portion of amplified unwanted DNA within the total amplified DNA fragment (theoretically 1%) will not interfere with the interpretation of the sequencing gels when direct sequencing of the amplified PCR product is performed. Accordingly, a sample from a lymph node of a patient positive for human immunodeficiency virus and infected with "*M. genavense*," which showed an extremely large number of acid-fast bacteria, was used as source of nucleic acids and subjected to broad-range primer amplification. Direct sequencing of the amplified product revealed the characteristic signature sequence of "*M. genavense*."

8. Using broad-range primers and direct sequence determination, one is limited to sterile biopsy specimens which microscopically show the organism of interest only and no other bacteria. Autopsy samples, which are frequently contaminated by members of the *Enterobacteriaceae*, are not suitable for this approach, since direct sequence determination will yield numerous ambiguous sequence positions, indicating a mixed bacterial population. In this instance the amplified PCR fragment has to be cloned and the determined 16S rRNA sequence of the target organism verified by in situ hybridization (1).
9. Details on basic molecular techniques such as blotting and hybridization are given in reference 23.

## References


1. **Arnoldi, J., C. Schlüter, M. Duchrow, L. Hübner, M. Ernst, A. Teske, H. D. Flad, J. Gerdes, and E. C. Böttger.** 1992. Species-specific assessment of *Mycobacterium leprae* in skin biopsies by in-situ hybridization and polymerase chain reaction. *Lab. Invest.* **66:**618–623.
2. **Böddinghaus, B., T. Rogall, T. Flohr, H. Blöcker, and E. C. Böttger.** 1990. Detection and identification of mycobacteria by amplification of rRNA. *J. Clin. Microbiol.* **28:**1751–1759.
3. **Böddinghaus, B., J. Wolters, W. Heikens, and E. C. Böttger.** 1990. Phylogenetic analysis and identification of different serovars of *Mycobacterium intracellulare* at the molecular level. *FEMS Microbiol. Lett.* **70:**197–204.
4. **Boom, R., C. J. A. Sol, M. M. M. Salimans, C. L. Jansen, P. M. E. Wertheim-van Dillen, and J. van der Noordaa.** 1990. Rapid and simple method for purification of nucleic acids. *J. Clin. Microbiol.* **28:**495–503.
5. **Böttger, E. C.** 1990. Frequent contamination of Taq polymerase with DNA. *Clin. Chem.* **36:**1258.
6. **Böttger, E. C.** 1990. Infection with a novel, unidentified mycobacterium. *N. Engl. J. Med.* **323:**1635.
7. **Böttger, E. C., A. Teske, P. Kirschner, S. Boost, H. R. Chang, V. Beer, and B. Hirschel.** 1992. Disseaminated infections with "*Mycobacterium genavense*" in patients with AIDS. *Lancet* **340:**76–80.

8. **Brisson-Noel, A., B. Gicquel, D. Lecossier, V. Lévy-Frébault, X. Nassif, and A. J. Hance.** 1989. Rapid diagnosis of tuberculosis by amplification of mycobacterial DNA in clinical samples. *Lancet* **ii:**1069–1071.
9. **Edwards, U., T. Rogall, H. Blöcker, M. Emde, and E. C. Böttger.** 1989. Isolation and direct complete nucleotide determination of entire genes. Characterization of a gene coding for 16S ribosomal RNA. *Nucleic Acids Res.* **17:**7843–7853.
10. **Haas, W., P. Kirschner, S. Ziesing, H. J. Bremer, and E. C. Böttger.** 1993. Cervical lymphadenitis in a child caused by a previously unknown mycobacterium. *J. Infect. Dis.* **167:**237–240.
11. **Hirschel, B., H. R. Chang, N. Mach, P. F. Piguet, J. Cox, J. D. Piguet, M. T. Silva, L. L. Lassnon, P. R. Klatser, J. E. R. Thole, L. Rigouts, and F. Portaels.** 1990. Fatal infection with a novel unidentified mycobacterium in a man with the acquired immunodeficiency syndrome. *N. Engl. J. Med.* **323:**109–113.
12. **Hultman, T., S. Stahl, E. Hornes, and M. Uhlen.** 1989. Direct solid phase sequencing of genomic and plasmid DNA using magnetic beads as support. *Nucleic Acids Res.* **17:**4937–4946.
13. **Kirschner, P., and E. C. Böttger.** 1992. Microheterogeneity within rRNA of *Mycobacterium gordonae. J. Clin. Microbiol.* **30:**1049–1050.
14. **Kirschner, P., A. Teske, K. H. Schröder, R. M. Kroppenstedt, and E. C. Böttger.** 1992. *Mycobacterium confluentis* sp. nov. *Int. J. Syst. Bacteriol.* **42:**257–262.
15. **Meier, A., P. Kirschner, K. H. Schröder, J. Wolters, R. M. Kroppenstedt, and E. C. Böttger.** *Mycobacterium intermedium* sp. nov. *Int. J. Syst. Bacteriol.*, in press.
16. **Meier, A., D. H. Persing, M. Finken, and E. C. Böttger.** Elimination of contaminating DNA within PCR reagents—implications for a general approach to detect uncultured pathogens. *J. Clin. Microbiol.* in press.
17. **Pääbo, S., and A. C. Wilson.** 1988. Polymerase chain reaction reveals cloning artefacts. *Nature* (London) **334:**387–388.
18. **Rand, K. H., and H. Houck.** 1990. Taq polymerase contains bacterial DNA of unknown origin. *Mol. Cell. Probes* **4:**445–450.
19. **Relman, D. A., J. S. Loutit, T. M. Schmidt, S. Falkow, and L. S. Tompkins.** 1990. The agent of bacillary angiomatosis. *N. Engl. J. Med.* **323:**1573–1580.
20. **Roberts, G. D., E. W. Koneman, and Y. K. Kim.** 1991. *Mycobacterium*, p. 304–339. *In* A. Balows, W. J. Hausler, Jr., K. L. Herrmann, H. D. Isenberg, and H. J. Shadomy (ed.), *Manual of Clinical Microbiology*, 5th ed. American Society for Microbiology, Washington, D.C.
21. **Rogall, T., T. Flohr, and E. C. Böttger.** 1990. Differentiation of Mycobacterium species by direct sequencing of amplified DNA. *J. Gen. Microbiol.* **136:**1915–1920.
22. **Rogall, T., J. Wolters, T. Flohr, and E. C. Böttger.** 1990. Towards a phylogeny and definition of species at the molecular level within the genus *Mycobacterium. Int. J. Syst. Bacteriol.* **40:**323–330.
23. **Sambrook, J., E. F. Fritsch, and T. Maniatis.** 1989. *Molecular Cloning: a Laboratory Manual*, 2nd ed. Cold Spring Harbor Laboratory, Cold Spring Harbor, N.Y.
24. **Stahl, D. A., and J. W. Urbance.** 1990. The division between fast- and slow-growing species corresponds to natural relationships among the mycobacteria. *J. Bacteriol.* **172:**116–121.
25. **Teske, A., J. Wolters, and E. C. Böttger.** 1991. The 16S rRNA nucleotide sequence of Mycobacterium leprae: phylogenetic position and development of DNA probes. *FEMS Microbiol. Lett.* **80:**231–238.
26. **Wilson, K. H.** 1992. New vistas for bacteriologists. *ASM News* **58:**318–321.
27. **Woese, C. R.** 1987. Bacterial evolution. *Microbiol. Rev.* **51:**221–271.

**Acknowledgments.** We thank Tom Oystein Jonassen, Microbiology Department, Ullevaal Hospital, Oslo, Norway, for comments on the isopropanol precipitation. Our apologies to all the colleagues engaged in developing nucleic acid amplification-based techniques whose work could not be cited because of space limitations.

# PCR Detection of *Mycobacterium tuberculosis*

# 1.2

Kathleen D. Eisenach, M. Donald Cave, and
Jack T. Crawford

## Background

Diagnostic tests for tuberculosis are based on either detection of *Mycobacterium tuberculosis* in a clinical sample or recognition of the host response to *M. tuberculosis*. Acid-fast microscopy of sputum or other appropriate clinical specimens, followed by culture confirmation, remains the cornerstone of the diagnosis of tuberculosis. These traditional bacteriological methods are slow and expensive. Furthermore, their sensitivity is quite low when clinical samples that contain small numbers of organisms are analyzed. Thus, the rapid direct tests that detect *M. tuberculosis* nucleic acid are very attractive for the laboratory diagnosis of tuberculosis. Eventually nucleic acid amplification assays will replace traditional methods and provide a more timely diagnosis.

Previously we described the isolation of a putative insertion sequence designated IS*6110* from a clinical isolate of *M. tuberculosis* (2, 5). IS*6110* is specific for strains of the *M. tuberculosis* complex (*M. tuberculosis*, *M. bovis*, *M. africanum*, and *M. microti*). Generally, IS*6110* is present in high copy number (10 to 20 copies) in most strains of *M. tuberculosis* and low copy number (1 to 5 copies) in *M. bovis* strains (1). The specificity and repetitive nature of IS*6110* make it an ideal target for amplification by PCR (Fig. 1). We have developed a PCR assay for the detection of *M. tuberculosis* DNA which amplifies a 123-bp sequence in IS*6110* (Table 1) (2). We have constructed an internal control DNA that uses the same primers as the target sequence. Use of the internal control DNA allows assessment of the efficacy of each individual reaction and ensures that the reaction is not inhibited by interfering substances. The PCR product of the internal control DNA is 600 bp long and is easily distinguished from the 123-bp product on an ethidium bromide-stained gel. In a clinical trial we used the PCR to detect *M. tuberculosis* in sputum and other respiratory samples. When compared with bacteriologic and clinical data, results indicate that the assay is very sensitive and specific (3). We expect this assay to greatly facilitate the laboratory diagnosis of tuberculosis. We have developed a method for DNA fingerprinting of *M. tuberculosis* strains by using probes of IS*6110* (1). Chromosomal fingerprinting with IS*6110* is currently used for tracing the spread of strains in outbreaks of tuberculosis (4, 6).

## Materials and Reagents

1.5-ml screw-cap microcentrifuge tubes, caps with O-ring

Repetitive syringe dispenser (Step-pette; Oxford model 8100)

**Figure 1.** Positions of PCR primers for detection of the *M. tuberculosis* IS*6110* target (1,361 bp). GenBank accession no. M29899.

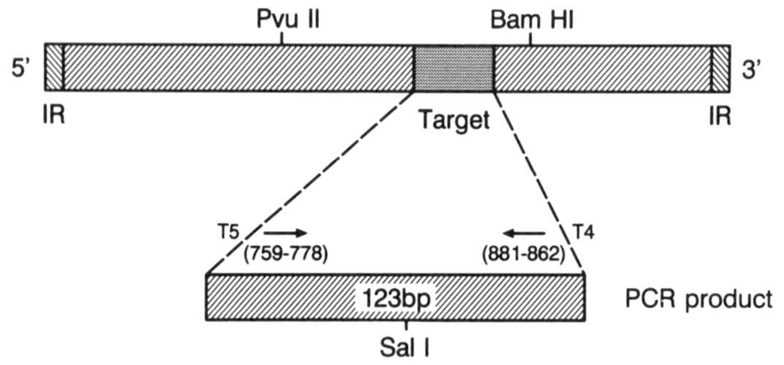


15-ml syringes (Oxford no. 8885-881055)

1.5-ml syringes and tips (Oxford no. 8885-881022)

Vacuum chamber

Lysozyme (chicken albumin)

2% Sodium dodecyl sulfate (SDS)

1 M HCl–1 M Tris-HCl

GeneClean Kit (Bio 101 Inc.)–6 M NaI, NEW concentrate (NaCl, Tris, EDTA), GLASSMILK

PCR reagents: 25 mM $MgCl_2$, 10× amplification buffer (100 mM Tris-HCl [pH 8.3], 500 mM KCl, 0.01% [wt/vol] gelatin), 10 mM each deoxynucleoside triphosphate, 50 μM each primer (T4, T5), 100 μg of internal control DNA per ml, 5 U of Ampli*Taq* polymerase (Perkin Elmer Cetus) per ml

12% polyacrylamide gels (acrylamide, ammonium persulfate, TBE buffer, TEMED)

Gel-loading dye: 50% glycerol, 75 mM EDTA, 4 mg of bromphenol blue per ml

## Sample Requirements

Collect sputum, bronchoalveolar lavage specimens, and bronchial washings by standard procedures. For sputum, collect a series of three to six single, early-morning specimens on successive days. Place the samples in sterile, disposable containers. Transport them to the laboratory as soon as possible. Refrigerate specimens if delivery is delayed.

## Methods

### Sample Preparation

*Note: All sample preparation is performed in a class II biological-safety cabinet. Sealed safety cups are used when centrifuging the clinical samples. The microcentrifuge is used inside the safety cabinet.*

**Table 1.** Primer sequences for detection of *M. tuberculosis*

| Target | Name | Nucleotide Sequence, 5′ to 3′ |
|---|---|---|
| IS*6110* | T4 | CCT GCG AGC GTA GGC GTC GG |
| IS*6110* | T5 | CTC GTC CAG CGC CGC TTC GG |

### *Lysis Procedure*

1. Treat specimens by the standard method with *N*-acetyl cysteine–NaOH. For sediments that remain after smear and culture, process them for PCR. Sediments can be stored at 4°C until tested.
2. Vortex the sample, and transfer it to a screw-cap microcentrifuge tube by using a disposable transfer pipette.
3. Spin at maximum speed in a microcentrifuge for 5 min. Carefully decant the supernatant; about 50 μl will remain in the bottom of the tube. *Note: Take care when removing the supernatant to avoid losing the cell pellet, which may not be visible and may not adhere to the wall of the tube.*
4. Prepare fresh lysozyme (20 mg/ml) in TE buffer by using a sterile serological pipette. Add 0.5 ml to each tube by using a Step-pette with a 15-ml syringe. Vortex.
5. Incubate at 37°C for at least 2 h.
6. Centrifuge for 3 min. Using a disposable Pasteur pipette, carefully aspirate the supernatant, leaving 0.1 ml of liquid in the tube. Avoid disturbing the pellet.
7. Mix equal volumes of 2% SDS and 1 N NaOH by using sterile serological pipettes. Add 0.1 ml to each tube by using a Step-pette with a 1.5-ml syringe. Vortex.
8. Set in a boiling water bath for 5 min.
9. Cool to room temperature. Add 50 μl of 1 M HCl–1 M Tris-HCl by using a Step-pette with a 1.5-ml syringe. Vortex. *Note: A precipitate will form, which disappears upon mixing. If this is not observed, check the pH with pH paper (should be 7.5 to 8.5) and adjust accordingly.*
10. If the samples have a lot of sediment or debris, centrifuge for 30 s prior to the GeneClean procedure. Transfer the supernatant to another screw-cap tube with a sterile transfer pipette.

### *DNA Extraction and Concentration Procedure (GeneClean)*

1. Mix NaI and GLASSMILK (750 and 5 μl, respectively, per sample) in a sterile 50-ml sterile tube by using a sterile serological pipette and a micropipettor with a sterile tip.
2. Add 750 μl to each tube by using a Step-pette with a 1.5-ml syringe. *Note: GLASSMILK must be kept in suspension during this step.*
3. Incubate at room temperature for 10 min, mixing occasionally.
4. Centrifuge in a microcentrifuge at 12,000 × *g* for 5 s. Decant the supernatant.
5. Aliquot the total amount of NEW wash needed into 50-ml sterile tubes; keep them on ice while washing glass pellet. Add 0.5 ml of NEW wash to each tube by using a Step-pette with a 15-ml syringe. Vortex.
6. Centrifuge for 5 s. Decant the supernatant.
7. Wash the glass pellet two more times with NEW wash.
8. After decanting the last wash, centrifuge again for 5 s. Tilt the tube at an angle, and drain the remaining NEW wash with a sterile micropipette. Wipe the wall of the tube dry with a sterile cotton-tip applicator. Alternatively, place tubes with caps loose in vacuum. *Note: It is necessary to remove all traces of NEW wash, which contains alcohol, to avoid inhibition of PCR.*

9. Add 10 μl of water with a positive-displacement micropipettor and sterile tip. Mix by tapping the tube with a finger.
10. Set at 55°C for 5 min.
11. Centrifuge for 30 s. Store at 4°C until PCR is performed.

### PCR Procedure

1. Prepare master mix in a biosafety cabinet which is located in a separate area. The volume prepared is based on the number of reaction tubes desired. For 15 reaction tubes, add to a 1.5-ml microcentrifuge tube the following: 480 μl of sterile distilled water, 120 μl of 25 mM $MgCl_2$, 82.5 μl of 10× amplification buffer, 15 μl of 10 mM dATP, 15 μl of 10 mM dCTP, 15 μl of 10 mM dGTP, 15 μl of 10 mM dTTP, 3 μl of primer T4, 3 μl of primer T5, 3 μl of internal control DNA (diluted to 1 pg/μl), and 3 μl of Ampli*Taq*. *Note: Maintain internal control DNA undiluted. Preparation of a large number of reaction tubes is convenient. Extra tubes can be stored at −20°C until used.*
2. Dispense 50 μl to each 0.5-ml microcentrifuge tube by using the Step-pette with the 1.5-ml syringe. Add 1 drop of mineral oil to each tube. *Note: Keep reaction mixture tubes on ice until used or freeze.*
3. Add 5 μl of DNA sample (or control sample) to the reaction tube with a positive-displacement pipettor, placing the sample below the oil.
4. Amplify in a programmable thermal cycler by using an initial denaturation step of 94°C for 5 min followed by a three-step profile: denaturation at 94°C for 2 min, annealing at 68°C for 2 min, and extension at 72°C for 2 min, for a total of 25 cycles. Increase the extension time by 5 s with each subsequent cycle. Start the thermal cycler, and place tubes in the block when the temperature reaches 70 to 90°C. Store tubes at 4°C at the end of the last cycle.
5. Remove the tubes from the block, and add 2 or 3 drops of chloroform. Store at 4°C until the reactions are run on a gel.

### Analysis of PCR Product

1. Prepare a 12% polyacrylamide gel in TBE buffer. Load the reaction mixture (25 μl per well), and electrophorese at 200 V until the bromphenol blue indicator is approximately 2 in. (ca. 5 cm) from the bottom of the gel (about 2 h). *Note: The reaction mixture is pipetted onto Parafilm and mixed with the gel loading dye. Contamination with oil can be visualized on Parafilm; slide the sample away from the oil droplet by using the pipette tip.*
2. Stain the gel with ethidium bromide, and visualize by UV transillumination. The sample is positive if the 123-bp fragment (*M. tuberculosis* target product) is detected. The sample is negative if the 123-bp fragment is not detected and the 600-bp fragment (internal control target product) is detected.
3. Photograph the gel if a permanent record is desired.

## Quality Control

### Positive Amplification Control

100 pg of *M. tuberculosis* DNA.

### Negative Amplification Control

Reaction mixture without sample.

### Internal Amplification Control

200 fg of internal control DNA.

### Positive Processing (Lysis) Control

$10^3$ *M. tuberculosis* bacilli.

### Negative Processing (Reagent) Controls

TE buffer in the lysis procedure and sterile distilled water in the GeneClean procedure.

## Comments

1. Triton X (2%) may be substituted for 2% SDS. The efficiency of lysis is comparable.
2. Certain clinical specimens may require a different processing procedure, especially samples which contain only a few organisms, e.g., smear-negative sputum, cerebrospinal fluid, pleural fluid, pericardial fluid, and blood.
3. When a large volume of reaction mixture is prepared or any reagent is used for the first time, subject the batch mixture to a quality control test. Specifically, test serial dilutions of *M. tuberculosis* DNA. The expected sensitivity should be 5 to 10 fg.
4. If the reaction is inhibited, as evidenced by the lack of amplification or decrease in amplification of the internal control DNA, test the remaining 5 μl of sample.
5. In the presence of a large amount of *M. tuberculosis* DNA, the following may be observed. Amplification of internal control DNA may be decreased. This is probably due to preferential amplification of the *M. tuberculosis* DNA, which is a smaller target. Additional larger fragments are observed on the gel. These are always the same size (about 200 to 400 bp) and will hybridize if probed with PCR product. They do not interfere with the interpretation of the results.
6. The PCR Carryover Prevention kit (Perkin Elmer Cetus) may be used with this assay. The number of cycles may have to be extended to 30 to obtain the same sensitivity.
7. There is a 1-bp discrepancy between the sequence in the GenBank data base (accession no. M29899) and the published sequences (2, 5). This occurs in the region of primer T4. The published sequence and the primer sequence in this chapter are correct. Note also that the sequence in reference 2 is the complement of the complete sequence of IS*6110* in reference 5.

## References

1. **Cave, M. D., K. D. Eisenach, P. F. McDermott, J. H. Bates, and J. T. Crawford.** 1991. IS6110: conservation of sequence in the *Mycobacterium tuberculosis* complex and its utilization in DNA fingerprinting. *Mol. Cell. Probes* **5:**73–80.
2. **Eisenach, K. D., M. D. Cave, J. H. Bates, and J. T. Crawford.** 1990. Polymerase chain reaction amplification of a repetitive DNA sequence specific for *Mycobacterium tuberculosis. J. Infect. Dis.* **161:**977–981.
3. **Eisenach, K. D., M. D. Sifford, M. D. Cave, J. H. Bates, and J. T. Crawford.** 1991. Detection of *Mycobacterium tuberculosis* in sputum samples using a polymerase chain reaction. *Am. Rev. Respir. Dis.* **144:**1160–1163.

4. **Mazurek, G. H., M. D. Cave, K. D. Eisenach, R. J. Wallace, Jr., J. H. Bates, and J. T. Crawford.** 1991. Chromosomal DNA fingerprint patterns produced with IS*6110* as strain-specific markers for epidemiologic study of tuberculosis. *J. Clin. Microbiol.* **29:**2030–2033.
5. **Thierry, D., M. D. Cave, K. D. Eisenach, J. T. Crawford, J. H. Bates, B. Gicquel, and J. L. Guesdon.** 1990. IS6110, an IS-like element of the *Mycobacterium tuberculosis* complex. *Nucleic Acids Res.* **18:**188.
6. **Zhang, Y., G. H. Mazurek, M. D. Cave, K. D. Eisenach, Y. Pang, D. T. Murphy, and R. J. Wallace, Jr.** DNA polymorphisms in strains of *Mycobacterium tuberculosis* analyzed by pulse-field gel electrophoresis: a tool for epidemiology. *J. Clin. Microbiol.* **30:**1551–1556.

# PCR Detection of *Rickettsia rickettsii* and *Ehrlichia chaffeensis*

**1.3**

Burt E. Anderson

## Background

Rocky Mountain spotted fever (RMSF) and human ehrlichiosis are the two tick-borne rickettsial diseases of greatest public health concern in the United States. The etiologic agent of RMSF is *Rickettsia rickettsii*, and the causative agent of human ehrlichiosis has recently been shown to be a new rickettsial agent, *Ehrlichia chaffeensis* (1). In terms of clinical presentations, RMSF and ehrlichiosis are similar diseases, often with overlapping symptoms (5). Patients with either RMSF or ehrlichiosis respond well to tetracycline or chloramphenicol, hence the need for early diagnosis and treatment.

The current definitive laboratory diagnosis of either RMSF or ehrlichiosis depends on the demonstration of a specific antibody response to these agents, although detectable levels of antibody may not be present during the early stages of these diseases (6). An early diagnosis for RMSF can sometimes be made by detecting *R. rickettsii* by the direct fluorescent antibody technique on skin lesions (9). However, not all RMSF patients have a visible rash, and this technique has not been used for diagnosing ehrlichiosis. We have reported PCR assays for the detection of *R. rickettsii* (based on the 17-kDa genus-common antigen gene sequence [3, 8]) (Fig. 1; Table 1) and *E. chaffeensis* (based on the 16S rRNA gene sequence [2]) (Fig. 2; Table 2) in blood specimens from the acute phase of infection. Our results indicate that the PCR assay for the detection of *E. chaffeensis* provides a sensitive and much-needed method for the early diagnosis of ehrlichiosis. The PCR assay for *R. rickettsii* appears to be slightly less sensitive, perhaps because of the low level of rickettsemia reported for RMSF (7), but it nevertheless provides an alternative means for the early diagnosis of this disease.

## Materials and Reagents

Sterile PCR diluent (10 mM Tris [pH 8.0], 10 mM NaCl)

Proteinase K (10 mg/ml in sterile water) (Sigma)

Buffer-saturated phenol (redistilled phenol saturated in PCR diluent) (Bethesda Research Laboratories, Gaithersburg, Md.)

Centricon 30 concentrators (Amicon, Beverly, Mass.)

High-speed centrifuge (e.g., Sorvall RC5C with SS34 rotor; Du Pont, Wilmington, Del.)

**Figure 1.** Diagram of PCR primers and probe for detection of *R. rickettsii*. *GenBank accession no. M28479.

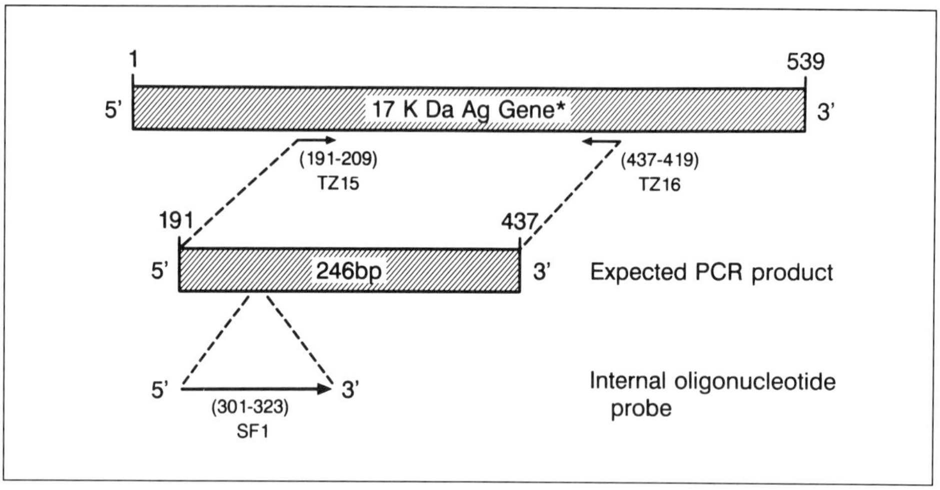


PCR master mix: 10 mM Tris (pH 8.3), 50 mM KCl, 1.5 mM $MgCl_2$, 0.001% gelatin, 200 μM each deoxynucleoside triphosphate, 2.5 U of *Taq* polymerase (Ampli*Taq*; Perkin-Elmer), 50 pmol of each primer (HE1 and HE3 for *E. chaffeensis* [Fig. 2; Table 2] or TZ15 and TZ16 for *R. rickettsii* [Fig. 1; Table 1])

Electrophoresis loading buffer: 33% glycerol, 5 mM Tris (pH 8.0), 0.07% bromphenol blue

TBE buffer (1×): 10.8 g of Tris base (Sigma) per liter, 5.5 g of boric acid (Sigma) per liter, 8 ml of 0.25 M EDTA (pH 8.0) per liter, 50 μl of ethidium bromide (10 mg/ml; Sigma) per liter

GeneScreen Plus nylon membrane (DuPont, Boston, Mass.)

E-link plus kit (Cambridge Research Biochemicals, Wilmington, Del.)

96-well filtration manifold

5× SSC (20× is 0.3 M sodium citrate [pH 7.0] and 3 M NaCl) with 0.5% SDS, 0.1% bovine serum albumin (BSA) (Sigma) fraction V, and 1× Denhardt's solution (50× is 10 g of Ficoll 400 [Pharmacia, Piscataway, N.J.] per liter, 10 g of polyvinylpyrrolidone [Sigma] per liter, and 10 g of BSA [Sigma] per liter)

1× SSC containing 0.5% SDS

0.25× SSC containing 0.5% SDS

# Methods

## Sample Collection

The preferred sample is 10 ml of whole blood collected in EDTA anticoagulant (purple top). Alternatively, clotted blood (red top) may be used by preparing

**Table 1.** Oligonucleotides for detection of *R. rickettsii*

| Function | Name | Nucleotide sequence, 5′ to 3′ |
|---|---|---|
| Primer | TZ15 | TTC TCA ATT CGG TAA GGC |
| Probe | SF1 | GAC TTG CAG AGC TTA CCT CAC AG |
| Primer | TZ16 | ATA TTG ACC AGT GCT ATT TC |

**Figure 2.** Diagram of PCR primers and probe for detection of *E. chaffeensis*. *GenBank accession no. M73222.

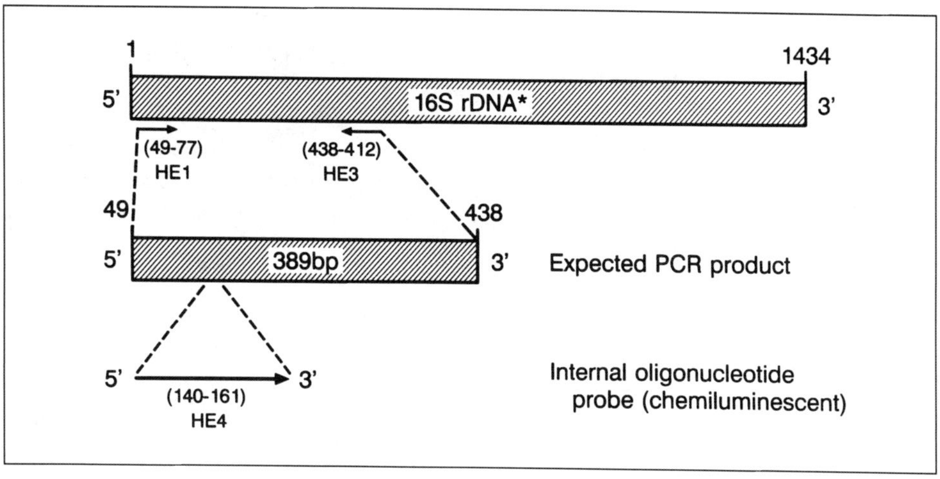


a 50% suspension in PCR diluent and vortexing with sterile glass beads to disperse the clot (if clots are used, begin at step 2 of the sample preparation protocol).

*Note: Because of the safety hazard in working with rickettsiae, handle suspect infected blood specimens in approved biosafety cabinets and take recommended safety precautions (4). Safety precautions for handling E. chaffeensis have not yet been established, but they may be assumed to be similar to those established for R. rickettsii (4).*

## Sample Preparation

For preparation of template DNA prior to amplification, prepare a total lysate (leukocytes and rickettsiae). Because of the small numbers of rickettsiae found in infected blood (7), do not attempt to purify the rickettsiae. *Note: Perform all pipetting operations for sample preparation and amplification set-up in a biosafety cabinet, and use positive-displacement pipettors.*

1. Add 150 μl of PCR diluent to 150 μl of blood in a 1.5-ml microcentrifuge tube.
2. Add 30 μl of 10% SDS and 5 μl of proteinase K, mix well by inverting, and place the tubes into a 55°C heating block for 1 h.
3. To the lysate add a mixture of 100 μl of buffer-saturated phenol and 100 μl of chloroform-isoamyl alcohol, vortex the mixture, and spin it for 3 min in a microcentrifuge at maximum speed to separate the layers. Repeat the extraction two or three times until no debris are visible at the interface.
4. Transfer the aqueous supernatant to a Centricon 30 concentrator. *Note: Take*

**Table 2.** Oligonucleotides for detection of *E. chaffeensis*

| Function | Name | Nucleotide sequence, 5′ to 3′ |
|---|---|---|
| Primer | HE1 | CAA TTG CTT ATA ACC TTT TGG TTA TAA AT |
| Probe | HE4 | TTA GAA ATG ATG GGT AAT ACT G |
| Primer | HE3 | TAT AGG TAC CGT CAT TAT CTT CCC TAT |

*care not to transfer any of the organic layer to the concentrator, which is not stable in the presence of these organic solvents (see Centricon 30 instructions).*

5. Dilute the supernatant with PCR diluent to 2 ml, and spin at 4,800 × *g* for 30 min. Perform the filtration twice more by adding an additional 2 ml of PCR diluent to the retentate and centrifuging as before. *Note: Centricon 30 filtration apparently removes compounds in blood which inhibit PCR (e.g., hematin).*
6. Collect the final Centricon 30 retentate by inverting the filter and centrifuging briefly (3 min) at 50 × *g* in a high-speed centrifuge. *Note: See Centricon instructions for details. Typical final retentate volumes are 30 to 50 μl, which contain the leukocyte and rickettsial nucleic acids.*

## PCR Amplification

1. Prepare the PCR master mix according to the recipe listed above, using either primer pair HE1 and HE3 for detecting *E. chaffeensis* or primer pair TZ15 and TZ16 for detecting *R. rickettsii*. Allow at least 100 μl of master mix for each reaction, and divide 90 μl equally among 0.5-ml microcentrifuge tubes. Add 10 μl of Centricon 30 retentate containing the target DNA.

2a. For *E. chaffeensis* PCR: Amplify (for a total of 40 cycles) in a programmable thermal cycler, using two three-step profiles coupled together. Profile 1 is three cycles at 94°C for 1 min, 55°C for 2 min, and 70°C for 1.5 min. Profile 2 is 37 cycles at 88°C for 1 min, 55°C for 2 min, and 70°C for 1.5 min. Use the fastest possible cycling time between the three temperatures. *Note: The first three cycles are performed at a higher denaturation temperature to allow strand separation of genomic DNA. The remaining cycles use a lower temperature to denature the initial PCR products for further amplification. This method presumably allows completion of more cycles before the Taq polymerase activity diminishes.*

2b. For *R. rickettsii* PCR: Amplify (for a total of 40 cycles) by using two three-step profiles coupled together. Profile 1 is three cycles at 94°C for 1 min, 52°C for 2 min, and 66°C for 1.5 min. Profile 2 is 37 cycles at 88°C for 1 min, 52°C for 2 min, and 66°C for 1.5 min.

3a. For *E. chaffeensis* PCR: Prepare a 1.4% Ultrapure agarose gel in TBE buffer. Electrophoretically separate the reaction products (10 μl per well) in TBE buffer; the bromphenol blue tracking dye should migrate approximately two-thirds the length of the gel.

3b. For *R. rickettsii* PCR: Prepare a 2% Nusieve GTG–1.0% Ultrapure agarose gel in TBE buffer. Electrophorese 10 μl of the PCR products as in step 3a.

4a. For *E. chaffeensis* PCR: Visualize the 389-bp amplification product by using a UV transilluminator.

4b. For *R. rickettsii* PCR: Visualize the 246-bp amplification product by using a UV transilluminator.

## Chemiluminescent Detection

1. Label oligonucleotide probe HE4 (for *E. chaffeensis* PCR) or probe SF1 (for *R. rickettsii* PCR) at the 5′ terminus with alkaline phosphatase by using the E-link plus system as specified by the manufacturer.
2. Heat 10 μl of each PCR product at 95°C for 5 min and filter slowly through a GeneScreen Plus membrane by using a 96-well filtration manifold.

3. Prehybridize the membrane filter, using a shaking water bath, for 90 min at 50°C in a solution of 5× SSC with 0.5% SDS, 0.1% BSA, and 1× Denhardt's solution. *Note: Heat the prehybridization solution for 30 min at 80°C to inactivate any inherent alkaline phosphatase activity.*

4a. For *E. chaffeensis* PCR: Add alkaline phosphatase-labeled HE4 probe to the prehybridization solution to a final concentration of 1 nM, and hybridize for 20 min at 50°C.

4b. For *R. rickettsii* PCR: Add alkaline phosphatase-labeled SF1 probe to the prehybridization solution, and hybridize as described in step 4a.

5. Wash the hybridized membrane filter twice for 5 min at 50°C in 1× SSC–0.5% SDS and twice for 5 min at 50°C in 0.25× SSC–0.5% SDS.

6. Wash the resulting filter briefly in 1× SSC, and spray the moist filter with Lumi-Phos chemiluminescent substrate solution (provided in the E-link plus kit). Cover the moist filter with Saran wrap, and expose it to Kodak X-Omat film for 1 to 2 h.

## Quality Control

### Positive Amplification Control

1 ng and 10 pg of *E. chaffeensis* (ATCC CRL10679) DNA or 1 ng and 10 pg of *R. rickettsii* DNA.

### Negative Amplification Control

10 μl of PCR diluent.

### Positive Patient Control (optional)

Whole-blood specimen from an *E. chaffeensis*-infected patient or from an *R. rickettsii*-infected patient.

### Negative Patient Control

Whole-blood specimen from an uninfected patient.

### Specificity Control

For *E. chaffeensis* PCR, 100 ng of DNA from *Ehrlichia canis.*
For *R. rickettsii* PCR, 100 ng of DNA from *Rickettsia prowazekii.*

## Comments

1. The primer pair HE1 and HE3 is highly specific under these conditions, reacting only with *E. chaffeensis*; thus, the specificity lies with amplification. Accordingly, hybridization of the PCR products with probe HE4 is strictly confirmatory and may be considered optional if the characteristic 389-bp *E. chaffeensis*-specific PCR product is observed.
2. The primer pair TZ15 and TZ16 amplifies DNA from both *R. rickettsii* and *R. conorii*. Likewise, the oligonucleotide probe SF1 hybridizes with both species under the conditions described in this report.
3. We routinely "sterilize" all water used for reagent preparation, specimen

preparation, and PCR amplification by gamma irradiation with $10^6$ rads from a $^{60}Co$ source. Although this method may not be widely available, it appears to be an effective means of reducing PCR contamination in the laboratory.

## References

1. **Anderson, B. E., J. E. Dawson, D. C. Jones, and K. H. Wilson.** 1991. *Ehrlichia chaffeensis*, a new species associated with human ehrlichiosis. *J. Clin. Microbiol.* **29:**2838–2842.
2. **Anderson, B. E., J. W. Sumner, J. E. Dawson, T. Tzianabos, C. R. Greene, J. G. Olson, D. B. Fishbein, M. Olsen-Rasmussen, B. P. Holloway, E. H. George, and A. F. Azad.** Detection of the etiologic agent of human ehrlichiosis by polymerase chain reaction. *J. Clin. Microbiol.* **30:**775–780.
3. **Anderson, B. E., and T. Tzianabos.** 1989. Comparative sequence analysis of a genus-common rickettsial antigen gene. *J. Bacteriol.* **171:**5199–5201.
4. **Centers for Disease Control.** 1988. Rickettsial agents, p. 69–72. *In* J. H. Richardson and W. E. Barkley (ed.), *CDC-NIH Biosafety in Microbiological and Biomedical Laboratories.* U.S. Dept. of Health and Human Services, Washington, D.C.
5. **Centers for Disease Control.** 1990. Rocky Mountain spotted fever and human ehrlichiosis—United States, 1989. *Morbid. Mortal. Weekly Rep.* **39:**281–284.
6. **Kaplan, J. E., and L. B. Schonberger.** 1986. The sensitivity of various serologic tests in the diagnosis of Rocky Mountain spotted fever. *Am. J. Trop. Med. Hyg.* **35:**840–844.
7. **Kaplowitz, L. G., J. V. Lange, J. J. Fischer, and D. H. Walker.** 1983. Correlation of rickettsial titers, circulating endotoxin, and clinical features in Rocky Mountain spotted fever. *Arch. Intern. Med.* **143:**1149–1151.
8. **Tzianabos, T., B. E. Anderson, and J. E. McDade.** 1989. Detection of *Rickettsia rickettsii* DNA in clinical specimens by using polymerase chain reaction technology. *J. Clin. Microbiol.* **27:**2866–2868.
9. **Walker, D. H., and B. G. Cain.** 1978. A method for specific diagnosis of Rocky Mountain spotted fever on fixed, paraffin-embedded tissue by immunofluorescence. *J. Infect. Dis.* **137:**206–209.

# PCR Detection of *Borrelia burgdorferi*

# 1.4

Paul N. Rys

## Background

The spirochete *Borrelia burgdorferi* is the causative agent of Lyme disease (9) and is transmitted to humans primarily through the bite of infected *Ixodes* ticks. The resulting arthritic, neurologic, and rheumatic symptoms vary in severity, are disease stage dependent, and often mimic other conditions (9). This multifaceted presentation often delays and confuses clinical diagnosis. Bacterial culture and serologic methods are currently used in diagnosis (9). However, limited sensitivity and specificity and lack of test standardization between laboratories have hindered the interpretation of results. Direct detection of *B. burgdorferi*-specific target DNA sequences by using PCR may be a promising adjunct to existing diagnostic tests.

Our laboratory (6, 7) and others (1, 8) have chosen to target specific DNA sequences encoding the outer surface protein A (OSPA) present on the 49-kb plasmid in *B. burgdorferi* B31, along with genomic sequences encoding the flagellin and 16S rDNA genes (Table 1). The range and quality of specimen types and collection and transport conditions have confounded attempts to design a single efficient standard processing technique. The varied physical characteristics and DNA content of the specimens, ranging from bacterial cultures, tick extracts, whole blood, serum, joint fluid, urine, and cerebrospinal fluid (CSF), require that protocols be developed for each specimen type to obtain a satisfactory yield of target DNA free from inhibitors that are often present. Our laboratory has evaluated several specimen-processing techniques, including simple dilution of crude extracts (bacterial cultures, ticks), conventional solvent extraction (archival ticks, CSF, joint fluids, serum, tissue digests), and commercially available extraction kits (whole blood, CSF, joint fluids, urine).

## Materials and Reagents

BSK H medium (Sigma)

"K" buffer/proteinase K: 10 mM Tris (pH 8.3), 50 mM KCl, 1.75 mM $MgCl_2$, 0.01% bovine serum albumin (BSA), 0.45% Tween 20, 0.45% Nonidet P-40, 100 μg of proteinase K per ml (International Biotechnologies, Inc.)

0.1-mm glass beads (BioSpec Products, Bartlesville, Okla.)

Serum separator tubes (Becton Dickinson)

**Table 1.** Oligonucleotides for PCR detection of *B. burgdorferi*

| Target | Name | Function | Nucleotide sequence, 5′ to 3′ |
|---|---|---|---|
| OSPA | OSPA2 | Primer | GTT TTG TAA TTT CAA CTG CTG ACC |
| | OSPA4 | Primer | CTG CAG CTT GGA ATT CAG GCA CTT C |
| | OSPA3 | Probe | GCC ATT TGA GTC GTA TTG TTG TAC TG |
| OSPA | OSPA149 | Primer | TTA TGA AAA AAT ATT TAT TGG GAA T |
| | OSPA319 | Primer | CTT TAA GCT CAA GCT TGT CTA CTG T |
| | OSPA6 | Probe | ATT GGG AAT AGG TCT AAT ATT AGC CT |
| Flagellin | FLA1 | Primer | GAT GAT GCT GCT GGC ATG GGA GTT TCT... GG |
| | FLA3 | Primer | CTG TCT GCA TCT GAA TAT GTG CCG TTA... CCT G |
| | FLA2 | Probe | ATT CAG ACA ACA GAA GGG AAT TTA AAT... GAA GTA G |
| 16S rDNA | DD02 | Primer | CCC TCA CTA AAC ATA CCT |
| | DD06 | Primer | ATC TGT TAC CAG CAT GTA AT |
| | DD04 | Probe | CCC GTA AGG GAG GAA GGT AT |

TE buffer

1 M dithiothreitol (DTT)

0.25 M EDTA (pH 8)

IsoQuick Nucleic Acid Extraction Kit (MicroProbe Corp., Bothell, Wash.)

PCR master mix: 10 mM Tris HCl (pH 8.3), 50 mM KCl, 1.75 mM $MgCl_2$, 0.01% BSA, 200 μM each deoxynucleoside triphosphate, 50 pmol of each primer per 50-μl reaction, 10% glycerol, 100 μg of isopsoralen compound 10 (HRI, Inc., Concord, Calif.) per ml, 1.25 U of Ampli*Taq* polymerase (Perkin-Elmer Corp.)

Sterile water for irrigation USP (Baxter Healthcare)

HRI-100 UV Photochemical Reaction Chamber for post-PCR sterilization of amplicon (HRI, Inc.)

Electrophoresis gel loading buffer: 27 ml of glycerol, 3 ml of 10× TBE, 1 ml of 10% sodium dodecyl sulfate (SDS), 10 ml 0.5 M EDTA (pH 8.0), 10 mg of bromphenol blue, $H_2O$ to 100 ml

20× TBE electrophoresis buffer: 216 g of Tris base (Sigma), 110 g of boric acid (Sigma), 80 ml of 0.5 M EDTA (pH 8.0; Sigma), $H_2O$ to 1 liter

Denaturation solution: 0.5 M NaOH, 1.5 M NaCl

Neutralization solution: 0.5 M Tris (pH 7.5; Sigma), 1.5 M NaCl

Nylon membrane (Hybond N; Amersham Corp., Arlington Heights, Ill.)

Prehybridization/hybridization solution: 5× Denhardt's solution, 5× SSPE, 0.5% SDS, 100 μg of denatured salmon sperm DNA (Sigma) per ml

Hybridization wash 1: 2× SSC (1× SSC is 0.15 M NaCl plus 0.015 M sodium citrate), 0.1% SDS

Hybridization wash 2: 1× SSC, 0.1% SDS

# Methods

## Sample Collection and Storage

***Bacterial Culture***

Grow *Borrelia* isolates in 7 ml of BSK H medium (Sigma) in screw-cap tubes (13 by 100 mm) at 32°C.

***Ticks***

Identify *Ixodes* ticks, and store them until processing as follows. Live ticks are suitable for all test methods. Prevent desiccation until testing. Store dead ticks in 70% alcohol prior to testing, although desiccated specimens may also be used.

***CSF***

Use 2 ml or more in sterile tubes. Store aliquots at −20°C until use. Avoid multiple freeze-thaw cycles.

***Joint Fluid***

Use 2 ml or greater in a red-top tube (no anticoagulant) or a purple-top tube (EDTA additive). Do not use heparin (green-top tube), which inhibits PCR. Store aliquots at −20°C until use. Avoid multiple freeze-thaw cycles.

***Whole Blood***

Use 5 to 10 ml in a purple-top tube (EDTA additive). Store at 4°C. Process or freeze aliquots as soon as possible after receipt.

***Serum***

Use 1 to 3 ml in a sterile tube. Store aliquots at −20°C until use. Avoid multiple freeze-thaw cycles.

***Urine***

Use 15 ml, preferably collected prior to antibiotic therapy. An equal volume of 95% ethanol is added prior to storage in some protocols.

***Tissue***

Use a single standard 3- to 5-mm skin punch biopsy or equivalent-sized tissue specimen stored in 70% ethanol at room temperature.

*Note: It is imperative that all clinical samples avoid contact with all sources of* Borrelia *DNA during collection, aliquot handling, and storage, or false-positive samples may occur (5). Avoid work areas where* Borrelia *cultures, antigen preparations, or amplified DNAs have been manipulated. Samples are best sealed at the patient's bedside and sent directly to the PCR laboratory, where aliquots for serologic testing can be made.*

## Sample Preparation

The diverse characteristics and DNA content of clinical samples require the use of different preparation methods. Specimens may contain nucleases (especially whole blood, serum, and urine) or *Taq* polymerase inhibitors (especially blood and urine), which disallow direct use of the specimen for PCR. Each method described is suitable for several specimen types.

***Cell Pellets***

Harvest the cell pellet from 1 ml of BSK culture containing $1 \times 10^9$ to $5 \times 10^9$ organisms per ml. Add 50 μl of TE buffer, and boil for 10 min in a temperature block. Prepare a 10-fold dilution series in TE buffer, and use 1- to 5-μl aliquots directly for PCR.

*Tick Extracts*

Process whole ticks as described previously (6, 7).

1. Remove ticks from alcohol and air dry on filter paper disks for 5 min. Put each whole tick into a 1.5-ml microcentrifuge tube, cover it with 20 μl of K buffer (without proteinase K) together with a small volume of BSA-coated glass beads, and crush it with a disposable plastic pestle or pipette tip. *Note: The buffer becomes slightly turbid when the tick contents are released.*
2. Heat tubes in a 95°C heating block for 10 min, and then immediately chill them on ice. Test 5 μl of supernatant by PCR. *Note: Additional buffer may be added to large specimens.*

*Tissue*

1. Immerse tissue (i.e., skin biopsy specimen, BB-sized to pea-sized tissue biopsy specimen) in 50 to 100 μl of K buffer-proteinase K. Incubate at 55°C in a temperature block until digestion is complete. *Note: The amount and type of tissue determine the required incubation time, which typically ranges from 1 to 24 h. Briefly vortexing sample several times during incubation will speed tissue digestion.*
2. Heat samples at 95°C for 10 min to inactivate proteinase K and immediately chill on ice. *Note: Excessive heating will decrease the DNA yield.* Test 1 to 5 μl of digest by PCR. Store processed samples at −20°C.

## Solvent Extraction

1. Add 500 μl of urine to each 1.5-ml microcentrifuge tube. Centrifuge at 16,000 × *g* for 5 min. Remove and discard the supernatant (the pellet may be stored at −20°C).
2. Add 246.2 μl of TE, 5 μl of 1 M DTT, 20 μl of 0.25 M EDTA (pH 8), 25 μl of 10% SDS, and 3.8 μl of 20-mg/ml proteinase K, and incubate the resuspended pellet at 37°C for 1 h.
3. Transfer the digest (approximately 300 μl) to pediatric serum separator tubes. *Note: This product greatly simplifies solvent extraction, allowing multiple extractions to be performed in the same tube.* Alternatively, the clinical specimen may be extracted directly by adding the specimen to the tube without digestion.
4. Extract with 150 μl of phenol-chloroform-isoamyl alcohol (25:24:1). Mix by inversion or vortexing, and then centrifuge at 16,000 × *g* for 30 s. *Note: The aqueous layer is above the inert plug and the organic solvent layer.* Repeat extraction in the same tube.
5. Extract with 150 μl of chloroform-isoamyl alcohol (24:1) in the same tube.
6. Transfer the upper aqueous layer containing nucleic acids to a 1.5-ml microcentrifuge tube. Add 30 μl of 3 M sodium acetate and 300 μl of isopropanol. Gently invert the tube to mix.
7. Centrifuge at 16,000 × *g* for 15 min at 4°C. Carefully remove the supernatant by using a fine-tip pipette without disturbing the pellet.
8. Wash the pellet with 500 μl of 70% ethanol. Centrifuge at 16,000 × *g* for 15 min at 4°C. Remove all but 20 μl of ethanol, and evaporate to dryness in a Speed

Vac centrifuge for 5 min. *Note: Air drying of the pellet is an alternative method.*

9. Add 50 μl of TE to resuspend the pellet. Store overnight at 4°C to resolubilize the nucleic acid. Use 2.5 μl in a 100-μl PCR reaction.

### Chaotropic Lysis/Solvent Extraction

The IsoQuick Nucleic Acid Extraction Kit employs chaotropic lysis to disrupt cells and stabilize the nucleic acid contained in a variety of clinical samples, including bacteria and whole blood (4). We have used this kit to successfully isolate DNA from 100- to 200-μl volumes of joint fluid, CSF, plasma, serum, urine, and whole blood. The kit uses a modified solvent extraction method, which, in brief, includes guanidine thiocyanate lysis, partitioning of nucleic acid into an aqueous phase, and isopropanol precipitation of nucleic acid. The resulting nucleic acid is dissolved in 20 to 30 μl of RNase-free water, and 5 μl is used for PCR. *Note: Addition of 20 μg of glycogen per sample during alcohol precipitation improves the recovery of small quantities of nucleic acid.*

### PCR Amplification

1. Prepare PCR master mix by using the recipe listed above. Substitute the recipe for each primer pair used as follows: OSPA2/OSP4, no change; OSPA149/OSP319, substitute 2.5 mM $MgCl_2$; FLA1/FLA3, no change; DD02/DD06, substitute 62.5 mM each dNTP, 50 μg of isopsoralen compound 10 per ml, and 2.5 U of Ampli*Taq* polymerase. Add 5 μl of processed DNA and 1 or 2 drops of mineral oil. *Note: Perform all pipetting operations with positive-displacement pipettors. It is useful to overestimate the volume of required master mix by one or two reactions to allow for pipetting inaccuracies and other volume losses during setup.*
2. Amplify in a DNA thermal cycler by using a three-step protocol for OSPA2/OSP4 and FLA1/FLA3 targets: denature at 94°C for 45 s, anneal at 55°C for 45 s, and extend at 72°C for 1 min for a total of 45 cycles. For the OSPA149/OSP319 target, denature at 94°C for 45 s, anneal at 55°C for 45 s, and extend at 72°C for 45 s for a total of 50 cycles. For the DD02/DD06 target use a two-step protocol: denature at 94°C for 25 s and anneal at 60°C for 25 s for a total of 50 cycles, and then heat to 72°C for a 5-min final extension.
3. Remove reaction tubes from the DNA cycler. UV irradiate tubes for 15 min at 4°C in an HRI-100 UV photochemical reaction chamber. *Note: Postamplification inactivation of amplified DNA is optional. However, routine use significantly reduces the risk of laboratory contamination with amplified DNA.*
4. Prepare a 3% NuSieve–1% SeaKem agarose gel in 1× TBE buffer. Electrophorese the PCR products (5 μl of product to 2 μl of gel-loading buffer) until the bromphenol blue indicator migrates at least two-thirds the length of gel.
5. Visualize the amplification products after staining with ethidium bromide. *Note: Use of isopsoralen will increase the apparent molecular weight of amplification products 10 to 25%.*
6. Denature and neutralize gel by soaking in denaturation and neutralization buffers for 45 and 15 min, respectively. Transfer reaction products to nylon membrane by Southern blot by using 10× SSC transfer buffer.

### DNA Probe Hybridization

1. Prepare the internal oligonucleotide probe (180 ng per hybridization) by $^{32}P$ end labeling.

2. Prehybridize blots for 1 h at 55°C in 10 ml of hybridization solution. Add probe, and hybridize for 3 h at 55°C. *Note: All prehybridization, hybridization, and wash steps can be performed in glass hybridization bottles incubated in a hybridization oven.*
3. Decant the probe solution, and wash the blots twice for 10 min with 150 ml of hybridization wash 1; perform a third high-stringency wash for 30 min at 55°C with 150 ml of prewarmed hybridization wash 2. *Note: Increased stringency may be achieved by repeating wash 3 or by adding a 15-min wash with 150 ml of hybridization wash 2 containing 2% sodium* $PP_i$.
4. Expose the Saran-wrapped blot to Kodak X-Omat film in an exposure cassette overnight for up to 48 h at −70°C.

## Quality Control

### Positive Amplification and Sensitivity Control

Use DNA prepared from a boiled *B. burgdorferi* B31 (ATCC 35210) bacterial culture. Use of both high-positive and low-positive reactions provides useful quality control information. Test DNA concentrations can be determined as follows. Prepare a 10-fold dilution series of *B. burgdorferi* DNA as described above. Amplify 5-μl aliquots of each dilution. Choose two of the last DNA dilutions yielding positive amplification reactions for routine controls.

### Specificity Control

1 ng each of *Borrelia hermsii* ATCC 35209 and *Treponema pallidum* ATCC 27087.

### Negative Amplification Control

5 μl of water added directly to the PCR master mix (triplicate).

## Comments

1. In serial dilution experiments with known numbers of organisms, the genomic and plasmid targets provide roughly comparable sensitivity (Fig. 1). In synovial fluid specimens, however, our most consistent positive results are obtained by amplification of the *B. burgdorferi* plasmid target (OSPA). For unknown reasons, genomic targets are often negative in these samples. We refer to this phenomenon as "target imbalance." When using the OSPA 149/319 primer pair, the sensitivity of PCR for detection of *B. burgdorferi* in synovial fluids from patients with untreated Lyme arthritis approaches 80%.
2. A discrete nonspecific amplification product of primers OSPA2/OSP4 migrating at approximately 350 bp may be seen when excess nonborrelial DNA (i.e., human lymphocyte) is present in processed specimens. This product also hybridizes to the internal probe, OSPA3, and can be used as an internal control in many specimens. Primers OSPA149/OSPA319 and DD02/DD06 do not produce this product. Positive results can be confirmed by using multiple independent DNA target sequences including OSPA (7), flagellin (6), and rDNA (3). The choice of target sequences will determine the sensitivity and specificity of the assay and may not yield equivalent results for all specimen types.

**Figure 1.** Primers and probes for detection of *B. burgdorferi.* GenBank accession numbers for each target are M57248 for OSPA, X57404 for 16S, and X16833 for flagellin.

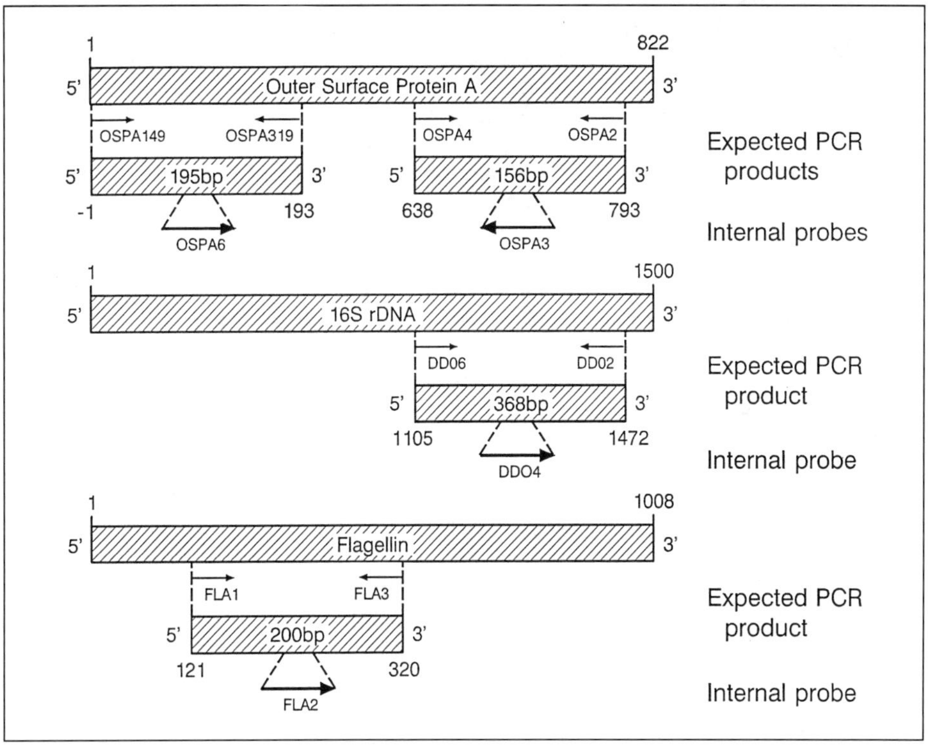


3. The routine use of isopsoralen compound 10 to inactivate amplified DNA is one of several methods now available to significantly reduce the risk of false-positive PCR results in the laboratory (chapter 5). The apparent shift in molecular weight and slight decrease in the intensity of ethidium bromide-stained bands on agarose gels are easily accommodated and are consistent from run to run. The sensitivity of probe hybridization results is not significantly compromised.
4. Consistent isolation of target DNA from a variety of clinical specimens is critical for reliable results. We have often observed inconsistent results when using small volumes (0.1 ml or less) of CSF, urine, and plasma. We suspect that this relates to the low copy number of the DNA target and/or to the stability of the target in the sample. Increasing the sample volume substantially may lead to better performance. The introduction of conveniently packaged kits for the extraction of nucleic acids will simplify processing protocols and facilitate the processing of large numbers of clinical samples, and further improvements in specimen collection techniques, storage, and processing will increase the reliability and consistency of test results.

## References

1. **Goodman, S. L., P. Jurkovich, S. M. Kramber, and R. C. Johnson.** 1991. Molecular detection of persistent *Borrelia burgdorferi* in the urine of patients with active Lyme disease. *Infect. Immun.* **59:**269–278.
2. **Innis, T. J., D. H. Gelfand, S. S. Sninsky, and T. J. White (ed.).** 1990. *PCR Protocols: A Guide to Methods and Applications*. Academic Press, Inc., San Diego, Calif.

3. **Malawista, S. E., R. T. Schoen, T. L. Moore, D. E. Dodge, T. J. White, and D. H. Persing.** 1992. Failure of multitarget detection of *Borrelia burgdorferi*-associated DNA sequences in synovial fluids of patients with juvenile rheumatoid arthritis: a cautionary note. *Arthritis Rheum.* **35:**246–247.
4. **MicroProbe Corp.** 1991. *IsoQuick Nucleic Acid Extraction Kit, Technical Insert.* MicroProbe Corp., Concord, Calif.
5. **Persing, D. H., S. W. Barthold, S. E. Malawista.** Molecular detection of *Borrelia burgdorferi. In* S. E. Schutzer (ed.), *Lyme Disease: Molecular and Immunologic Approaches.* Cold Spring Harbor Laboratory, Cold Spring Harbor, N.Y., in press.
6. **Persing, D. H., S. R. Telford III, P. N. Rys, D. E. Dodge, T. J. White, S. E. Malawista, and A. S. Spielman.** 1990. Detection of *Borrelia burgdorferi* DNA in museum specimens of *Ixodes dammini* ticks. *Science* **249:**1420–1423.
7. **Persing, D. H., S. R. Telford III, A. S. Spielman, and S. W. Barthold.** 1990. Detection of *Borrelia burgdorferi* infection in *Ixodes dammini* ticks with the polymerase chain reaction. *J. Clin. Microbiol.* **28:**566–572.
8. **Rosa, P. A., and T. G. Schwan.** 1989. A specific and sensitive assay for the Lyme disease spirochete *Borrelia burgdorferi* using the polymerase chain reaction. *J. Infect. Dis.* **160:**1018–1029.
9. **Steere, A. C.** 1989. Lyme disease. *N. Engl. J. Med.* **321:**568–596.
10. **White, T. J., R. Madej, and D. H. Persing.** 1992. The polymerase chain reaction: clinical applications. *Adv. Clin. Chem.* **29:**161–196.

# Rapid Ribosequencing for the Detection and Classification of Tick-Borne *Borrelia burgdorferi* and Related Species

**1.5**

Ulf B. Göbel, Barbara Graf, and Thomas Adam

## Background

Lyme borreliosis, the most common arthropod-borne disease in the United States, is caused by the spirochete *Borrelia burgdorferi*, an organism first isolated from *Ixodes dammini*, the deer tick. It is a multisystemic disorder involving the skin, heart, joints, and nervous system and presents with numerous symptoms and signs. Some of these, e.g., acrodermatitis chronica atrophicans (ACA), erythema migrans (EM), and meningoradiculitis, were reported in Europe as early as 1902, 1909, and 1922, respectively, whereas other manifestations, e.g., arthritis, were first described in the United States as Lyme disease in the late 1970s (12). The marked phenotypic and genotypic heterogeneity of strains from different geographic and biological origins led to an intensive search to delineate phylogenetic relationships among these isolates (1, 2, 5, 8–10). Regardless of the methods used, all isolates looked at so far have clustered into three major phylogenetic groups, represented by strains B31, 20047, and VS461. On the basis of DNA-DNA hybridization, rRNA restriction patterns, multilocus enzyme electrophoresis, the molecular weight of major outer proteins, and monoclonal antibody reactivity, Baranton et al. proposed three genospecies: *B. burgdorferi* sensu strictu (type strain B31), *B. garinii* sp. nov. (type strain 20047), and a third genospecies known as the VS461 group (4). Hence, the existence of more than one species offers a possible explanation for the differences in the clinical appearance observed in various geographic areas. One might speculate, though, that certain manifestations reflect infections with distinct species, rather than stages of a disease developing after initial infection with *B. burgdorferi* sensu strictu. Recent findings of a rather close association of group VS461 borreliae with skin lesions of Lyme borreliosis seem to point in this direction (2, 7, 13). However, a significantly larger number of isolates and clinical samples must be analyzed before these observations can be substantiated. We therefore developed a method which allows not only for the rapid phylogenetic classification of Lyme disease-associated borreliae but also for the sensitive detection of these organisms in clinical samples. Our approach is based on the analysis of 16S rRNA signature sequences representing the three genospecies. Appropriate fragments are amplified after reverse transcription of bulk RNA by using universal (eubacterial) or borrelia-specific primers. The amplification products are identified by subsequent hybridization with specific probes or by direct sequencing.

## Materials and Reagents

RNAzol B (Cinna/Biotech, Houston, Tex.)

Pestles (Walter Ebel, Hamburg, Germany)

Mini-Bead Beater (BioSpec Products Inc.)

Zirconium beads 0.1 and 0.5 mm in diameter (BioSpec Products Inc.)

RNase-free DNase I (Promega)

Avian myeloblastosis virus reverse transcriptase (Stratagene)

10× RT buffer: 500 mM KCl, 100 mM Tris-HCl (pH 8.3), 15 mM $MgCl_2$, 0.01% (wt/vol) gelatin

PCR master mix: 50 mM KCl, 10 mM Tris-HCl (pH 8.3), 0.01% (wt/vol) gelatin, dNTPs (200 μM each), 2.5 U of *Taq* DNA polymerase (Cetus), 60 pmol of TP primer per 80 μl of master mix

5× TBE buffer: 450 mM Tris base, 450 mM boric acid, 10 mM EDTA (pH 8.0)

Denaturation solution (Southern blots): 1.5 M NaCl, 0.5 M NaOH

Neutralization solution (Southern blots): 1.5 M NaCl, 1 M Tris-HCl (pH 8.0)

Uncharged nylon membranes (Biodyne A; pore size, 0.2 μm; Pall)

Nonisotopic-labeling kit (ECL; Amersham)

ECL hybridization buffer: 5× SSC (1× SSC is 0.15 M NaCl plus 0.015 M sodium citrate), 0.5% (wt/vol) blocking reagent (ECL kit), 0.1% (wt/vol) *N*-lauryl-sarcosine sodium salt, 0.02% (wt/vol) sodium dodecyl sulfate (SDS)

Electroelution (Elutrap; Schleicher & Schuell)

5× forward reaction buffer: 50 mM $MgCl_2$, 75 mM mercaptoethanol, 300 mM Tris-HCl (pH 7.8)

[$\gamma$-$^{32}$P]ATP (3,000 Ci/mmol; Amersham)

Herring sperm DNA (Boehringer Mannheim)

Prehybridization solution: 5× SSC, 0.1% (wt/vol) bovine serum albumin (BSA), 0.1% (wt/vol) polyvinylpyrrolidone, 0.1% (wt/vol) Ficoll, 0.2% (wt/vol) SDS, 1 mg of heat-denatured herring sperm DNA per ml

Hybridization solution: 5× SSC, 0.1% BSA, 0.1% (wt/vol) polyvinylpyrrolidone, 0.1% (wt/vol) Ficoll, 0.2% (wt/vol) SDS, 250 μg of yeast tRNA per ml

Binding buffer: 10 mM Tris-HCl (pH 7.5), 100 mM NaCl, 1 mM EDTA, 0.15% (vol/vol) Triton X-100

20× SSPE: 3 M NaCl, 0.2 M $NaH_2PO_4$, 0.02 M EDTA (pH 7.4)

Streptavidin-coated Dynabeads M-280 (Dynal)

Magnetic particle concentrator for microtubes of the Eppendorf type (Dynal)

Fluorescence chromatography plate (silica gel plates F 245, 0.2 mm; no. 5738, Merck)

Denaturation solution for single-strand sequencing: 0.1 M NaOH, 0.4 mM EDTA

Sequenase kit (United States Biochemical Corp.)

[$\alpha$-$^{35}$S]dATP (1,000 Ci/mmol; Amersham)

## Methods

### Collection of Clinical Samples

1. For EM lesions, biopsy specimens should be taken from the seemingly healthy sites just outside of the erythema, whereas for ACA lesions the specimens should be collected from within the lesion.

2. Shock-freeze samples in liquid nitrogen and store them at −80°C if not analyzed immediately. Ship on dry ice if samples are to be processed in a different laboratory. *Note: There are no systematic studies on the collection, transport, and processing of biopsy specimens from skin lesions. Our recommendations are based on our personal experience. So far we have been analyzing samples from skin lesions exclusively, but the same transport and processing procedures should apply for other biopsy specimens as well.*

### RNA Extraction

1. Biopsy specimens are disintegrated effectively by macerating the tissue with a pestle in an Eppendorf tube containing a few zirconium beads in the presence of RNAzol. Alternatively, disrupt the tissue by placing the sample in a tightly sealed (screw-cap) 2.0-ml microcentrifuge tube filled with RNAzol and a total of 600 ml of zirconium beads (0.1 and 0.5 mm in diameter, 300 ml each) on a Mini-Bead Beater and shaking it at room temperature for 5 min. *Note: RNAzol is used to extract bulk RNA from both clinical samples and culture isolates as recommended by the manufacturer. This solution contains guanidinium thiocyanate, a chaotropic agent which effectively inactivates RNases. Use a sufficient amount of RNAzol (a minimum of 2 ml of RNAzol is required for 100 mg of tissue or $10^7$ cells).* Culture-grown borreliae can be lysed in RNAzol without mechanical disruption.
2. Precipitate bulk RNA with ethanol at −20°C overnight. *Note: Wash the precipitate twice in 75% ethanol DEPC–$H_2O$ to remove substances which could adversely affect the reverse transcription.* Resuspend RNA in 10 μl of DEPC-treated $H_2O$.

### DNase I Treatment

1. Use 2 μl of your RNA preparation plus DEPC-$H_2O$, RNase-free DNase (1 U), and 10× RT buffer to make a final reaction volume of 10 μl of 1× RT. Incubate at 37°C for 10 min. *Note: RNase-free DNase I can be purchased from commercial suppliers (e.g., Promega) or prepared by following a protocol described in manuals of molecular biological techniques, such as Sambrook et al. (11) or Berger and Kimmel (5). It is crucial to check the DNase I and the RT buffer for absence of RNase activity prior to use!*
2. Inactivate the enzyme by heating at 95°C for 5 min, spin down any condensate, and place the tube on ice.

### Reverse Transcription

1. Add 5 U of avian myeloblastosis virus reverse transcriptase (Stratagene), 2 μl of the respective RT primer (30 μM), 1 μl of 10× RT buffer, and 4 nmol each dATP, dCTP, dGTP, and dTTP (final reaction volume, 20 μl), and incubate at 40°C for 30 min.
2. Place the tube in a heating block at 95°C for 5 min to inactivate the enzyme and to denature the DNA-RNA heteroduplexes. Place on ice immediately.

### Amplification and Analysis of Amplification Products

1. Spin down the condensate in a microcentrifuge at maximum speed for 15 s. Bring the volume up to 100 μl by adding 80 μl of the PCR master mix. *Note: Positive-displacement pipettors or plugged tips should be used throughout the procedure. It is advisable to work under a fume hood.*
2. Amplify the target cDNA by using a three-step profile (93, 50, and 72°C for 1 min each) in a programmable thermocycler for a total of 30 cycles.

## Electrophoresis

1. Prepare a 1% agarose gel in 1× TBE buffer. Add 10 μl of each reaction per well, and electrophorese until the bromophenol blue indicator has migrated about two-thirds of the length of the gel. *Note: Add appropriately sized markers, e.g., a 123-bp ladder, in a separate well to estimate the length of your amplification product.*
2. Stain with ethidium bromide to visualize the amplification products.
3. Soak gel in denaturation and neutralization solutions with gentle agitation for 20 min each.
4. Transfer the reaction products onto a nylon membrane as described previously (11).
5. Immobilize the DNA by baking the filters at 80°C for 2 h. *Note: Alternatively you can use a UV crosslinker (Stratagene).*

## Hybridization

### *Chemiluminescent Detection*

1. Label the oligonucleotide BBU30 (containing a hexylthiol linker at the 5′ end) with horseradish peroxidase by using the ECL system.
2. Purify the labeled oligonucleotide by electrophoretic separation on a preparative 20% polyacrylamide gel. Identify the labeled product by UV shadowing on fluorescent chromatography plates, and electroelute it with an Elutrap electroelutor or comparable device. *Note: Labeling of the oligonucleotide results in a reduction of electrophoretic mobility and thus in an increase of the apparent length.*
3. Determine the concentration by measuring the $A_{260}$ in a spectrophotometer.
4. Prehybridize filters in ECL hybridization buffer for 1 h at 42°C. Add labeled and purified oligonucleotide (15 ng/ml) and hybridize for 2 h at 42°C.
5. Wash the membranes in 3× SSC containing 0.1% SDS for 15 min at 42°C, discard the buffer, and wash them for 5 min at the same temperature in 3× SSC with 0.1% SDS and 3.4 M urea. For a final wash, transfer the membranes into 2× SSC at room temperature for 5 min.
6. Immerse the membranes in the signal generation solutions, wrap with plastic wrap, and expose to Kodak X-Omat film for 3 min to 3 h.

### *Detection with Radiolabeled Probe*

1. Label the oligonucleotide probe BBU30 (10 pmol) at the 5′-end by adding 10 U of T4 polynucleotide kinase and 50 μCi (10 μmol) of [$\gamma$-$^{32}$P]ATP in the forward reaction buffer (final volume, 20 μl) and incubating at 37°C for 1 h. To inactivate the enzyme, heat the tube at 68°C for 10 min.
2. Purify the labeled oligonucleotide by electrophoresis on a 20% polyacrylamide gel. Localize the product by autoradiography (10 min), cut it out, and electroelute (Elutrap).
3. Prehybridize membranes in prehybridization solution at 60°C for 60 min. Remove the prehybridization solution, and add hybridization solution with 1.5 $\times 10^6$ cpm of the radiolabeled probe. Hybridize at 60°C for 2 h with occasional agitation. Wash three times (300 ml for 2 min each) in 5× SSC with 0.5% (wt/vol) SDS at 60°C. Place the filter on a paper towel to remove excess buffer.
4. Cover with plastic wrap, and expose on Kodak X-Omat film at −70°C for 12 to 48 h.

### Direct Single-Strand Sequencing

1. Wash streptavidin-coated magnetic beads (40 μl) twice in binding buffer, and resuspend the beads in 40 μl of binding buffer. *Note: The magnetic particle concentrator (Dynal) was used throughout all washing steps during sequencing to pellet the particles.*
2. Take 40 μl of the PCR product, incubate at 65°C for 10 min, and add 40 μl of washed beads together with 21 μl of 20× SSPE buffer.
3. Incubate for 30 min at 37°C by gently resuspending sedimented beads every 5 min.
4. Wash three times in prewarmed binding buffer at 37°C.
5. Denature DNA by adding 40 μl of freshly prepared denaturation solution. Mix well, and keep at room temperature for 5 min.
6. Carefully take off the supernatant fluid containing the nonbiotinylated strand of the amplification product.
7. Wash beads three times with 100 μl of binding buffer. Take care to completely remove excess fluid with a cloth. The solid-phase bound biotinylated strand of the PCR product is now ready for sequencing by using the Sequenase 2.0 kit.
8. Add 2 μl of annealing buffer (provided with the kit) and 1 μl of sequencing primer (30 μM). Incubate for 2 min at 65°C. Cool to room temperature over 30 min by resuspending sedimented beads every 5 min. All reactions are done as described by the manufacturer with [$\alpha$-$^{35}$S]dATP as labeling substrate.
9. Heat tubes at 80°C for 5 min, and apply 3 μl from each tube onto an 8% polyacrylamide gel.
10. Dry the gel, and expose to Kodak X-Omat for 12 to 36 h.

## Quality Control

### Positive Amplification Control

1 pg and 100 pg of *B. burgdorferi* DNA in 20 μl of 1× RT buffer.

### Negative Amplification Control

14 μl of sterile distilled water in 1× RT buffer.

### Positive System Control

Bulk RNA prepared from 50 to 1,000 borreliae (as determined by dark-field microscopy) from a liquid culture of *B. burgdorferi* B31.

### Negative System Control

RNA preparation, reverse transcription, and amplification by processing 200 μl of sterile distilled water instead of biological material.

### Negative Internal Control

PCR amplification of part of the bulk RNA without prior reverse transcription. *Note: This provides an excellent control to detect contamination by external DNA and carryover of amplification products.*

### Specificity of Amplification Primers

10 ng of bulk RNA prepared from inflamed periodontal pocket material, containing more than 50% spirochetes per high-power field as determined by dark-field or phase contrast microscopy.

**Table 1.** Primer/probe sequences for amplification/identification of *B. burgdorferi* 16S rRNA

| Function | Name | Nucleotide sequence, 5′ to 3′ | Location[a] |
|---|---|---|---|
| Universal primer | TPU1 | AGA GTT TGA TC[C,A] TGG CTC AG | 9–28 |
| Universal primer | RTU3 | G[T,A]A TTA CCG CGG C[G,T] CTG | 510–527 |
| Specific primer | TP6 | TAC CAC AGC TCA ACT GTG GA[A,C,G] CTA TG | 600–625 |
| Specific primer | RT1 | TCG GTA CTA ACT TTT AGT TAA CA | 820–842 |
| Probe | BBU30 | AAC TTC CTC TAT CAG ACT CTA GAC ATA TAG | 633–662 |

[a] Numbers correspond to appropriate positions of the published *B. burgdorferi* B31 sequence (GenBank accession number X57404.

## Comments

1. Utmost care should be taken to avoid any RNase contamination of buffers and enzymes. Use DEPC-treated distilled water only. A more detailed discussion of this issue is given in reference 5.
2. We use stock solutions of gel-purified oligonucleotides adjusted to 30 μM with DEPC-treated water and stored at −20°C. Forward and reverse primers are designated TP (*Taq* polymerase) and RT (reverse transcriptase), respectively. RT primers are used for reverse transcription and amplification. TP primers are added to the PCR master mix. The sequences are given in Table 1. *Borrelia*-specific oligonucleotides TP6 and RT1 amplify a 244-bp fragment of the variable region V4 of *Borrelia* 16S rRNA. We recommend using biotinylated TP6 for amplification and immobilization of the amplified product. RT1 should be used for reverse transcription and sequencing. The amplified fragment contains the target sequence for the internal oligonucleotide probe BBU30, which recognizes all genospecies associated with Lyme borreliosis but none of the relapsing-fever borreliae. By direct sequencing of the 244-bp fragment, signatures of all three genospecies can be identified (Table 2). It should be noted that a few *B. garinii* strains are not differentiated by sequencing V4. However, this should be of minor importance for clinical purposes, as these strains have so far been isolated from ticks exclusively. If this distinction is required, it is necessary to amplify and sequence variable regions V1 and V2 by using TPU1 and RTU3 as primers with RTU3 as the biotinylated primer and TPU1 for sequencing. Since these primers are not genus specific, they are susceptible to contamination with DNA and RNA from other bacteria. Thus, these primers are recommended for the identification of pure cultures only. For this latter purpose, it may be more convenient to prepare DNA from pure cultures and subsequently amplify and sequence regions V1 and V2 of the 16S rRNA gene (3).
3. PCR products obtained from clinical samples should be identified by hybridization to the internal oligonucleotide BBU30 or by direct sequencing. The ECL system is about 1 order of magnitude less sensitive than the radiolabeled probe. Visible bands in the ethidium bromide-stained gels can be used immediately for direct sequencing. If a positive hybridization signal is obtained, corresponding to a weak or invisible band in the ethidium bromide-stained agarose gel, the amplification product can be purified by using Strataclean or phenol-chloroform, and part of the purified DNA (6 μl) or dilutions thereof can be reamplified by adding 90 μl of PCR master mix (without primer!) and TP and RT primers (60 pmol each).
4. Because of the natural amplification of rRNA (about $10^3$ to $10^4$ copies per

**Table 2.** 16S rRNA V4 signature sequences

| Species | Nucleotide sequence, 5′ to 3′ |
|---|---|
| *B. garinii* | GTG GAC CTA TGT TGG AAA CTA TAT GTC TAG |
| *B. burgdorferi* | GTG GAG CTA TGT TGG AAA CTA TAT GTC TAG |
| Group VS461 | GTG GAA CTA TGT TGG AAA CTA TAT ATC TAG |

bacterium), high sensitivity can be achieved after reverse transcription, even if the peroxidase-labeled probe is used to identify the amplification product (2).

5. With primers TP6 and RT1, some nonspecific amplification may occur in samples containing *B. hermsii* or *B. parkeri*. However, these products will not hybridize to the internal probe BBU30 (2).

## References


1. **Adam, T., G. S. Gassmann, C. Rasiah, and U. B. Göbel.** 1991. Phenotypic and genotypic analysis of *Borrelia burgdorferi* isolates from various sources. *Infect. Immun.* **59:**2579–2589.
2. **Adam, T., B. Graf, U. Neubert, and U. B. Göbel.** 1992. Detection and classification of *Borrelia burgdorferi* by direct sequencing of 16S rRNA amplified after reverse transcription. *Med. Microbiol. Lett.* **1:**120–126.
3. **Adam, T., U. Neubert, and U. B. Göbel.** Direct sequencing of amplified 16S rRNA genes identifies a distinct *Borrelia burgdorferi* genotype in skin manifestations of Lyme borreliosis. Submitted for publication.
4. **Baranton, G., D. Postic, I. Saint-Girons, P. Boerlin, J. C. Pifaretti, M. Assous, and P. A. D. Grimont.** Delineation of *Borrelia burgdorferi* sensu strictu, *Borrelia garinii* sp. nov., and group VS461 associated with Lyme borreliosis. *Int. J. Syst. Bacteriol.* **42:**378–383.
5. **Berger, S. L., and A. R. Kimmel.** 1987. Guide to molecular cloning techniques. *Methods Enzymol.* **152:**1–811.
6. **Boerlin, P., O. Peter, A. G. Bretz, D. Postic, G. Baranton, and J. C. Piffaretti.** 1992. Population genetic analysis of *Borrelia burgdorferi* isolates by multilocus enzyme electrophoresis. *Infect. Immun.* **60:**1677–1683.
7. **Graf, B., and U. B. Göbel.** Unpublished observations.
8. **Marconi, R. T., and C. F. Garon.** 1992. Phylogenetic analysis of the genus *Borrelia*: a comparison of North American and European isolates of *Borrelia burgdorferi. J. Bacteriol.* **174:**241–244.
9. **Marconi, R. T., and C. F. Garon.** 1992. Identification of a third genomic group of *Borrelia burgdorferi* through signature nucleotide analysis and 16S rRNA sequence determination. *J. Gen. Microbiol.* **138:**533–536.
10. **Paster, B. J., F. E. Dewhirst, W. G. Weisburg, L. A. Tordoff, G. J. Fraser, R. B. Hespell, T. B. Stanton, L. Zablen, L. Mandelco, and C. R. Woese.** 1991. Phylogenetic analysis of spirochetes. *J. Bacteriol.* **173:**6101–6109.
11. **Sambrook, J., E. F. Fritsch, and T. Maniatis.** 1989. *Molecular Cloning: A Laboratory Manual*, 2nd ed. Cold Spring Harbor Laboratory, Cold Spring Harbor, N.Y.
12. **Steere, A. C.** 1989. Lyme disease. *N. Engl. J. Med.* **321:**586–596.
13. **Wilske, B., V. Preac-Mursic, U. B. Göbel, B. Graf, S. Janris, E. Soutschek, E. Schwab, and G. Zumstein.** 1993. An OspA serotyping system for *Borrelia burgdorferi* based on reactivity with monoclonal antibodies and OspA sequence analysis. *J. Clin. Microbiol.* **31:**340–350.

# PCR Detection of *Yersinia pestis* in Fleas

# 1.6

J. Hinnebusch and Tom G. Schwan

## Background

*Yersinia pestis*, the causative agent of plague, is widely distributed throughout the world in endemic foci involving a diversity of mammalian hosts and flea vectors (5, 6). The plague bacillus was discovered in 1894 by Yersin (11) and Kitasato (3), who were working independently to identify the cause of a plague epidemic killing thousands of people in Hong Kong and neighboring China. In 1897 while working in Formosa, Ogata determined that rat fleas were involved in transmitting the bacterium (6), although another 6 years lapsed before this species of flea, now called *Xenopsylla cheopis*, was described and named by N. C. Rothschild on the basis of specimens collected from the Sudan. For several years following Ogata's publication and another supporting study by Simond (6), the scientific community was slow to accept the importance of fleas as vectors of bubonic plague. When the first epidemic in North America began in San Francisco's Chinatown in 1900, no attention was given at all to fleas associated with the domestic rats also dying of plague. By 1905, however, epidemiologists and other public health workers finally realized the risk of acquiring plague from rat fleas, and monitoring and killing fleas of urban rats became an integral part of plague control operations. By 1910, pioneering work on plague in California's ground squirrels and their fleas also demonstrated that controlling fleas in wild habitats was another important aspect of reducing the number of humans succumbing to plague.

In the western United States, most human cases of plague result from the bite of infected fleas associated with wild rodents such as ground squirrels, chipmunks, and prairie dogs (5). For this reason, laboratory testing to detect infection with *Y. pestis* in such fleas collected from natural habitats has been a primary component of county, state, and federal health programs responsible for surveillance and control of plague. The laboratory test currently used to detect *Y. pestis* in fleas is the same one used by Yersin and Kitasato 100 years ago in Hong Kong when testing tissues from human corpses: experimental animal inoculations, subsequent reisolations from the sacrificed animal tissues, and identification by standard bacteriological procedures. This method is expensive, relatively insensitive when testing individual fleas, philosophically objectionable for some workers, and slow. Often decisions about controlling fleas in natural habitats and closing recreational areas to the public must be made faster than results can be obtained from experimental animal inoculations in the laboratory. A few years ago, a DNA hybridization probe that detected *Y. pestis* in individuals of three species of fleas having fed on experimentally infected laboratory mice was developed (9). The DNA probe method was relatively insensitive, however, and was unable to detect less than $10^5$ to $10^6$ bacteria. Recently we have developed a sensitive and specific PCR assay that is able to detect as few as 10 *Y. pestis* cells in experimentally infected oriental rat fleas, *X. cheopis* (2). This assay should greatly facilitate the testing of fleas to determine the prevalence and distribution of *Y. pestis* in wild-animal populations to better assess the potential risk of transmission to humans.

## Materials and Reagents

Brain heart infusion (BHI) broth (Difco, Detroit, Mich.)

PCR dilution buffer: 10 mM Tris (pH 8.3), 100 mM KCl

PCR "bottom-layer" master mix (2×): 10 mM Tris (pH 8.3), 3 mM $MgCl_2$, 0.6 μM each Yp1 and Yp2 (see Table 1), 400 μM each deoxyribonucleoside triphosphates

Ampli*Taq* DNA polymerase (Perkin-Elmer Corp., Norwalk, Conn.)

AmpliWax PCR Gem paraffin beads (Perkin-Elmer)

Southern blot wash solution 1: 2× SSC (1× SSC is 0.15 M NaCl plus 0.015 M sodium citrate), 0.1% sodium dodecyl sulfate (SDS), 1 mM EDTA

Southern blot wash solution 2: 0.1× SSC, 0.1% SDS, 1 mM EDTA

## Methods

### Sample Collection

Collect fleas from anesthetized wild rodents by brushing the fleas from the hair into a white enamel pan, or collect fleas from rodent nest material. Immerse the fleas in 70% ethanol and keep them at room temperature, or immerse them in sterile saline and keep them at 4°C until testing. Alternatively, put them into dry tubes, freeze them, and keep them at −20°C until testing.

### Sample Preparation

1. Place each flea to be analyzed into an individual sterile 0.5-ml microcentrifuge tube. Allow ethanol-stored fleas to air dry.
2. With a micropipettor, take up 50 μl of BHI broth into a sterile disposable pipette tip. Before expelling the BHI broth into the microcentrifuge tube, macerate the flea with the end of the pipette tip. Repeat for each sample. *Note: We have pooled fleas in one tube, but the concentration of 50 μl of BHI broth per flea should be maintained (e.g., use 250 μl of BHI for a pool of five fleas).*
3. Place the tubes containing the triturated fleas in BHI broth into a boiling water bath for 10 min.
4. Remove the tubes from the boiling water bath, and place on ice. Centrifuge the tubes for 5 to 10 s at maximum speed in a microcentrifuge, and keep them on ice before proceeding with the PCR.

### PCR Amplification

1. Label a separate 0.5-ml PCR tube for each sample, and dispense 50 μl of 2× PCR "bottom-layer" master mix into each tube.

2. Add one paraffin bead (AmpliWax PCR Gem) to each tube, close the tubes, and heat them in an 80°C heating block for 5 min.
3. Remove the tubes from the heating block, and leave them at room temperature for 5 min or longer to allow the wax barrier to solidify.
4. Prepare the template solution. Using a sterile 0.5-ml microcentrifuge tube for each sample, mix 2 μl of the flea-BHI broth preparation and 43 μl of PCR dilution buffer (total volume, 45 μl).
5. Prepare the enzyme solution. Mix 0.5 μl of Ampli*Taq* DNA polymerase and 4.5 μl of cold PCR dilution buffer per reaction. Prepare just before use, and hold on ice. It is advisable to prepare a slight excess of enzyme solution to compensate for losses during pipetting; e.g., for nine samples, mix 5 μl of DNA polymerase and 45 μl of PCR dilution buffer.
6. After the wax barrier has formed in the PCR tubes, add 5 μl of enzyme solution to each tube.
7. Lastly, add the 45 μl of sample template to the proper tube (to make a total reaction volume of 100 μl). Close the lids tightly, and perform PCR with a programmable thermal cycler.
8. The PCR conditions are as follows: for the first two cycles, denaturation at 95°C for 2 min, annealing at 51°C for 1 min, and extension at 72°C for 1 min; for the next 38 cycles, denaturation at 94°C for 1 min, annealing at 51°C for 1 min, and extension at 72°C for 1 min; for the final incubation, 72°C for 10 min.

## Sample Analysis

1. Prepare a 2% NuSieve agarose gel in TBE buffer. Add 0.1 μg of ethidium bromide per ml to the molten agarose before pouring the gel.
2. Transfer 25 μl of each PCR to a clean microcentrifuge tube (the wax seal in the PCR tube is easily pierced by a pipette tip), and add 5 μl of 6× electrophoresis loading buffer. Prepare a separate tube containing DNA size standards.
3. Load the samples into the wells of the agarose gel, and separate the amplification products by electrophoresis in TBE buffer containing 0.1 μg of ethidium bromide per ml.
4. When the bromophenol blue dye is approximately three-fourths of the way down the gel, stop the electrophoresis and place the gel on a UV transilluminator to visualize the ethidium bromide-stained DNA fragments. Photograph the gel. The *Y. pestis*-specific amplification product is a 478-bp fragment.
5. If desired, perform a Southern blot of the gel by standard methods (7). Hybridize the Southern blot membrane with a labeled DNA probe. A 269-bp DNA fragment can be generated with the primers Yp2 and Yp3 (Fig. 1) in a PCR with *Y. pestis* DNA as the template. The resulting amplification product is then gel purified and labeled for use as a specific probe. An oligonucleotide whose binding site lies within the specific PCR-amplified fragment, such as Yp3 (Table 1), may also be labeled for use as an internal probe.
6. Following hybridization with the specific probe, wash the Southern blot membrane at high stringency. For example, when using the radiolabeled fragment amplified from *Y. pestis* DNA by primers Yp2 and Yp3, wash the membrane once at room temperature for 15 min in wash solution 1 and then twice at 65°C for 15 min in wash solution 2. Rinse the membrane at room

**Figure 1.** The 9.5-kb pesticin plasmid of *Y. pestis* and location of the binding sites for PCR primers Yp1 and Yp2 and internal primer Yp3. The *Bam*HI-*Hind*III fragment indicated has been shown to hybridize specifically to *Y. pestis* DNA (4). Numbers refer to nucleotide positions in the plasminogen activator/coagulase gene (*pla*) (8).

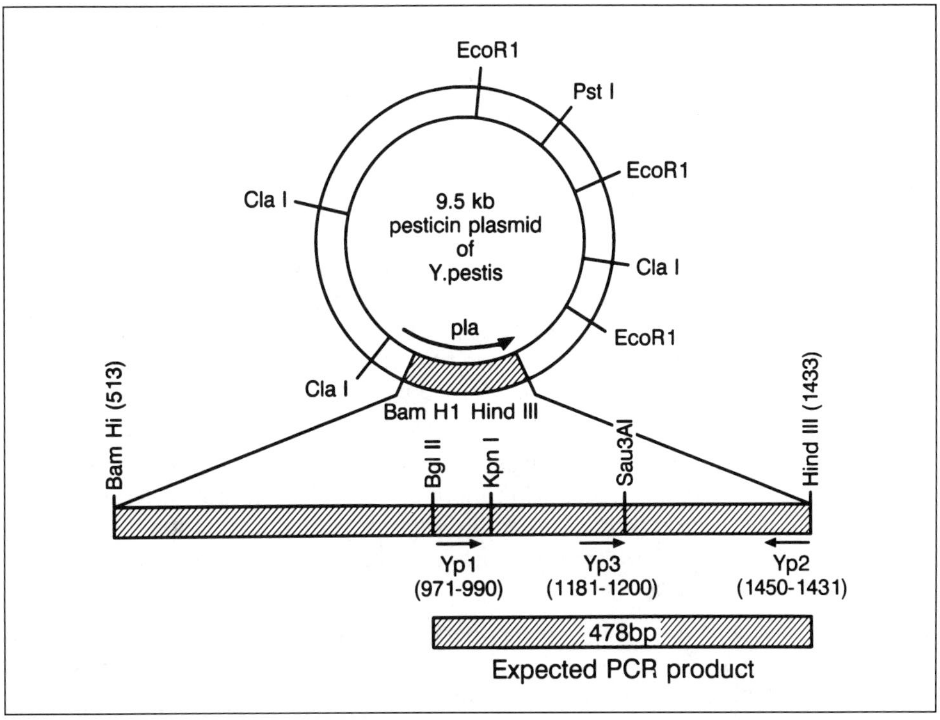


temperature in 0.1× SSC, seal it in plastic wrap, and expose it to autoradiographic film.

# Quality Control

## Positive Amplification Control

10 pg of *Y. pestis* DNA (ATCC 23053).

## Negative Amplification Control

45 μl of PCR dilution buffer added as template.

## Positive Flea Control

One uninfected *X. cheopis* flea (or other species) triturated in 50 μl of BHI broth and 100 *Y. pestis* cells.

**Table 1.** Oligonucleotides for detection of *Y. pestis* in fleas

| Function | Name | Nucleotide sequence, 5′ to 3′ |
|---|---|---|
| Primer | Yp1 | ATC TTA CTT TCC GTG AGA AG |
| Primer | Yp2 | CTT GGA TGT TGA GCT TCC TA |
| Primer | Yp3 | ATA CTG TGA CGG CGG GTC TG |

### Negative Flea Control

One uninfected *X. cheopis* flea (or other species) triturated in 50 μl of BHI broth.

### Specificity Controls

100 ng of DNA from *Y. pseudotuberculosis* (ATCC 29833); 100 ng of DNA from *Rickettsia typhi* (ATCC VR 738).

## Comments

1. Fleas infected with *Y. pestis* are indicated by the presence of a 478-bp amplified fragment. The PCR protocol utilizes the "hot-start" technique of Chou et al. (1) and is designed to optimize sensitivity. Fragments smaller than 478 bp, which stain weakly with ethidium bromide, may also be seen. These secondary fragments most likely result from the accumulation of single-stranded DNA amplification products.
2. Preparation of the flea samples was based on a procedure described by Webb et al. (10). This simple preparation method resulted in a more sensitive PCR assay than did more elaborate procedures that involved incubation of triturated fleas in cell lysis buffers, with or without subsequent phenol and chloroform extraction.
3. *Y. pestis* contains a 9.5-kb virulence-associated plasmid that is not present in other *Yersinia* species (4, 8) (Table 1) and contains the target sequence for this assay. This plasmid encodes pesticin (a bacteriocin) and the plasminogen activator/coagulase gene of *Y. pestis* (8). The PCR primers Yp1 and Yp2 (Fig. 1) amplify nucleotides 971 to 1450 of the plasminogen activator/coagulase gene and were selected from the published sequence of Sodeinde and Goguen (8) from a region shown previously by DNA hybridization to be specific for *Y. pestis* (4).
4. In trials with 10 to 50 or more *Y. pestis* cells mixed with negative *X. cheopis* fleas, the 478-bp amplification product was visible on the ethidium bromide-stained agarose gel. With the specific probe, 1 to 10 *Y. pestis* cells per flea were detectable (2).
5. We prepare the 269-bp DNA probe by random-primer labeling, with [$\alpha$-$^{32}$P]dATP and a commercial kit (Boehringer Mannheim, Indianapolis, Ind.). Other labeling methods, including nonisotopic labeling, should also be effective.

## References


1. **Chou, Q., M. Russel, D. E. Birch, J. Raymond, and W. Bloch.** 1992. Prevention of pre-PCR mis-priming and primer dimerization improves low-copy-number amplifications. *Nucleic Acids Res.* **20:**1717–1723.
2. **Hinnebusch, J., and T. G. Schwan.** 1993. A new method for plague surveillance using the polymerase chain reaction to detect *Yersinia pestis* in fleas. *J. Clin. Microbiol.* **31,** in press.
3. **Kitasato, S.** 1894. The bacillus of bubonic plague. *Lancet* **ii:**428–430.
4. **McDonough, K. A., T. G. Schwan, R. E. Thomas, and S. F. Falkow.** 1988. Identification of a *Yersinia pestis*-specific DNA probe with potential for use in plague surveillance. *J. Clin. Microbiol.* **26:**2515–2519.

5. **Poland, J. D., and A. M. Barnes.** 1979. Plague, p. 515–558, *In* J. H. Steele (ed.), *CRC Handbook Series in Zoonoses, Section A: Bacterial, Rickettsial, and Mycotic Diseases*, vol. 1. CRC Press, Inc., Boca Raton, Fla.
6. **Pollitzer, R.** 1954. Plague. *W.H.O. Monogr. Ser.* **22:**1–698.
7. **Sambrook, J., E. F. Fritsch, and T. Maniatis.** 1989. *Molecular Cloning: A Laboratory Manual*, 2nd ed. Cold Spring Harbor Laboratory, Cold Spring Harbor, N.Y.
8. **Sodeinde, O. A., and J. D. Goguen.** 1989. Nucleotide sequence of the plasminogen activator gene of *Yersinia pestis*: relationship to *ompT* of *Escherichia coli* and gene *E* of *Salmonella typhimurium. Infect. Immun.* **57:**1517–1523.
9. **Thomas, R. E., K. A. McDonough, and T. G. Schwan.** 1989. Use of DNA hybridization probes for detection of the plague bacillus (*Yersinia pestis*) in fleas (Siphonaptera: Pulicidae and Ceratophyllidae). *J. Med. Entomol.* **26:**342–348.
10. **Webb, L., M. Carl, D. C. Malloy, G. A. Dasch, and A. F. Azad.** 1990. Detection of murine typhus infection in fleas by using the polymerase chain reaction. *J. Clin. Microbiol.* **28:**530–534.
11. **Yersin, A.** 1894. La peste bubonique a Hong-Kong. *Ann. Inst. Pasteur* (Paris) **8:**662–667.

# PCR Detection of *Treponema pallidum*

1.7

Justin D. Radolf

## Background

The genus *Treponema* within the family *Spirochaetaceae* comprises one species of animal pathogen (*Treponema paraluiscuniculi*), two species pathogenic for humans (*T. pallidum* and *T. carateum*), and a number of nonpathogenic species. In humans, *T. pallidum* subspp. *pallidum*, *pertenue*, and *endemicum* cause venereal syphilis, yaws, and endemic syphilis, respectively, and *T. carateum* causes pinta; these four pathogens are virtually indistinguishable on the basis of morphological, antigenic, and genetic criteria (6). Venereal syphilis is the most important of the human treponematoses and, despite the availability of highly effective antimicrobial therapy, remains a major global public health problem. In fact, epidemic increases in venereal syphilis were reported in the United States throughout much of the 1980s (8), and genital ulcers caused by syphilis also have been identified as cofactors for the transmission of the human immunodeficiency virus (11).

Diagnosis of venereal syphilis is complicated because *T. pallidum* subsp. *pallidum* is one of the few major pathogens of humans that cannot be propagated on artificial media. Current methods for the direct detection of the syphilis spirochete in clinical specimens are either insensitive, as in the case of dark-field microscopy, or impractical, as with rabbit intratesticular inoculation (rabbit infectivity testing [RIT]) (5, 10). Although serological tests for syphilis have become the mainstays of laboratory diagnosis, they lack sensitivity, specificity, or both for many forms of the disease (e.g., neurosyphilis, congenital syphilis) (5). To circumvent these problems, we used the PCR to develop a highly sensitive and specific assay for detection of *T. pallidum* based on amplification of a portion of the gene encoding the *T. pallidum*-specific 47-kDa major membrane immunogen (Fig. 1; Table 1) (1). More recently, we used RIT, the current "gold standard" for detection of the spirochete in clinical materials, to validate our PCR results (2, 9). Our studies indicate that detection of *T. pallidum* by PCR promises to become a significant addition to the diagnostic armamentarium for syphilis.

## Materials and Reagents

Superspeed centrifuge

10- and 25-ml serological pipettes (individually wrapped)

1-ml syringes

**Figure 1.** PCR primers and probe for detection of *T. pallidum* by using the 47-kDa gene target. GenBank accession no. M88769 and M27493.

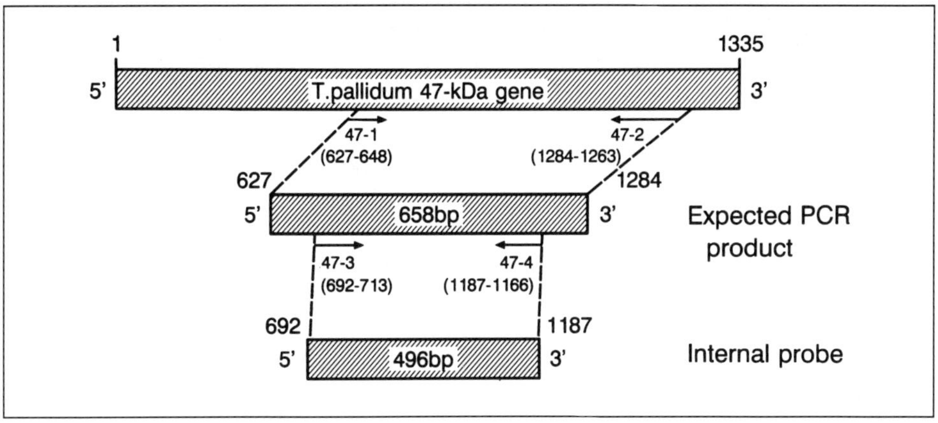


Sterile water (Baxter-Travenol)

Chelex-100 (Bio-Rad)

For alkaline lysis: 5 M NaCl, 2 M NaOH, 20% sodium dodecyl sulfate (SDS), phenol, isopropanol, chloroform-isoamyl alcohol (24:1)

Sterile TE buffer: 10 mM Tris (pH 7.4), 1 mM EDTA

Neutralization buffer (for alkaline lysis): 0.5 M Tris HCl (pH 7.0)

10× PCR buffer: 100 mM Tris (pH 8.3), 500 mM KCl, 30 mM $MgCl_2$, 1 mg of gelatin per ml

1.25 mM deoxynucleoside triphosphates (dNTPs) (Pharmacia) in water

47-1 and 47-2 primers (0.5 μg/μl in water)

Bg-1 and Bg-2 primers (0.25 μg/μl in water)

1 μg of human placental DNA (Sigma) per μl

PCR master mix: 11 μl of sterile water, 10 μl of 10× PCR buffer, 10 μl of dNTPs, 1 μl of primer 47-1, 1 μl of primer 47-2, 1 μl of primer Bg-1, 1 μl of primer Bg-2, 1 μl of *Taq* (2.5 U), and 0.5 μl of human DNA (total volume of master mix, 50 μl per sample)

Electrophoresis running buffer: 1× TAE (5a) plus 0.012% ethidium bromide

Electrophoresis loading buffer: 15% Ficoll, 0.25% bromophenol blue

**Table 1.** Sequences of oligonucleotide primers

| Primer | Nucleotide sequence, 5′ to 3′ | Coding strand | Location[a] |
|---|---|---|---|
| 47-1 | GAC AAT GCT CAC TGA GGA TAG T | + | 627 to 648 |
| 47-2 | ACG CAC AGA ACC GAA TTC CTT G | − | 1263 to 1284 |
| 47-3 | TTG TGG TAG ACA CGG TGG GTA C | + | 692 to 713 |
| 47-4 | TGA TCG CTG ACA AGC TTA GGC T | − | 1166 to 1187 |
| Bg-1[b] | CAA CTT CAT CCA CGT TCA CC | − | 54 to 73 |
| Bg-2[b] | GAA GAG CCA AGG ACA GGT AC | + | −195 to −176 |

[a] Nucleotides.
[b] Human β-globin gene.

Denaturation solution: 1.5 M NaCl, 0.5 M NaOH

Neutralization solution: 1 M Tris HCl (pH 8.0), 1.5 M NaCl

Nitrocellulose (Schleicher & Schuell)

0.2× SSC (1× SSC is 0.15 M NaCl plus 0.015 M sodium citrate)–0.01 M SDS solution

2× SSC–0.01 M SDS solution

Hybridization solution: 1 M NaCl, 1% SDS, 10% dextran sulfate, 100 μg of salmon sperm DNA per ml

*T. pallidum* recombinant 47-kDa plasmid (4)

Random hexanucleotide primer labeling kit (Boehringer Mannheim)

[α-$^{32}$P]dATP

## Methods

### Sample Collection

Use blood, cerebrospinal fluid (CSF), amniotic fluid, or urine as samples. To obtain serum, centrifuge blood in a clinical centrifuge.

### Sample Preparation

Prepare sera and amniotic fluids for PCR by using the alkaline lysis-DNA extraction procedure described below.

Prepare CSF and urine by "boil" and "spin" procedures described below.

Sample preparation protocols are designed such that the final volume for addition to PCR mixture is 50 μl.

***Alkaline Lysis Extraction*** (Sera and amniotic fluids)

*Note: When extraction is performed in a 1.5-ml microcentrifuge (as is customary), the maximum sample volume is 108 μl.*

1. Starting with 5 M NaCl, 2 M NaOH, and 20% SDS, prepare the lysis solution in microcentrifuge tubes such that when NaCl, NaOH, and SDS are added to the sample or specimen, their final concentrations are 1 M, 0.1 M, and 0.5%, respectively.
2. Boil for 1 min, and place on ice for 5 min.
3. Neutralize with 4 volumes of 0.5 M Tris (pH 7).
4. Phenol extract with 0.5 volume for approximately 30 s; centrifuge in a microcentrifuge for 3 min.
5. Transfer the aqueous phase by using a P1000 Pipetman and Aeroseal tip to another microcentrifuge tube containing 0.5 volume $CHCl_3$-isoamyl alcohol (24:1); extract for 30 s, and microcentrifuge for 3 min.
6. Using a P1000 Pipetman and Aeroseal tip, transfer the aqueous phase to a tube containing 0.6 to 1 volume of isopropanol. Incubate at −70°C for at least 2 h or −20°C overnight.
7. Microcentrifuge the samples at 4°C for 15 to 30 min.
8. Decant the samples onto paper towels. Dry inverted for about 1 h.
9. Suspend the pellet in 50 μl of water.
10. Prepare for PCR as described below.

*"Boil-Only" Method*

1. Aliquot the sample of CSF or serum into microcentrifuge tubes (up to 50 μl).
2. Add 150 μl of 5% Chelex-100 to the sample.
3. Boil the sample for 10 min, and then place it directly on ice for 5 min; microcentrifuge for 1 min.
4. Add sample (50 μl) to the PCR mixture.

*Spin Method*

1. Aliquot as much sample as possible into microcentrifuge tubes: top off each tube to 1 ml (if necessary) with PBS.
2. Centrifuge samples in a superspeed centrifuge for 1 h at 20,000 × *g* at 4°C.
3. Remove the supernatant, leaving a wet pellet of approximately 50 μl.
4. Add 150 μl of 5% Chelex-100 to the wet pellet.
5. Boil samples for 10 min, and place directly on ice for another 5 min; microcentrifuge for 1 min.
6. Add 50 μl of supernatant to the PCR mixture.

### PCR Amplification

1. Prepare the master mix from the recipe provided.
2. Vortex the master mix.
3. Pipette 50 μl of the master mix into 0.5-ml microcentrifuge tubes.
4. Overlay each tube with 80 μl of mineral oil.
5. Place the samples on ice.
6. Add the sample, processed as described above.
7. Perform PCR under the following conditions for 40 cycles: 94°C for 1 min 15 s (denaturing), 58°C for 1 min 15 s (annealing), and 72°C for 1 min (extension).
8. After the 40th cycle, incubate for 10 min at 72°C; store at 4°C.

### Analysis of Sample

1. Electrophoresis on 1.5% agarose to visualize PCR products (most importantly the 269-bp β-globin fragment).
2. Slot or dot blot by Southern hybridization with an internal probe derived from primers 47-3 and 47-4 labeled with $^{32}P$ by random hexanucleotide priming (Boehringer Mannheim).
3. Subject to autoradiography.

## Quality Controls

### Alkaline Lysis

*Negative Control*

108 μl of normal serum, amniotic fluid, or CSF.

*Positive Control*

108 μl of normal serum or amniotic fluid "spiked" with $10^3$ *T. pallidum* organisms.

### "Boil Only"

***Negative Control***

50 μl of normal CSF or urine.

***Positive Control***

50 μl of normal CSF or urine "spiked" with $10^3$ *T. pallidum* organisms.

### Spin Method

***Negative Control***

200 μl of normal CSF or urine.

***Positive Control***

200 μl of normal CSF or urine "spiked" with $10^3$ *T. pallidum*.

### PCR Reagent Controls

***Negative Control***

No DNA (50 μl of water).

***Positive Control***

2 ng of plasmid containing the cloned gene encoding the *T. pallidum* 47-kDa membrane protein in 50 μl of water or *T. pallidum* isolated from inoculated rabbit testes.

## Comments

1. This PCR protocol results in amplification from *T. pallidum* DNA of a 658-bp product. Many specimens are positive only by DNA-DNA hybridization. We have found that the use of a large internal probe is much more sensitive than the use of a synthetic oligonucleotide and also improves the specificity of the assay.
2. Our PCR assay is not capable of distinguishing among the different subspecies of *T. pallidum*. However, the assay is *T. pallidum* specific in that it does not amplify sequences from other spirochetal pathogens (including *B. burgdorferi*, the agent of Lyme disease, other venereal disease pathogens, or commensal flora) (1, 7).
3. Our anticontamination procedures consist of meticulous technique and handling of all specimens in a dedicated *T. pallidum*-free facility. We have not found it necessary to employ "sterilization" procedures to prevent "amplicon build-up."
4. Inhibition of the PCR by substances present in clinical specimens is, in our experience, a major cause of false-negative results and one which is difficult to detect. We have circumvented this problem by including in each reaction mixture an internal control in which a 269-bp fragment of the human β-globin gene is amplified. A small amount (2 ng) of human DNA (isolated from human placenta) is added to each reaction mixture to ensure sufficient β-globin PCR product to be visualized on an agarose gel stained with ethidium bromide. The diminution in the sensitivity of the assay for *T. pallidum* is negligible.
5. We have found that freezing and thawing diminish the PCR signal intensity. Whenever possible, we work up specimens immediately.

6. Because of the relatively large amounts of protein in sera and amniotic fluids, we have found that these specimens must be processed by extraction of DNA.
7. Other PCR assays for the detection of *T. pallidum* DNA in clinical specimens have been published. These include a report by Hay et al. (3) in which sequences from the *tmpA* and 4D genes were used as separate targets and that of Noordhoek et al. (7) in which sequences from the *bmp* gene were amplified.

## References

1. **Burstain, J. M., E. Grimprel, S. A. Lukehart, M. V. Norgard, and J. D. Radolf.** 1991. Sensitive detection of *Treponema pallidum* by using the polymerase chain reaction. *J. Clin. Microbiol.* **29:**62–69.
2. **Grimprel, E., P. J. Sánchez, G. D. Wendel, J. M. Burstain, G. H. McCracken, Jr., J. D. Radolf, and M. V. Norgard.** 1991. Polymerase chain reaction and rabbit infectivity to detect *Treponema pallidum* in amniotic fluids, fetal and neonatal sera, and cerebrospinal fluids. *J. Clin. Microbiol.* **29:**1711–1718.
3. **Hay, P. E., J. R. Clark, R. A. Strugnell, D. Taylor-Robinson, and D. Goldmeier.** 1990. Use of the polymerase chain reaction to detect DNA sequences specific to pathogenic treponemes in cerebrospinal fluid. *FEMS Microbiol. Lett.* **68:**233–238.
4. **Hsu, P.-L., N. R. Chamberlain, K. Orth, C. Moomaw, L.-Q. Zhang, C. A. Slaughter, J. D. Radolf, S. Sell, and M. V. Norgard.** 1989. Sequence analysis of the 47-kilodalton major integral membrane immunogen of *Treponema pallidum. Infect. Immun.* **57:**196–203.
5. **Lukehart, S. A., and K. K. Holmes.** 1991. Syphilis, p. 651–661. *In* E. Braunwald, K. J. Isselbacher, R. G. Petersdorf, J. D. Wilson, J. B. Martin, and A. S. Fauci (ed.), *Harrison's Principles of Internal Medicine*. McGraw-Hill Book Co., New York.

5a. **Maniatis, T., E. F. Fritsch, and J. Sambrook.** 1982. *Molecular Cloning: A Laboratory Manual.* Cold Spring Harbor Laboratory, Cold Spring Harbor, N.Y.

6. **Miao, R. M., and A. H. Fiedsteel.** 1980. Genetic relationship between *Treponema pallidum* and *Treponema pertenue*, two noncultivable human pathogens. *J. Bacteriol.* **141:**427–429.
7. **Noordhoek, G. T., E. C. Wolters, M. E. J. DeJonge, and J. D. A. Van Embden.** 1991. Detection by polymerase chain reaction of *Treponema pallidum* DNA in cerebrospinal fluid from neurosyphilis patients before and after antibiotic treatment. *J. Clin. Microbiol.* **29:**1967–1984.
8. **Rolfs, R. T., and A. K. Nakashima.** 1990. Epidemiology of primary and secondary syphilis in the United States, 1981 through 1989. *JAMA* **264:**1432–1437.
9. **Sánchez, P. J., G. D. Wendel, Jr., E. Grimprel, M. Goldberg, M. Hall, O. Arencibia-Mireles, J. D. Radolf, and M. V. Norgard.** 1993. Evaluation of molecular methodologies and rabbit infectivity testing for the diagnosis of congenital syphilis and neonatal central nervous system invasion by Treponema pallidum. *J. Infect. Dis.* **167:**148–157.
10. **Turner, T. B., P. H. Hardy, and B. Newman.** 1969. Infectivity tests in syphilis. *Br. J. Vener. Dis.* **45:**183–196.
11. **Wasserheit, J. N.** 1992. Epidemiological synergy. Interrelationships between human immunodeficiency virus infection and other sexually transmitted diseases. *Sex. Transm. Dis.* **19:**61–77.

# Molecular Detection of *Treponema pallidum* via a 16S Target

**1.8**

JEFFREY S. SARTIN

## Background

*Treponema pallidum* subsp. *pallidum*, the causative agent of syphilis, represents one of the oldest and most notorious human pathogens. Indeed, 8,000-year-old bones from well-preserved human specimens have shown evidence of treponemal infection; more recently, for five centuries following Columbus, the dreaded Great Pox engulfed Europe and swayed the course of philosophy and public attitudes, as well as medicine (4). That syphilis continues to be a wily foe is attested to by the fact that in 1989 the United States experienced a 40-year high in reported cases (an incidence of 18.4/100,000) (2). Among certain subpopulations, the incidence has proven even more dramatic; e.g., that for black men is 147.4/100,000.

The introduction of the Wasserman test for syphilis in 1906 represented a revolutionary leap in the diagnosis of infectious diseases. In the decades following that innovation, the diagnosis of syphilis has become refined by use of dark-field microscopy, treponemal tests such as the fluorescent treponemal antibody absorption (FTA-ABS) test, and reaginic tests such as the rapid plasma reagin (RPR) test and the Venereal Disease Research Laboratory (VDRL) test. However, significant shortcomings remain, particularly in the diagnosis of neurosyphilis and syphilis in patients with human immunodeficiency virus (HIV) infection (5, 7). Both false-negative and false-positive results with current diagnostic strategies make the development of a sensitive, specific test highly desirable.

PCR amplification has enabled the detection of minute amounts of DNA in a broad range of specimens. Using primers directed at the gene encoding the 16S ribosomal subunit, we have amplified a highly specific DNA sequence in cerebrospinal fluid specimens, which is confirmed by use of an internal probe (Table 1).

## Materials and Reagents

IsoQuick DNA/RNA extraction kit (Microprobe Corp., Bothell, Wash.)

UV photochemical reaction chamber (e.g., HRI-100; HRI, Walnut Creek, Calif.) for post-PCR sterilization

ECL kit (Amersham) or $^{32}$P isotope for hybridization

10× PCR buffer: 10 mM Tris (pH 8.3), 50 mM KCl, 1.5 mM $MgCl_2$, 0.001% bovine serum albumin fraction V (50 mg/20 ml of buffer)

K buffer: 1 volume of 10× PCR buffer, 0.5% Nonidet P-40, 0.5% Tween 20, 100

**Table 1.** TPr primer sequences

| Name | Nucleotide sequence, 5′ to 3′ |
|---|---|
| TPr-3 | CTC AGA GAT GAG CCT GCG ACC CAT T |
| TPr-4 | GCA TTC CCT CCC GTC CTC ATT CTT C |
| TPr-6 | AGC TTG TTG GTG GGG TAA TGG |
| TPr-7 | GTC GGC AAA AGA ACT TTA CAA T |

μg of proteinase K per ml (detergents and protease obtained from Sigma Chemical Co., St. Louis, Mo.)

TE buffer: 10 mM Tris-HCl, 0.1 mM EDTA (pH 8.0)

PCR master mix: 0.1 volume of 10× PCR buffer, 0.2 mM each deoxynucleoside triphosphate, 0.5 U of *Taq* polymerase (Perkin-Elmer Corp.), 50 μg of isopsoralen compound 10 (HRI) per ml, 10% glycerol, 50 pmol each of primer 1 and primer 2 (Table 1)

Denaturation solution: 1.5 M NaCl, 0.5 M NaOH

Neutralization solution: 1 M Tris-HCl (Sigma) (pH 8.0), 1.5 M NaCl

## Methods

### Sample Collection

Collect 1 to 2 ml of cerebrospinal fluid (CSF) from lumbar puncture by the usual technique; if it is not used immediately, store it at −70°C.

### Sample Preparation

Preparation of CSF for *T. pallidum* PCR involves lysis of bacterial cells, with removal of inhibitory proteins and recovery of target DNA and RNA. You may

**Figure 1.** PCR primers and probe for detection of *T. pallidum*. *GenBank accession no. M88721.

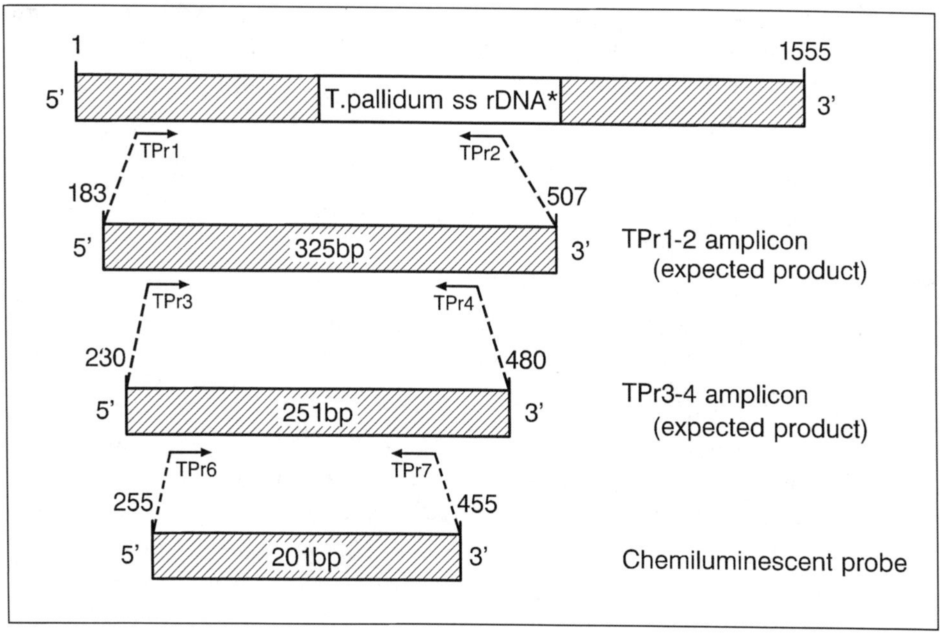

use the IsoQuick extraction kit; follow the procedure for total nucleic acid extraction. Alternatively, use the following alkaline lysis procedure, as described by Grimprel et al. (1).

1. Add 100 μl of sample to 100 μl of lysing solution (1 M NaCl, 1 N NaOH, 0.1% sodium dodecyl sulfate [SDS]), and boil for 1.5 min.
2. Neutralize with 400 μl of 0.5 M Tris-HCl (pH 8.0).
3. Extract successively with 150 μl of phenol-chloroform-isoamyl alcohol (25: 24:1) and then with 150 μl of chloroform-isoamyl alcohol (24:1).
4. Add 30 μl of 3 M sodium acetate and 300 μl of isopropanol, invert gently to mix, and spin for 15 min in a 4°C refrigerated centrifuge at 16,000 × *g*.
5. Remove the supernatant, and wash with 500 μg of 70% ethanol; spin for 15 min at 16,000 × *g* as above.
6. Remove the supernatant, and dry by evaporation (e.g., with a Speed-Vac).
7. Resuspend the pellet in 50 μl of TE buffer, and refrigerate at 4°C.

## PCR Amplification

1. Prepare the master mix by using the recipe above. Aliquot 40 μl of the master mix per tube (with 10 μl of sample). *Note: Because of pipetting error, it is preferable to prepare two or three tubes more than calculated, to ensure enough reagent. Positive-displacement pipettes or filter-tip pipettes are necessary for all pipetting steps.* Add 10 μl of sample.
2. Use the following amplification protocol: time delay cycle, 4 min at 95°C; step cycle, denaturation for 1 min at 95°C and annealing/elongation for 1 min at 58°C; time delay cycle (terminal extension), 7 min at 72°C; soak cycle, hold at 4°C. Run for a total of 40 cycles.

## Gel Electrophoresis Protocol

1. Prepare 2% SeaKem agarose gel in TBE buffer. Using constant voltage, separate the reaction products by electrophoresis; e.g., run at 120 V for 2 h.
2. Stain with 1 μg of ethidium bromide per ml for 30 to 45 min. *Note: Destain for 10 to 15 min with distilled water if desired.*
3. Visualize the 251-bp reaction product after staining with ethidium bromide; photograph if desired.

## Southern Blot Protocol

1. Denature and neutralize in denaturation and neutralization buffers for 30 and 15 min, respectively.
2. Transfer product onto nylon filters by using SSC, as described previously (6).
3. Couple the reaction products to the membrane by exposure to a UV cross-linker at a preset level of 1.2 × 105 μJ.

## Chemiluminescent Detection

1. To prepare the chemiluminescent internal probe, add 10 μl of the gel-purified TPr-3,4 amplification product to the master mix containing nested primers TPr-6 and TPr-7; amplify by PCR as described above for 20 cycles.
2. Gel purify the 201-bp internal product, determine its concentration by spectrophotometry, and label 10 ng as specified by the manufacturer of the enhanced chemiluminescence (ECL) gene detection system (Amersham).

3. Prehybridize in blocking buffer from the kit for 30 min (20 ml per membrane), add 10 ng of labeled probe per ml, and incubate the filter for 3 to 16 h at 42°C in a hybridization oven with rolling tubes or in trays with gentle agitation.
4. Wash filters twice for 20 min in ECL wash buffer consisting of 0.1× SSC with 0.17 M urea and 0.01 M SDS and then twice for 5 min in 2× SSC.
5. Cover moist filters with ECL detection reagents (20 ml per membrane) for exactly 1 min; blot dry, place between two pieces of plastic wrap, and immediately expose to Kodak X-Omat film for 1 min to overnight (see Comments).

## Quality Control

### Positive Amplification Control

1 ng and 10 pg of *T. pallidum* DNA (approximately $10^4$ and $10^2$ organisms, respectively), one reaction each.

### Negative Amplification Control

10 μl of distilled water added to PCR, five reactions.

### Positive Patient Control

10 μl of CSF from a patient with positive CSF-VDRL.

### Negative Patient Control

10 μl of CSF from a patient with negative CSF- (and serum) FTA.

### Specificity Control

1 μg of DNA from *T. bryantii*, *T. denticola*, *T. pectinovorum*, *Borrelia burgdorferi*, and representative CSF pathogens (*Neisseria meningitidis*, etc.).

## Comments

1. Although our experience with the IsoQuick kit is limited, it appears to provide recovery of DNA and RNA comparable to that of the alkaline lysis procedure or similar techniques. It offers the benefit of superior elimination of inhibitory proteins usually found in CSF and recovery of RNA, which then can be reverse transcribed and amplified by using an RNA PCR technique, e.g., Gene Amp Thermostable rTth Reverse Transcriptase RNA PCR kit (Perkin-Elmer Corp.).
2. A study by Noordhoek et al. suggests that up to 99% of intact DNA in specimens may be lysed and therefore unsuitable for standard PCR by repeated cycles of freezing and thawing (3). Therefore, if specimens or processed DNA are to be stored for a long period, we recommend storage at −70°C and aliquoting to prevent repeated thawing of a specimen.
3. A convenient method for performing phenol extractions involves use of serum separator tubes (Becton Dickinson and Co.). When samples and reagents are spun at 16,000 × *g* for 30 s, the aqueous layer will remain above the inert plug; extractions can be repeated in the same tube.
4. This two-step cycle allows extension during the transition from annealing to denaturation temperatures; the cycle can be set for a temperature step-up as

rapidly as possible. Our experience indicates that sensitivity is not compromised, and specificity may even be enhanced through the elimination of nonspecific reactants.

5. Use of isopsoralen for sterilization will result in a noticeable shift in the molecular weight of the product relative to the kilobase ladder.
6. Alternatively, comparable results may be obtained with a $^{32}$P-end-labeled oligonucleotide probe, e.g., TPr-7, a 22-bp primer. This procedure is similar to others described in reference 6.
7. The photoreactivity of the ECL probe may be extremely high, so that only 1 to 2 min is needed for optimal exposure. We generally take exposures at 1 and 10 min and adjust the time for longer exposures if necessary. The ECL probe may be synthesized in advance and stored for up to 6 months in 50% glycerol.
8. Previous experiments indicate that the specificity of the reaction depends greatly on the stringency of the conditions; i.e., higher temperatures favor greater specificity. Use of an annealing temperature of 50°C or less has produced cross-reactivity with other treponemal species; this is eliminated at higher temperatures.

## References

1. **Grimprel, E., P. J. Sanchez, G. D. Wendel, J. M. Burstain, G. H. McCracken, J. D. Radolf, and M. V. Norgard.** 1991. Use of polymerase chain reaction and rabbit infectivity testing to detect *Treponema pallidum* in amniotic fluid, fetal and neonatal sera, and cerebrospinal fluid. *J. Clin. Microbiol.* **29:**1711–1718.
2. **Lukehart, S. A.** 1991. Syphilis: issues for the 1990's. *Clin. Microbiol. Newsl.* **13:**117–120.
3. **Noordhoek, G. T., E. C. Wolters, M. E. J. de Jonge, and J. D. A. van Embden.** 1991. Detection by polymerase chain reaction of *Treponema pallidum* DNA in cerebrospinal fluid from neurosyphilis patients before and after antibiotic treatment. *J. Clin. Microbiol.* **29:**1976–1984.
4. **Quetel, C.** 1990. *History of Syphilis*. Polity Press, Cambridge.
5. **Romanowski, R., R. Sutherland, G. H. Fick, D. Mooney, and E. J. Love.** 1991. Serologic response to treatment of infectious syphilis. *Ann. Intern. Med.* **114:**1005–1009.
6. **Sambrook, J., E. F. Fritsch, and T. Maniatis.** 1989. *Molecular Cloning: A Laboratory Manual*, 2nd ed. Cold Spring Harbor Laboratory, Cold Spring Harbor, N.Y.
7. **Zenker, P. N., and R. T. Rolfs.** 1990. Treatment of syphilis, 1989. *Rev. Infect. Dis.* **12**(Suppl.)**:**S590–S609.

# PCR Detection of *Chlamydia trachomatis*

# 1.9

Linda D. Bobo

## Background

*Chlamydia trachomatis* is an obligate intracellular bacterium composed of 18 serovars which infect mucosal surfaces of the genitourinary tract, rectum, nasopharynx, or conjunctiva. Serious sequelae of infection include infertility, ectopic pregnancy, acute epididymitis, proctitis, and blindness. However, many chlamydial infections are mild or asymptomatic. If unrecognized infection occurs during pregnancy, the neonate is at risk for conjunctivitis or pneumonia. *C. trachomatis* is the leading cause of sexually transmitted diseases in the United States, with an estimated 4 million new cases reported annually (10). In heterosexual males, this organism accounts for 40 to 50% of nongonoccal urethritis (8). The lymphogranuloma venereum biovars, L1 to L3, may infect the regional inguinal lymph nodes, causing systemic manifestations. The trachoma biovars, A to C, are the leading cause of infectious blindness in the world; 500 million people are infected, of whom 7 million are blind (3).

Definitive laboratory diagnosis of *C. trachomatis* infection is made by demonstrating inclusion bodies grown in cell cultures. Other commonly used tests are direct immunofluorescence for microscopic demonstration of elementary bodies and enzyme immunoassay (EIA). Both use antibodies directed against the major outer membrane protein (MOMP) or lipopolysaccharide (LPS) antigens of the organism. However, both culture and antigen detection methods may be only 70% sensitive on a single endocervical swab and 40% or less sensitive for patients with clinical follicular or inflammatory trachoma (4, 6, 7, 9). The PCR has been reported recently for detection of *C. trachomatis* from genital or conjunctival specimens by using primers in a conserved area of the MOMP gene (1, 2). We have coupled PCR with liquid, nonisotopic hybridization and an EIA for detection of biotinylated RNA-DNA hybrids. The quantitative nature of PCR-EIA has allowed us to monitor the response to antibiotic therapy and to relate the chlamydial DNA load to the intensity of clinical disease (2). In addition, we have described a culture-negative, nucleic acid-positive state in an animal model for trachoma (5).

## Materials and Reagents

Lysis buffer: 100 μg of proteinase K per ml, 0.5% (vol/vol) Nonidet P-40–Tween 20, 100 pg of salmon sperm DNA per ml

Millipore Ultrafree UFC filters

PCR holding buffer: 10 mM Tris (pH 8.3), 1 mM EDTA, 1% (vol/vol) Nonidet P-40–Tween 20

PCR master mix cocktail: 1× PCR buffer (10 mM Tris [pH 8.3], 50 mM KCl, 2.5 mM $MgCl_2$, 0.01% gelatin), 0.5 μM each primer (Ct.0005/06), 0.2 mM each dNTP, 2.5 U of *Taq* polymerase per reaction

Biotinylated RNA probe: transcription of DNA made with primers Ct.03T7/04 with 2.5 mM biotin 11-UTP (Enzo), 20 U of T7 RNA polymerase, 2.5 mM each rATP, rCTP, and rGTP, 1× transcription buffer (40 mM Tris [pH 7.5], 6 mM $MgCl_2$, 2.5 mM spermidine, 12 mM NaCl, 70 to 100 U of RNasin [Promega])

1 U of RQ1 DNase (Promega)

Sephadex G-25 columns (NAP-5; Pharmacia)

0.1% Diethylpyrocarbonate water (DEPC)

DEPC–0.5% sodium dodecyl sulfate (SDS)

2× hybridization buffer: 600 mM NaCl, 60 mM sodium citrate (pH 7), 20 mM HEPES (*N*-2-hydroxyethylpiperazine-*N*′-2-ethanesulfonic acid; pH 7.4), 4 mM EDTA, 0.5% SDS

10% Triton X-100

Black microtiter plates (Microfluor "B"; Dynatech)

Monoclonal goat anti-biotin (final concentration, 2 μg/ml; Zymed)

0.06 M carbonate buffer (pH 9.6)

Plate wash (10 mM phosphate buffer [pH 7.2], 150 mM NaCl, 0.05% Tween 20 [PBST])

Monoclonal antibody–β-galactosidase conjugate (184 ng/ml; gift from Miles Lab)

PBST plus 0.5% gelatin (PBSTG)

Normal mouse serum (final concentration, 0.5%)

Substrate solution (1 mM 4-methylumbelliferyl-β-D-galactosidase and 10 mM phosphate buffer [pH 7], 100 mM NaCl, 1 mM $MgCl_2$, 50 μg of bovine serum albumin per ml)

Fluorometer (detection wavelength, 365 nm; emission wavelength, 450 nm) (Microfluor; Dynatech)

## Methods

### Sample Preparation

Collect genitourinary or conjunctival specimens with Dacron-tipped swabs, and place them in chlamydia transport medium or PCR holding buffer, respectively. Lyse eye samples by adding proteinase K and incubating at 55°C for 1 h. Lyse specimens in chlamydia transport medium with entire lysis buffer, and incubate them as described. Inactivate enzyme by boiling for 5 min or at 95°C for 10 min. Snap-chill specimens at −70°C for at least 1 h. Microdialyze thawed samples three times with water containing 100 pg of sonicated salmon sperm DNA per ml.

### PCR Amplification

Amplify each sample in duplicate with 5- and 50-μl portions mixed with PCR master mix to a final volume of 100 μl. Add two drops of sterile mineral oil to all

**Table 1.** Primers for detection of *C. trachomatis*

| Name | Nucleotide sequence,[a] 5′ to 3′ |
|---|---|
| Ct.0005 | GAT AGC GAG CAC AAA GAG AGC TAA |
| Ct.06 | TTC ACA TCT GTT TGC AAA ACA CGG TCG AAA ACA AAG |
| Ct.03T7 | TTA ATA CGA CTC ACT ATA GGG TCT GCT TCC TCC TTG... CAA GCA AGT CTG CC |
| Ct.04 | CCA TAG TAA CCC ATA CGC ATG CTG |

[a] Underlined area denotes T7 promoter.

tubes, followed by the PCR master mix and then the sample. Perform amplification in a thermocycler for 30 cycles as follows: denaturation at 94°C for 1 min, annealing at 55°C for 1 min, and extension at 72°C for 1 min.

### Probe Preparation

Amplify DNA by using stock *C. trachomatis* L2 and primers Ct.03T7 and Ct.04 (Table 1). Use ca. 500 ng in a 100-μl transcription reaction at 37°C for 1.5 h. Perform RQ1 DNase digestion of the DNA template for 1 h, and inactivate the enzyme by boiling for 5 min. Remove unincorporated biotin 11-UTP by using column chromatography, and store the probe in 20-μl aliquots at −70°C. Determine the titer of the probe at 1:500, 1:1,000 and 1:2,000 in 2× hybridization buffer with stock strain L2 diluted to 100, 10, and 0 inclusion-forming units (IFU) and amplified with Ct.0005 and Ct.06 primers. Also, probe alone is included to check for background.

### Liquid Hybridization

Perform all subsequent steps in an area separate from the sample-processing laboratory. Mix 60 μl of PCR product with 60 μl of probe, boil for 3 min, and incubate at 78°C for 0.5 h. Add Triton X-100 to each tube to 1%, and mix well.

### EIA

Add 50 μl of hybridization mixture in duplicate to anti-biotin-coated wells of a microtiter plate, and incubate overnight at 4°C or for 1 h at 37°C. Wash plates three times with PBST, add 50 μl of a 1:2,000 dilution of monoclonal antibody–β-galactosidase conjugate to DNA-RNA hybrids in PBSTG, and incubate plates at 37°C for 1 h. Wash plates as described, and add 50 μl of 1 part substrate-2 parts substrate buffer to the wells. Let the reaction develop for 2 h, and read it on the fluorometer as fluorescence units. Calculate the positive cutoff from the mean and five standard deviations by using the five negative clinical controls. Schematics showing the location of the primers and the entire PCR-EIA assay are in Fig. 1 and 2, respectively.

## Quality Control

### Negative Controls

We use water blanks as every fifth specimen during processing. We also test five negative clinical controls on each run in addition to a PCR buffer blank. All the water blanks used in sample preparation are tested in sequence with the clinical specimens.

**Figure 1.** Location of the outer and nested primers within the MOMP gene. The numbering system is derived from sequence information from GenBank accession no. M14738 (1).

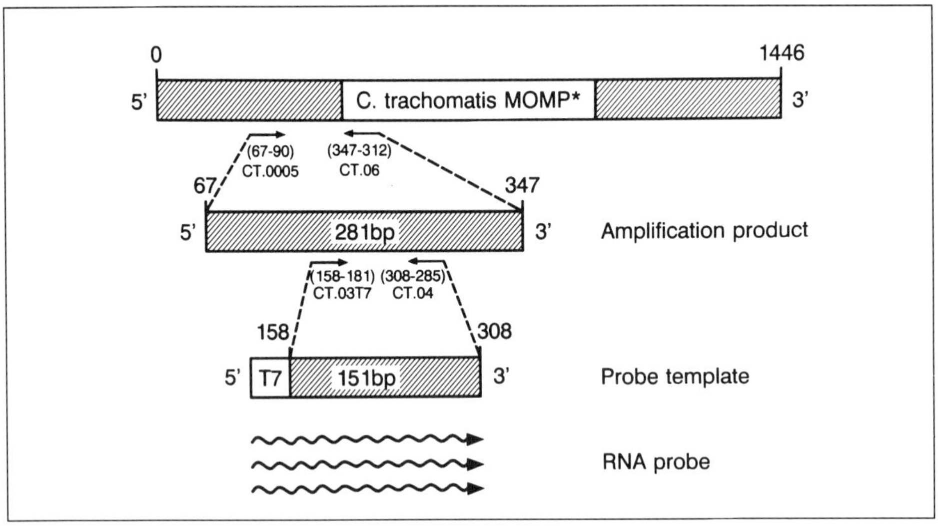


## Positive Controls

We use *C. trachomatis* L2 (ATCC 434) dilutions containing 100, 10, 1, and 0 inclusion-forming units (IFU) (linear portion of the assay from 100 to 1 IFU). We use a probe control consisting of probe alone with no target to check for background in the EIA.

# Comments

1. Methods used to minimize contamination were as follows. Surfaces and equipment were decontaminated with 1 N HCl, gloves were changed frequently, and filter-block tips were used on pipettors. In addition, specimen processing was performed in a room separate from the hybridization and EIA.
2. Concentrations of target DNA ($\geq 10^{12}$ copies) which is $>10^{-9}$ M *Taq* polymerase in the reaction can be inhibitory to amplification. Dilution of the sample into the correct range would thus eliminate the plateau effect. In addition, as mentioned above, antigenic excess of DNA-RNA in the EIA will result in a lower value than that for a more dilute sample. It has been helpful to use a 5-μl sample in addition to 25 or 50 μl of target.
3. Infrequently, the 50-μl split was PCR-EIA negative and the 5-μl sample was positive. This usually indicated the presence of an inhibitor to the PCR, because spiking the negative sample with *C. trachomatis* DNA resulted in no amplification.
4. The primers used to generate the target are specific for *C. trachomatis*, as is the internal sequence corresponding to the probe.
5. Detection of amplified DNA by EIA is 10- to 100-fold more sensitive than gel detection but equally as sensitive as $^{32}P$ hybridization of a Southern blot. Quantitative detection of *C. trachomatis* DNA can be influenced by the presence of inhibitors or stimulators (salts or metal ions) in the specimen material. Sample adequacy can be controlled by coamplification of β-globin,

**Figure 2.** Schematic of the PCR-EIA amplification detection system. The DNA target from a clinical sample and the biotinylated probe were generated in two separate procedures, which converged to yield duplexes in a solution hybridization reaction. Biotinylated RNA-DNA hybrids captured in wells of microtiter plates coated with goat anti-biotin antibody were reacted with a β-D-galactosidase-linked monoclonal antibody which recognizes DNA-RNA hybrids. The fluorogenic substrate was cleaved by the enzyme to yield a quantifiable product, 4-methylumbelliferone (1).

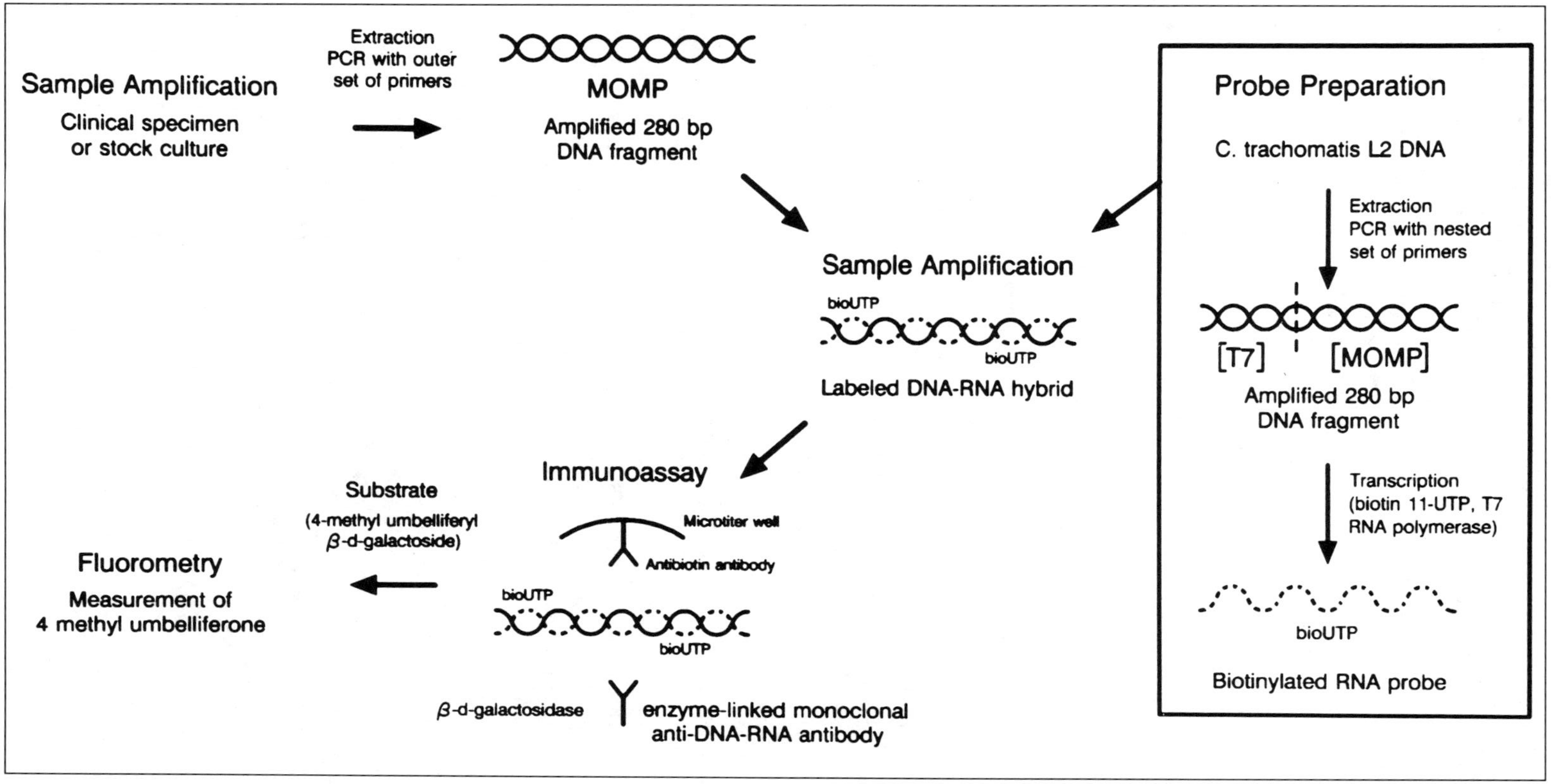

human leukocyte antigen DHQα (HLA-DHQα), or β-actin gene sequences along with the target *C. trachomatis* DNA. We have controlled for inhibitors or stimulators by using a primer set constructed of sequences for a plasmid at the 5′ end followed by a sequence for the target gene. The sample is "spiked" with a known amount of control plasmid and then amplified with the chimeric primers. The chimeric primers will amplify both types of DNA, and the products are analyzed separately with a probe for the gene of interest and one for the plasmid DNA fragment. An additional amplification of the same amount of plasmid alone is performed. The ratio [(plasmid alone amplification value/plasmid value with sample amplification) × sample value] yields a correction for inhibition or stimulation of amplification. A linear regression from a standard line generated by known amounts of amplified stock DNA can be used in combination with the above controls to quantitate copies of DNA (11).

6. Detection of less than 1 IFU in the positive control is possible because of noninfectious forms (reticulate bodies) which can be present in the sample.
7. In our experience, a probe control background of 70 (or greater) fluorescence units could be indicative of incomplete digestion of the DNA template by RQ1 DNase in the transcription reaction. Backgrounds of 100 (or greater) fluorescence units could indicate contamination of the PBST wash solution with bacteria.
8. Conjunctival specimens from a country where trachoma was endemic were stored and transported in PCR holding buffer without maintenance of a cold chain.
9. Since we are using a highly sensitive assay (compared with more traditional tests), we have tried to determine the significance of positive PCR results as follows. Laboratory verification included a repeat positive with a different primer set. In addition, we designed our studies so that we could correlate our results with clinical findings and compare sequence type in sexual partners who were linked epidemiologically in a transmission study. Moreover, we have compared fluorescence signals on sequential samples from patients before and after treatment and have compared PCR-EIA results with a $^{32}P$ probe hybridization for rRNA in an animal model for trachoma. In the transmission study, sexual partners who were PCR positive and culture negative had the same gene type by PCR sequencing. In studies in a trachoma-endemic area, only 3% of pretreatment PCR-positive subjects with trachoma who were mass treated for 1 month with topical tetracycline were PCR positive. The subjects who were PCR positive throughout treatment and 1 month after treatment stopped had the same gene type at all time points. In an animal model for trachoma, high PCR and RNA values were found in 2 of 6 culture-negative monkeys 1 to 4 weeks after a second challenge with *C. trachomatis*.

## References

1. **Bobo, L., F. Coutlee, R. Yolken, T. Quinn, and R. Viscidi.** 1990. Diagnosis of *Chlamydia trachomatis* cervical infection by detection of amplified DNA with an enzyme immunoassay. *J. Clin. Microbiol.* **28:**1968–1973.
2. **Bobo, L., B. Munoz, R. Viscidi, T. Quinn, H. Mkocha, and S. West.** 1991. Diagnosis of *Chlamydia trachomatis* eye infection in Tanzania by polymerase chain reaction/enzyme immunoassay. *Lancet* **338:**847–850.
3. **Dawson, C. R., B. R. Jones, and M. L. Tarizzo.** 1981. Guide to trachoma control in programmes for the prevention of blindness. World Health Organization, Geneva.

4. **Dunlop, E. M. C., B. T. Goh, S. Darougar, and R. Woodland.** 1985. Triple-culture diagnostic tests for the diagnosis of chlamydial infection of the female genital tract. *Sex. Transm. Dis.* **12:**68–71.
5. **Holland, S., A. Hudson, L. Bobo, J. Whittum-Hudson, R. Viscidi, T. Quinn, and H. R. Taylor.** 1992. Demonstration of chlamydial RNA and DNA during a culture-negative state. *Infect. Immun.* **60:**2040–2047.
6. **Mabey, D. W. C., J. N. Robertson, and M. E. Ward.** 1987. Detection of *Chlamydia trachomatis* by enzyme immunoassay in patients with trachoma. Lancet **ii:**1491–1492.
7. **Schachter, J.** 1985. Immunodiagnosis of sexually transmitted diseases. *Yale J. Biol. Med.* **58:**443–452.
8. **Stamm, W.** 1988. Diagnosis of *Chlamydia trachomatis* genitourinary infection. *Ann. Intern. Med.* **108:**710–717.
9. **Taylor, H. R., P. A. Rapoza, S. West, et al.** 1989. The epidemiology of infection in trachoma. *Invest. Ophthalmol. Visual Sci.* **30:**1823–1833.
10. **Washington, A. E., R. E. Johnson, and L. L. Sanders.** 1987. *Chlamydia trachomatis* infections in the U.S.: what are they costing us? *JAMA* **257:**2070–2072.
11. **Yang, B., R. Yolken, and R. Viscidi.** 1993. Quantitative polymerase chain reaction by monitoring enzymatic activity of DNA polymerase. *Anal. Biochem.* **208:**110–116.

# 3SR Detection of *Chlamydia trachomatis*

# 1.10

PAUL V. HAYDOCK AND SHARON A. KOCHIK

## Background

Diagnosis and treatment of *Chlamydia trachomatis* infections is a critical challenge to health care professionals because of the prevalence and insidious nature of the pathogen. Chlamydial infections can cause a variety of pathologic conditions including nongonococcal urethritis, cervicitis, lymphogranuloma venereum, conjunctivitis, prostatitis, and trachoma. Untreated *C. trachomatis* infections can lead to more serious complications such as salpingitis and pelvic inflammatory disease, which could result in infertility or ectopic pregnancy (2, 8). Since chlamydial infections are frequently asymptomatic, accurate and sensitive detection methods are required. The "gold standard" *Chlamydia* assay is cell culture of patient samples in McCoy cells followed by fluorescent-antibody staining; in this assay a positive sample reveals bright fluorescent cytoplasmic inclusions (5, 9). A variety of enzyme-linked immunosorbent assays and DNA probe-based detection methods have been developed to avoid the need for cell culture of patient samples (1, 10). These methods are very sensitive, but they do not allow determination of viability as does cell culture. An advantage of cell culture is that viability is tested. Enzyme immunoassay and PCR will detect both viable and nonviable organisms. Thus, enzyme immunoassay and PCR will be positive even though the organisms causing an active infection have been killed by antibiotics. The advantage to the system presented here is that a positive result will be obtained only from a viable specimen.

We have developed a method for the detection of *Chlamydia* species based on amplification of target RNA by 3SR (Fig. 1) (3). After culture of patient samples in McCoy cells, adherent cells are lysed and the lysates are amplified by using the 3SR followed by hybridization analysis. The combination of cell culture and RNA amplification allows very sensitive detection of chlamydial infections.

## Materials and Reagents

5× 3SR buffer: 200 mM Tris (pH 8.1), 150 mM $MgCl_2$, 100 mM KCl, 50 mM dithiothreitol, 5 mM spermidine · 3HCl.

5× nucleotide mix: 35 mM each ribonucleotide (ICN), 5 mM each deoxyribonucleotide (Sigma), 50 mM Tris (pH 8.1).

Primers CT220 and CT447 at 50 ng/μl each (Table 1)

Distilled water treated with diethylpyrocarbonate (DEPC)

**Figure 1.** Primers and probe for 3SR detection of *C. trachomatis.*

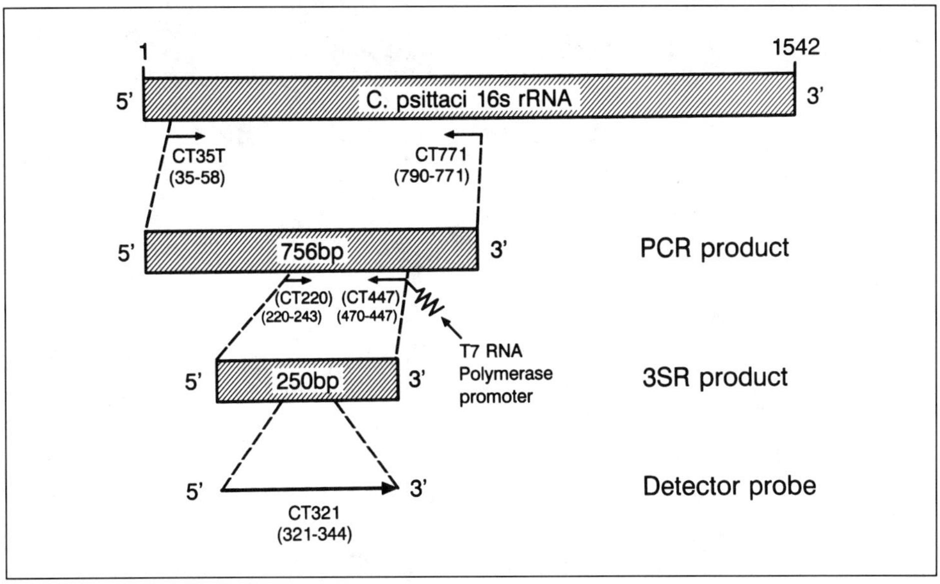


3SR enzymes: 20 U of avian myeloblastosis virus reverse transcriptase per μl, 100 U of T7 RNA polymerase per μl, 1 U of *Escherichia coli* RNase H per μl

Detector oligonucleotide CT321 at 60 ng/μl

TSF: 7.4% formaldehyde, 10× SSC (1× SSC is 0.15 M NaCl plus 0.015 M sodium citrate)

Nitrocellulose sheet cut to fit slot blot apparatus

Hybridization mix: 0.5% bovine serum albumin, 0.5% polyvinylpyrrolidine, 5× SSC, 1% sodium dodecyl sulfate (SDS)

Wash solution: 1× SSC, 1% SDS

Chlamydial transport medium (e.g., from Baxter Diagnostics Inc., Bartels Division, Issaquah, Wash.)

Sterile Dacron swab

Cell culture for chlamydia isolation (McCoy cells grown in 1-dram (3.697-ml) vials, or equivalent cell culture) (e.g., from Baxter Diagnostics Inc., Bartels Division)

Chlamydia isolation medium with 1 μg of cycloheximide per ml (e.g., from Baxter Diagnostics Inc., Bartels Division)

Lysis mix A: 0.05% SDS

Lysis mix B: sterile distilled water

Centrifuge equiped with rotor to fit culture vials and attain 3,000 × *g*

## Methods

### Sample Collection

Collect samples on endocervical swabs and transfer to chlamydia transport medium. Store samples at 2 to 8°C after collection. It is best to perform assay within 2 days of collecting the sample. Store samples at −70°C or colder.

**Table 1.** Oligonucleotides for detection of *C. trachomatis*

| Function | Name | Nucleotide sequence, 5′ to 3′ |
|---|---|---|
| 3SR primers | CT220 | ACC TTT CGG TTG AGG GAG AGT CTA |
| | CT447 | AAT TTA ATA CGA CTC ACT ATA GGG ACC AAT. . . TCT TAT TCC CAA GCG A |
| Detector | CT321 | AAC ACT GGG ACT GAG ACA CTG CCC A |
| PCR primers | CT35T | AAT TTA ATA CGA CTC ACT ATA GGG AGC TGG. . . CGG CGT GGA TGA GGC |
| | CT771 | CCT GTT TGC TCC CTT GCT TTC GC |

### Sample Preparation

1. Extract swab in transport medium, and press the swab against the inside of the tube to expel the sample suspension. Vortex the sample for 30 to 60 s.
2. Remove any media from the McCoy vials. Wash once with 1 ml of phosphate-buffered saline (PBS). Inoculate 0.1 to 0.2 ml of the specimen into the McCoy cell vials. Centrifuge the tubes at 3,000 × *g* for 30 to 60 min at room temperature.
3. Aeseptically add 1 ml of fresh, prewarmed chlamydia isolation medium with cycloheximide to each vial. Incubate vials at 37°C for 24 to 48 h. *Note: Incubation for 48 h may yield more consistent results. Visible inclusions may be evident.*

### Processing Samples for Amplification

1. Remove the medium from each vial, and rinse the monolayer with 1 ml of PBS.
2. Add 20 μl of lysis solution A, and vortex to disperse the reagent and lyse the cells.
3. Add 180 μl of lysis solution B. Assay samples within 1 h of performing the lysis step.

### 3SR Amplification

1. Prepare a 3SR master mix by mixing 20 μl of 5× buffer, 20 μl of 5× nucleotides, 5 μl each of primers CT220 and CT447 (Table 1), and 40 μl of $H_2O$ for each reaction to be performed. *Note: Prepare extra master mix to allow for pipetting error.* Aliquot 90 μl of the reaction mix into fresh sterile tubes.
2. Add 10 μl of the sample to be assayed, and vortex to mix. Hold at 42°C. *Note: Do not exceed 10 μl of sample, as the SDS may inhibit the 3SR reaction.*
3. Prepare an enzyme mix by combining on ice 1.5 μl of AMV reverse transcriptase, 1 μl of T7 RNA polymerase, and 1 μl of RNase H for each reaction. *Note: The enzyme mix may be prepared and stored on ice for up to 2 h.* Add 3.5 μl of enzyme mix to each reaction mix. Incubate at 42°C for 1 h. *Note: The reaction may be incubated for up to 2 h.*

### Detection of Amplification Products

1. Mix 5 μl of each amplification reaction with 95 μl of TSF. Heat to 65°C for 10 min.
2. Soak nitrocellulose first in water and then in 10× SSC. Blot samples onto nitrocellulose by using the slot blot apparatus. Bake the nitrocellulose for at least 20 min at 80°C.

3. Mix 1.5 μl of 10× kinase buffer, 2.5 μl of CT321 probe, and 10 μl of [$^{32}$P]ATP. Add 1 μl of T4 kinase, and incubate the reaction at 37°C for 30 min. Stop the reaction by adding 85 μl of hybridization mix.
4. Hybridize the blot in 5 ml of hybridization mix with 20 μl of labeled CT321 probe for 30 to 60 min at 55°C.
5. Wash the blot four times in wash solution at room temperature. Dry the blot, and expose to X-ray film.

## Quality Control

### Positive Amplification Control

5 amol of in vitro-generated transcript. The transcript is synthesized from a PCR product generated from *Chlamydia* RNA by using primers CT35T (which contains a T7 transcription initiation site) and CT771 in an RNA PCR. Exact details of RNA PCR are presented in reference 6. An RT reaction is performed with CT771 and total RNA purified from *C. trachomatis*-infected McCoy cells (the amounts of template RNA and primer CT771 are determined empirically). The RT products are then PCR amplified with CT35T and CT771. The 756-bp PCR product, which also contains a T7 RNA polymerase binding site, is purified by ethanol precipitation and G-50 spin column chromatography according to the manufacturer's protocol (e.g., BMB). The 756-bp fragment is transcribed in vitro to yield a positive-control RNA template (7). The RNA template (positive amplification control) is purified by spin column chromatography and is accurately quantitated spectrophotometrically.

### Negative Amplification Control

10 μl of distilled water.

### Positive Sample Control

10 pg of total RNA from *C. trachomatis*-infected McCoy cells.

### Negative Sample Control

10 pg of uninfected McCoy cell RNA.

### Positive Hybridization Control

1, 10, and 100 fmol of RT-PCR product as described above.

## Comments

1. The presence of *C. trachomatis* RNA is indicated by the amplification of a 250-base single-stranded RNA derived from 16S rRNA. On slot blot analysis, positive samples are heavily labeled and yield dense signals after short autoradiographic exposures. The primers are directed toward divergent regions of the 16S rRNA, which prevents cross-reaction with unrelated species. This procedure is rapid after the culture step and can be used to process multiple samples with a high degree of sensitivity. Standard fluorescent staining can be tedious with large numbers of patient samples. This assay also determines the presence of viable organisms only (only viable organisms will

be cultured). (Some of our data suggest that 3SR will detect only viable organisms.)

2. Specific amplification has been obtained from all serovars of *C. trachomatis* tested to date.
3. When the products of the 3SR amplification are run on nondenaturing gels, the specific product is obscured by many nonspecific products. These products are apparently due to some unknown interaction between the three enzymes, since the products can be obtained by incubation of the enzymes in buffer and nucleotides without primers and template. Specific products can be visualized by Northern (RNA) blot analysis (7) with CT321 as a probe.
4. Background contamination control for the amplification reaction is judged by examining the signal obtained from the water blank. If a significant signal is obtained, it is necessary to decontaminate working areas and pipettors with bleach before repeating the experiment.
5. If further confirmation is required, duplicate vials may be prepared and the second one may be stained with a fluorescent antibody.
6. 3SR products can also be detected by nonradioactive methods such as described in reference 4.

## References


1. **Bobo, L., F. Coutlee, R. H. Yolken, T. Quinn, and R. P. Viscidi.** 1990. Diagnosis of *Chlamydia trachomatis* cervical infection by detection of amplified DNA with an enzyme immunoassay. *J. Clin. Microbiol.* **28:**1968–1973.
2. **Fraiz, J., and R. B. Jones.** 1988. Chlamydial infections. *Annu. Rev. Med.* **39:**357–370.
3. **Guatelli, J. C., D. Y. Whitfield, D. Y. Kwoh, J. Barringer, D. D. Richman, and T. J. Gingeras.** 1990. Isothermal, in vitro amplification of nucleic acids by a multi-enzyme reaction modeled after retroviral replication. *Proc. Natl. Acad. Sci. USA* **87:**1874–1878.
4. **Haydock, P. V., J. T. Brown, and E. W. Radany.** 1992. Amplification and detection of rare RNAs using the self-sustained sequence replication reaction, p. 139–146. *In* R. Greenshields (ed.), *Biomedical Technology*. Century Press, London.
5. **Hipp, S. S., Y. Han, and D. Murphy.** 1987. Assessment of enzyme immunoassay and immunofluorescence tests for detection of *Chlamydia trachomatis. J. Clin. Microbiol.* **25:**1938–1942.
6. **Kawasaki, E. S.** 1990. Amplification of RNA, p. 21–27. *In* M. Innis, D. Gelfand, J. Sninsky, and T. White (ed.), *PCR Protocols: A Guide to Methods and Applications.* Academic Press, Inc., San Diego, Calif.
7. **Sambrook, J., E. F. Fritsch, and T. Maniatis.** 1989. *Molecular Cloning: A Laboratory Manual*, 2nd ed. Cold Spring Harbor Laboratory, Cold Spring Harbor, N.Y.
8. **Schachter, J.** 1988. The intracellular life of *Chlamydia. Curr. Top. Microbiol. Immunol.* **138:**109–139.
9. **Thomas, B. J., R. T. Evans, D. A. Hawkins, and D. Taylor-Robinson.** 1984. Sensitivity of detecting *Chlamydia trachomatis* elementary bodies in smears by use of fluorescein-labeled monoclonal antibody: comparison with conventional chlamydial isolation. *J. Clin. Pathol.* **37:**812–816.
10. **Woods, G. L., A. Young, J. C. Scott, T. M. H. Blair, and A. M. Johnson.** 1990. Evaluation of a non-isotopic probe for detection of *Chlamydia trachomatis* in endocervical specimens. *J. Clin. Microbiol.* **28:**370–372.

# PCR Detection of *Chlamydia pneumoniae*

# 1.11

Lee Ann Campbell

## Background

*Chlamydia pneumoniae* is a major cause of acute respiratory disease in humans and is responsible for approximately 10% of cases of community-acquired pneumonia (4). This organism has also been associated with other clinical manifestations including coronary artery disease, asthma, and sarcoidosis (5–7, 9). Antibody prevalence studies have shown that 50 to 60% of the adult population in the diverse geographical places studied have serological evidence of *C. pneumoniae* infection, making this one of the most prevalent infectious agents (4). The two serodiagnostic tests that have been used routinely for *C. pneumoniae* are the microimmunofluorescence test and the complement fixation test (4, 10). The microimmunofluorescence test is specific for *C. pneumoniae*; the complement fixation test measures cross-reactive antibodies against the genus *Chlamydia*. Despite the identification of cell lines that have improved culture of *C. pneumoniae*, isolation is often difficult, especially in patients who have begun antibiotic therapy before diagnosis or who have developed a high immunoglobulin G antibody titer by the time the specimens are obtained. Moreover, the best source of specimens for routine culturing has not yet been defined. The development of better direct detection methods is necessary to facilitate studies on transmission, host range, tissue tropism, and association of the organism with specific diseases.

In previous reports, we described the isolation of several cloned *C. pneumoniae Pst*I restriction fragments that reacted with all *C. pneumoniae* isolates tested, differentiated *C. pneumoniae* from the other chlamydial species, and were specific for *C. pneumoniae* (1). One of these clones, a 474-bp *Pst*I fragment, was sequenced, and primers were chosen on the basis of GenBank analysis, G+C ratio, and lack of secondary structure (Fig. 1; Table 1) (2). These primers amplified all of the *C. pneumoniae* isolates tested and did not amplify the other chlamydial species or any of the other organisms tested (2). Our recent studies on throat swab specimens from seropositive patients from whom the organism was cultured have shown that PCR is as sensitive as culture. The detection of *C. pneumoniae* by PCR in seropositive patients from whom the organism was not cultured suggests that this may provide an alternative method for laboratory diagnosis of this ubiquitous pathogen. In addition, we have used this method for detection of *C. pneumoniae* in lung, coronary artery, and aorta tissue.

**Figure 1.** PCR primers and probe for detection of *C. pneumoniae*.

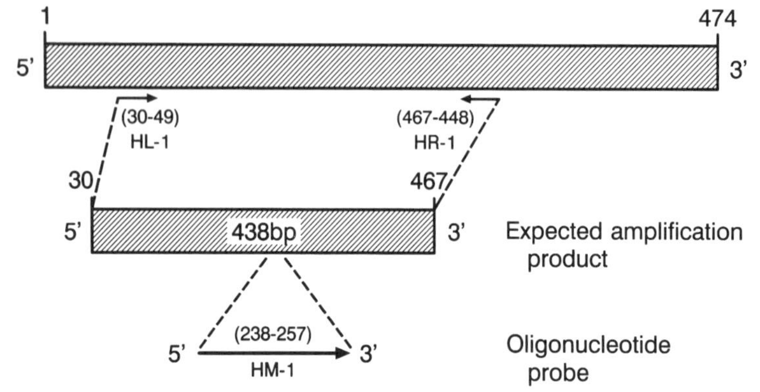


## Materials and Reagents

Sterile Tris (TE) buffer: 10 mM Tris (pH 7.4), 1 mM EDTA

PCR master mix: 50 mM Tris, 3 mM $MgCl_2$, 100 μM each deoxyribonucleoside triphosphate, 1.0 U of *Taq* polymerase (Perkin Elmer Cetus), 0.5 μM each primer. *Note: We routinely prepare master mix in 48-μl aliquots so that in the 100-μl PCR volume (after the addition of sample and sterile water), the final concentration is as listed above.*

Sample preparation buffer: proteinase K (100 μg/ml) (Boehringer Mannheim Biochemicals, Indianapolis, Ind.), 0.5% Tween 20, 0.5% Nonidet P-40

Electrophoresis loading buffer: 0.6 ml of 0.5 M EDTA, 0.5 ml of 0.5% bromphenol blue, 3.0 g of sucrose, distilled water to 10 ml

Nylon membrane (Nytran; Schleicher & Schuell)

Depurination solution: 0.25 M HCl

Denaturation solution: 0.5 M NaOH, 1 M NaCl

Neutralization solution: 0.5 M Tris, 3 M NaCl

If end-labeling oligonucleotide: [$\gamma$-$^{32}$P]ATP (DuPont NEN, Boston, Mass.), 10× kinase buffer (500 mM Tris-HCl [pH 7.5], 100 mM $MgCl_2$, 50 mM dithiothreitol, 1 mM spermidine HCl), T4 polynucleotide kinase (GIBCO BRL)

For detection by immunochemiluminescence: Genius DNA-labeling and detection kit (Boehringer Mannheim Biochemicals), Lumi-Phos 530 (Boehringer Mannheim Biochemicals)

Unless otherwise noted, all chemicals and detergents were obtained from Sigma.

**Table 1.** Primers for PCR detection of *C. pneumoniae*

| Name | Nucleotide sequence, 5′ to 3′ |
|---|---|
| HL-1 | GTT GTT CAT GAA GGC CTA CT |
| HM-1 | GTG TCA TTC GCC AAG GTT AA |
| HR-1 | TGC ATA ACC TAC GGT GTG TT |

# Methods

## Sample Collection

If *C. pneumoniae* culturing is being done, elute throat swab specimens in 1 ml of chlamydia transport medium (SPG), which contains (per liter) 75 g of sucrose, 0.52 g of $KH_2PO_4$, 1.22 g of $Na_2HPO_4$, and 0.72 g of glutamic acid. Store at −70°C. If culture is not being done, throat swab specimens can be eluted into an alternate buffer such as TES (10 mM Tris [pH 8.0], 1 mM EDTA, 150 mM NaCl).

If fresh tissue specimens are used, place them into 1 ml of SPG (or more if needed), refrigerate for 30 min to 4 h, and transfer to a −70°C freezer.

## Sample Preparation

1. Centrifuge a 200-μl aliquot of the throat swab specimen at 15,000 × *g* in a microcentrifuge for 30 min.
2. Remove the supernatant with a sterile pipette tip. Be careful to not disturb the pellet. Resuspend the pellet in sample preparation buffer at 60°C for 60 min. Boil the samples for 10 min, and immediately place them on ice to prevent reannealing of DNA.
3. Use a 15-μl aliquot for each PCR.
4. If tissue specimens are being tested, isolate DNA from tissues by the method described by Sambrook et al. (8). Resuspend purified DNA in 100 μl of TE buffer. Use a 1-μl aliquot per PCR.

## PCR Amplification

1. Prepare PCR master mix by using the recipe given above. For throat swab specimens, add 15 μl of specimen to the master mix and bring to 100 μl with sterile water. For DNA samples isolated from tissue, add 1 μl of sample to master mix and bring to a final volume of 100 μl with sterile water.
2. Amplify samples in a programmable thermocycler by using a three-step profile for 40 cycles. Each cycle consists of denaturation at 94°C for 1 min, annealing at 48°C for 1 min, and primer extension at 72°C for 1 min.
3. Prepare a 1.5% agarose gel in TBE buffer. Electrophoretically separate reaction products (10 μl per well) in TBE buffer. Molecular size standards (123-bp DNA ladder; GIBCO BRL) should be run simultaneously. Stain with ethidium bromide for 10 min to visualize the 437-bp amplification product.
4. Depurinate (10 min), denature (30 min), and neutralize (30 min) the gel by soaking in the appropriate buffers and transferring to Nytran by standard methods (8).

## Confirmation of Amplified Product

1. If the internal primer (HM-1) is used as the diagnostic probe, end label with [$\gamma$-$^{32}$P]ATP (8). Add 1.3 μl of 10× kinase buffer, 1 μl of HM-1 (100 ng/μl), 10 μl of [$\gamma$-$^{32}$P]ATP, and 1.0 μl of T4 polynucleotide kinase. Incubate at 37°C for 1 h and then at 65°C for 10 min. Follow hybridization conditions for end-labeled oligonucleotides as described by Sambrook et al. (8). *Note: We have found no difference in results between using the labeled internal primer and using the original 474-bp cloned C. pneumoniae PstI fragment labeled with digoxigenin-dUTP as the confirmatory probe. Nonspecific amplification products do not hybridize to either probe.*
2. If chemiluminescent detection is used, label the purified 474-bp fragment with

digoxigenin-dUTP with the Genius DNA-labeling and detection kit as specified by the manufacturer.

3. After a 1-h prehybridization of filter at 63°C, add 10 ng of labeled probe per ml. Incubate the filter overnight at 63°C.
4. Wash filters with 2× SSC–0.1% SDS (twice at room temperature) and with 0.1× SSC–0.1% SDS (twice at 68°C).
5. Detect hybrids by immunochemiluminescence with Lumi-Phos 530 as specified by the manufacturer. Expose film for 10 min to 2 h.

## Quality Control

### Positive Amplification Control

*C. pneumoniae* cells are grown in the HL human cell line (2), and the organisms are purified with a linear gradient of meglumine diatrizoate (Hypaque 76; Winthrop Breon Laboratories, New York, N.Y.) (1).

*C. pneumoniae* purified elementary bodies for which the titer (inclusion-forming units [IFUs] per milliliter) has been determined are concentrated, lysed, and treated in the same manner as the clinical specimens. In each test, 3 and 300 IFU (one reaction each) are amplified. The number of IFUs per milliliter should be determined by the method of Furness et al. (3).

Alternatively, 10 ng and 10 pg of *C. pneumoniae* DNA can be used. *C. pneumoniae* type strain TW-183 (ATCC VR2282) or isolate AR-39 (ATCC 53592) can be used.

### Negative Amplification Control

All components used in the reaction, with sterile water added instead of the sample.

### Specificity Control

100 ng of *Chlamydia trachomatis* DNA, any human serovar.

### Positive Patient Control

If available, a specimen from a culture-positive patient.

### Negative Patient Control

Throat swab specimen from an isolation-negative, seronegative patient who has been determined to be PCR negative.

## Comments

1. For consistency, all reagents are prepared in sterile water (Baxter, Health Care Corp.). We have tried several other methods of sample treatment, including addition of either sodium azide (0.04%, vol/vol) or EDTA (0.4 mM) or both, use of only one detergent, and proteinase K treatment without addition of detergent. All possible combinations of these reagents with proteinase K have been tried. No increase was observed in the sensitivity of detection compared with that obtained by using the buffer described above.
2. If possible, sample preparation, PCR amplification, and analysis of amplifica-

tion products should be done in separate rooms. The room in which we prepare samples consists of individual cubicles with sliding doors and germicidal UV bulbs. The cubicle used for sample preparation for PCR is used for this purpose only. Clinical specimens are stored in a different room, in which no aspect of sample preparation or analysis is done.

3. The aliquot of sample to be used is given as a guideline. In general practice, we test more than 1 volume (e.g., 5, 15, and 25 μl for throat swab specimens). The lower volume is used to dilute inhibitors of PCR, whereas the larger volume is used to increase the possibility that organisms are in the sample to be amplified.
4. For clinical specimens, to ensure that negative results are true negatives and not the result of inhibition of PCR by inhibitors present in the specimen, equal volumes are seeded with 25 IFU of purified *C. pneumoniae* elementary bodies. If there is no amplification product after PCR, it can be concluded that the sample contains inhibitory substances, which can either be removed by drop dialysis or decreased by using a smaller sample volume.
5. An annealing temperature of 48°C results in maximum product amplification without loss of specificity. However, in some clinical samples, an increase in spurious products is observed. This can be circumvented by adjusting the annealing temperature to 55°C.
6. In the course of our experiments, we have obtained our primers from different sources. Presumably because the percentage of complete primers may vary, we have found that the optimal deoxynucleotide concentrations may differ and should be standardized with each new batch of primers.

## References


1. **Campbell, L. A., C.-C. Kuo, and J. T. Grayston.** 1987. Characterization of the new *Chlamydia* agent, TWAR, as a unique organism by restriction endonuclease analysis and DNA-DNA hybridization. *J. Clin. Microbiol.* **25:**1911–1916.
2. **Campbell, L. A., M. Perez Melgosa, D. J. Hamilton, C.-C. Kuo, and J. T. Grayston.** 1992. Detection of *Chlamydia pneumoniae* by polymerase chain reaction. *J. Clin. Microbiol.* **30:**434–439.
3. **Furness, G., D. M. Graham, and P. Reeve.** 1960. The titration of trachoma and inclusion blenorrhoeae viruses in cell culture. *J. Gen. Microbiol.* **23:**613–619.
4. **Grayston, J. T., L. A. Campbell, C.-C. Kuo, C. Mordhorst, P. Saikku, D. H. Thom, and S.-P. Wang.** 1990. A new respiratory tract pathogen: *Chlamydia pneumoniae* strain TWAR. *J. Infect. Dis.* **161:**618–625.
5. **Gronhagen-Riska, C., P. Saikku, H. Riska, B. Froseth, and J. T. Grayston.** 1988. Antibodies to TWAR—a novel type of *Chlamydia*—in sarcoidosis, p. 297–301. *In* C. Grassi, G. Rizzato, and E. Pozzi (ed.), *Sarcoidosis and Other Granulomatous Disorders.* Excerpta Medica, New York.
6. **Hahn, D. L., R. W. Dodge, and R. Golubjatnikov.** 1991. Association of *Chlamydia pneumoniae* (strain TWAR) infection with wheezing, asthmatic bronchitis, and adult-onset asthma. *JAMA* **266:**225–230.
7. **Saikku, P., K. Mattila, and M. S. Nieminen.** 1988. Serological evidence of an association of a novel *Chlamydia*, TWAR, with chronic coronary heart disease and acute myocardial infarction. *Lancet* **ii:**983–986.
8. **Sambrook, J., E. F. Fritsch, and T. Maniatis.** 1989. *Molecular Cloning: A Laboratory Manual*, 2nd ed. Cold Spring Harbor Laboratory Cold Spring Harbor, N.Y.
9. **Thom, D. H., S.-P. Wang, J. T. Grayston, D. S. Siscovick, D. K. Stewart, R. A. Kronmal, and N. S. Weiss.** 1991. *Chlamydia pneumoniae* strain TWAR antibody and

angiographically demonstrated coronary artery disease. *Arterioscler. Thromb.* **11:**547–551.

10. **Wang, S.-P., and J. T. Grayston.** 1986. Microimmunofluorescence serological studies with the TWAR organism, p. 329–332. *In* J. D. Oriel, G. Ridgway, J. Schachter, D. Taylor-Robinson, and M. Ward (ed.), *Chlamydial Infections. Proceedings of the Sixth International Symposium on Human Chlamydial Infections*. Cambridge University Press, Cambridge.

# PCR Detection of *Mycoplasma pneumoniae*

# 1.12

Maurice Garret and Jacques Bonnet

## Background

*Mycoplasma pneumoniae* is generally confined to the respiratory tract and probably accounts for 20% of all community-acquired pneumonias (4). The development of a sensitive and rapid laboratory diagnostic method is important to ensure the correct identification of the organism and for epidemiological studies of *M. pneumoniae* infections. Two methods are currently used for diagnosis (1). The first is difficult and time-consuming since it requires the cultivation of *M. pneumoniae* for 3 weeks. The second, which is more widely used, is based on serological methods. Since the presence of cold agglutinins is neither constant nor specific, the only reliable immunological technique is the demonstration of a significant increase (over a 2-week period) of the level of specific antibodies. The measures are currently based on a complement-binding test with a glycolipidic antigen.

PCR is used for the rapid detection of a variety of pathogenic microorganisms (2, 9). We cloned a specific segment of *M. pneumoniae* from a genomic DNA library (3). This sequence was used as a target in the PCR detection of the presence of this pathogen in clinical samples. We showed that a set of *M. pneumoniae* reference strains consistently gave a positive signal. Human DNA, DNA from species generally found in the respiratory tract, and DNA from various *Mycoplasma* species different from *M. pneumoniae* gave negative results. These data show that this PCR approach provides a specific and sensitive tool for the detection of *M. pneumoniae* DNA (3, 6).

## Materials and Reagents

40% acrylamide solution: 38 g of acrylamide and 2 g of *N,N'*-methylenebisacrylamide (Sigma) per 100 ml stored in a dark bottle at 4°C

*N,N,N',N'*-Tetramethylethylenediamine (TEMED; Merck)

10% ammonium persulfate (Merck) solution in water (this solution should be freshly made)

6× Electrophoresis loading buffer: 0.25% bromophenol blue, 30% glycerol in water

DNA size markers: *Hae*III-digested pBR322 (Boehringer Mannheim)

Denaturing solution: 1.5 M NaCl, 0.5 M NaOH

Neutralizing solution: 1 M Tris-HCl (pH 8.0), 1.5 M NaCl, 1 mM EDTA

50× Denhardt's solution: 1 g of bovine serum albumin (Boehringer Mannheim), 1 g of Ficoll (Pharmacia), 1 g of polyvinylpyrrolidone (Sigma) per 100 ml

5 mg of salmon sperm DNA per ml in 10 mM Tris-HCl (pH 8.0)–1 mM EDTA

5× PCR master mix: 50 mM Tris-HCl (pH 8.3), 250 mM KCl, 12.5 mM $MgCl_2$, 1 mM each deoxyribonucleoside triphosphate (Boehringer Mannheim), 5 μM each primers MP-1 and MP-2, 100 U of *Taq* polymerase (Perkin Elmer Cetus) per ml

Vertical electrophoresis apparatus

Hybond nylon membrane (Amersham)

## Methods

### Sample Preparation and Storage

1. Centrifuge 10 ml of bronchoalveolar lavage specimen (5) for 20 min at 2,000 × *g*.
2. Remove and discard 9 ml of the supernatant fraction, and resuspend the pellet in the remaining supernatant. This sample can be stored at −70°C for further analysis.
3. Before PCR amplification, dilute 15 μl of thawed sample with 35 μl of $H_2O$, and then heat the sample in a boiling water bath for 10 min. This step allows lysis of the cells and denaturation of the DNA.
4. Transfer 40 μl of the sample to a 0.5-ml microcentrifuge tube.

### PCR Amplification and Analysis

1. Add 10 μl of the 5× PCR master mix prepared as indicated above. *Note: To avoid contamination, sample preparation and PCR amplification must be performed in two separate rooms with different sets of pipettes.*
2. Add 50 μl of mineral oil to each tube.
3. Amplify in a programmable thermal cycler with the following profile: 1 min of denaturation at 94°C, 1 min of annealing at 55°C, and 1 min of polymerization at 72°C, for a total of 30 cycles.
4. Prepare an 8% acrylamide solution in TBE buffer (8). *Note: A gel 10 to 15 cm long and 1 mm thick is convenient for this purpose.* Transfer 10 μl of PCR medium to a tube containing 2 μl of 6× loading buffer. *Note: It is not necessary to extract oil when pipetting is done carefully.* Load samples and 0.75 μg of DNA markers on the gel.
5. Apply 10 V/cm until the bromophenol blue dye has migrated approximately two-thirds the length of the gel.
6. At the end of the run, remove one glass plate and soak the gel for 15 min at room temperature in TBE buffer containing 1 μg of ethidium bromide per ml.
7. Photograph the gel under UV light (8).

### PCR Product Characterization

1. After ethidium bromide staining, denature and neutralize the gel by soaking in denaturation and neutralization solutions for 30 min each. Set up a capillary blot on a nylon filter as described by the manufacturer.

**Table 1.** Oligonucleotides for PCR detection of *M. pneumoniae*

| Function | Name | Nucleotide sequence, 5′ to 3′ |
|---|---|---|
| Primer | MP5-1 | GAA GCT TAT GGT ACA GGT TGG |
| Primer | MP5-2 | ATT ACC ATC CTT GTT GTA AG |
| Probe | MP5-4 | CGT AAG CTA TCA GCT ACA TGG AGG |

2. Label 20 pmol of oligonucleotide MP5-4 (Table 1) with T4 DNA polynucleotide kinase and 50 μCi of [γ-$^{32}$P]ATP (8).
3. Prehybridize for 1 h at 50°C in 5× SSPE–5× Denhardt's–0.5% sodium dodecyl sulfate (SDS)–0.1 μg of denatured salmon sperm DNA per ml. *Note: This prehybridization buffer may have to be slightly adapted to each particular nylon filter. Read manufacturer's instructions carefully.*
4. Hybridize overnight at 50°C in 10 ml of the prehybridization buffer described above containing 20 pmol of labeled oligonucleotide. *Note: This hybridization solution can be used several times for several filters. Between each hybridization, store the solution at −20°C for up to 2 weeks.*
5. Wash filters twice for 5 min in 5× SSPE–0.1% SDS at room temperature and once for 30 min at 50°C.
6. Autoradiograph with an intensifying screen for 2 h to overnight at −70°C.

# Quality Control

## Positive Amplification Control

1 ng of *M. pneumoniae* DNA (FH or PI 1428 strain).

## Negative Amplification Control

45 μl of distilled water added to the PCR master mix.

## Specificity Control

100 ng of *Escherichia coli* DNA and 100 ng of human DNA.

## Positive Control

Simulated positive clinical samples obtained by adding a serial dilution of an *M. pneumoniae* culture to a culture-negative PCR sample.

# Comments

1. The presence of *M. pneumoniae* is indicated by the amplification of a 144-bp segment of the *M. pneumoniae* genome.
2. In some experiments, nonspecific amplification products are seen on ethidium bromide-stained gels. These products will not hybridize with the specific MP5-4 oligonucleotide.
3. Biotin can be used as an alternative to the $^{32}$P labeling of the oligonucleotide probe (7).
4. Extraction of DNA from clinical specimens was generally not necessary.

**Figure 1.** Nucleotide sequence of the amplified *M. pneumoniae* DNA fragment. Sequences corresponding to oligonucleotides described in Table 1 are boxed. The *Dde*I restriction enzyme site is overlined.

5' [GAAGCTTATGGTACAGGTTGG] TTAATTGACTGGAAGGATGTTAAGCAGGACAA
MP5-1

Dde I
CAATTTACTGCTTA**CTTAG**CTACCAA [CCTCCATGTAGCTGATAGCTTACG] AA
MP5-4

ATAAAGACGATTACAAG [CCTTACAACAAGGATGGTAAT] 3'
MP5-2

However, if background is observed, the sample may be digested with proteinase K and then subjected to phenol extraction (3).

5. The amplified *M. pneumoniae* DNA fragment contains a *Dde*I restriction site (Fig. 1). This restriction site can be used for characterization of the PCR product.
6. With purified genomic *M. pneumoniae* DNA, we were able to detect the equivalent of one *Mycoplasma* organism after a 50-cycle amplification. With the protocol described in this paper and with 25 cycles of amplification we were able to detect 100 to 1,000 organisms in clinical samples.
7. Throat swab samples can be processed by following the same protocol with the primers described in this paper (6).

## References


1. **Bébéar, C., B. de Barbeyrac, C. Bernet, and H. Renaudin.** 1989. Méthode d'exploration des infections a mycoplasmes. *Ann. Biol. Clin.* **27:**2492–2496.
2. **Bej, A. K., M. H. Mahbubanl, and R. M. Atlas.** 1991. Amplification of nucleic acids by polymerase chain reaction (PCR) and other methods and their applications. *Crit. Rev. Biochem. Mol. Biol.* **26:**301–334.
3. **Bernet, C., M. Garret, B. de Barbeyrac, C. Bébéar, and J. Bonnet.** 1989. Detection of *Mycoplasma pneumoniae* by using the polymerase chain reaction. *J. Clin. Microbiol.* **27:**2492–2496.
4. **Cassel, G. H., and B. C. Cole.** 1981. Mycoplasmas as agents of human diseases. *N. Engl. J. Med.* **304:**80–89.
5. **Chastre, J., J. Y. Fagon, P. Soler, Y. Domart, J. Pierre, M. C. Dombret, C. Gibert, and A. J. Hance.** 1989. Quantification of BAL cells containing intracellular bacteria rapidly identifies ventilated patients with nosocomial pneumoniae. *Chest* **95:**190S–192S.
6. **De Barbeyrac, B., C. Bernet, R. Teyssou, F. Poutiers, and C. Bébéar.** Detection of human mycoplasmas by *in vitro* DNA amplification. *In* I. Kahane and A. Adoni (ed.), *Rapid Diagnosis of Mycoplasma*, in press. Plenum Publishing Corp., New York.
7. **Guitteny, A. S., B. Fouquet, C. Mougin, R. Theoule, and B. Bloch.** 1988. Histological detection of messenger RNA with biotinylated synthetic oligonucleotide probes. *J. Histochem. Cytochem.* **36:**563–571.
8. **Sambrook, J., E. F. Fritsch, and T. Maniatis.** 1989. *Molecular Cloning: A Laboratory Manual*, 2nd ed. Cold Spring Harbor Laboratory, Cold Spring Harbor, N.Y.
9. **Skov Jensen, J., S. A. Uldum, J. Sondergard-Andersen, J. Vuust, and K. Lind.** 1991. Polymerase chain reaction for detection of *Mycoplasma genitalium* in clinical samples. *J. Clin. Microbiol.* **29:**46–50.

# Rapid and Sensitive PCR Method for Identification of *Mycoplasma* Species in Tissue Culture

# 1.13

JOLENE G. WONG-LEE AND MICHAEL LOVETT

## Background

*Mycoplasma* spp., which belong to the class *Mollicutes*, are the smallest known free-living organisms (1). They constitute a common and pervasive tissue culture contaminant and are implicated in a number of human and veterinary disease states. The small size and lack of a cell wall make mycoplasmas extraordinarily difficult to detect by conventional methods (1). The effects of mycoplasma infection are significant, including changes in cell morphology, growth rates, and metabolic states (1, 4). These changes can profoundly affect the results obtained in any cell culture-based bioassay system. Mycoplasmas are also involved in a number of diseases; *Mycoplasma pneumoniae* is the major causative agent of community-acquired pneumonia (1, 2).

The use of DNA probes for the detection of *Mycoplasma* spp. has been shown to lack sensitivity (2). We have designed a method in which PCR is used to detect the eight most common *Mycoplasma* species: *M. hyorhinis*, *M. arginini*, *M. pneumoniae*, *M. fermentans*, *M. orale*, *M. pirum*, *Acholeplasma laidlawii*, and *Spiroplasma mirum* (4–6). By using published 16S rRNA sequences from the above species, we aligned the sequences to determine the selection of oligonucleotides (Table 1). The oligonucleotides selected for use in the PCR detect all of these species without cross-reacting with other common tissue culture contaminants such as bacteria or yeasts and without the need for cultivation of the organism as with earlier techniques (3).

A positive control that contains a lambda DNA insert within the target sequence has been developed by cloning of the target sequence from *M. arginini* into a plasmid vector followed by ligation of a 1.1-kb *Sac*II fragment of a lambda DNA digest to a single *Sac*II site within the target sequence (Fig. 1). This allows discrimination of the positive control product from the *Mycoplasma*-specific product on the basis of size (1.6 and 0.5 kb, respectively).

## Materials and Reagents

PCR master mix: 31.75 μl of water, 5 μl of reaction buffer (final concentrations: 10 mM Tris-HCl, 50 mM KCl, 1.5 mM $MgCl_2$, 0.001% [wt/vol] gelatin), 1 μl of each dNTP (final concentrations, 200 μM each), 0.25 μl of *Taq* polymerase (final concentration, 2.5 U), 2.5 μl of primer A (final concentration, 1.0 μM), 2.5 μl of primer B (final concentration, 1.0 μM), 2.0 μl of positive control

**Figure 1.** Positions of universal primers-RI for building the positive control. See the text for additional details. *GenBank accession code MYCRRNAAR.

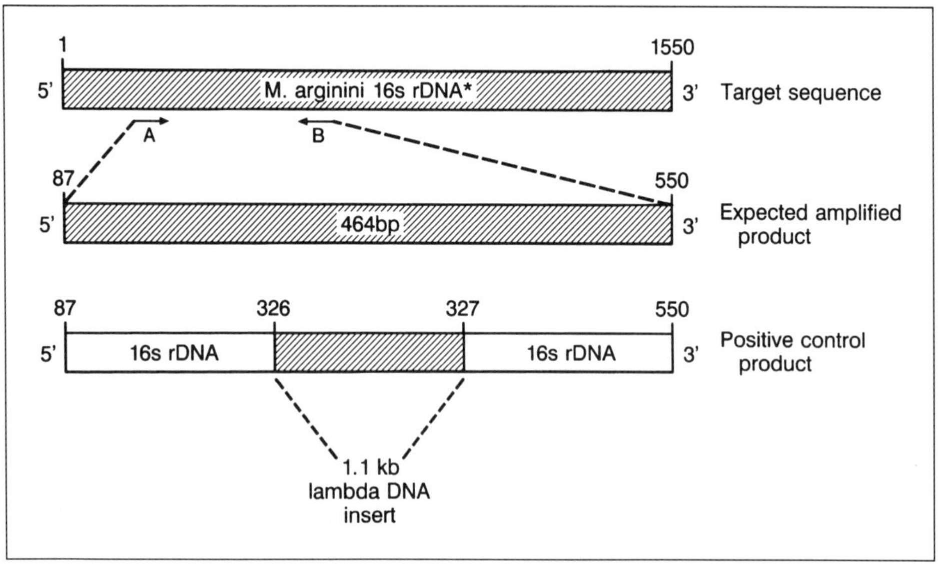


*Escherichia coli* HB101 competent cells

SOC medium (per liter): 20 g of Bacto tryptone, 5 g of Bacto yeast extract, 0.5 g of NaCl, 20 mM glucose

pBS$^+$ vector (Stratagene)

*Eco*RI (Promega)

*Sac*II (Promega)

*Bam*HI (Promega)

Lambda DNA (Stratagene)

Plasmid Midi and Maxi Prep (Quiagen)

NA 45 paper

T4 DNA ligase

STET buffer (sodium chloride, Tris, EDTA, Triton): 0.1 M NaCl, 10 mM Tris-Cl (pH 8.0), 1 mM EDTA (pH 8.0), 5% Triton X-100

**Table 1.** Oligonucleotides for PCR detection of *Mycoplasma* spp.[a]

| Function | Name | Nucleotide sequence, 5′ to 3′ |
|---|---|---|
| Universal *Mycoplasma* primer | Primer A | GGC GAA TGG GTG AGT AAC ACG |
| Universal *Mycoplasma* primer | Primer B | CGG ATA ACG CTT GCG ACC TAT G |
| Universal *Mycoplasma* primer-RI | Primer A-RI | **GAA TTC** GGC GAA TGG GTG AGT AAC ACG |
| Universal *Mycoplasma* primer-RI | Primer B-RI | CGG ATA ACG CTT GCG ACC TAT G**CT TAA G** |

[a] This set of universal primers is used to create the positive control. The nucleotides in boldface are the sequence in which the *Eco*RI enzyme will recognize and cut.

LB + ampicillin (50 μmol/ml) plates
PrimeErase Quik Kit (Stratagene)

## Methods

### Sample Collection

Remove 2 to 3 ml of spent tissue culture medium (medium should be at least 2 days old). Make sure that the medium is not turbid with bacteria or yeasts. Samples can be stored at 4°C for up to 2 weeks prior to analysis or at −20°C for 2 to 3 months.

### Sample Preparation

Run samples undiluted and at 1:100 dilution, because the degree of infection is unknown for each sample and a high degree of infection will inhibit the PCR and compete for the positive control.

1. Remove 10 μl of each sample, and add 990 μl of *Mycoplasma*-free water (1:100 dilution). *Note: Injection or irrigation grade water is Mycoplasma free.*
2. Mark 0.6-ml microcentrifuge tubes with each sample number and either A or B (A = undiluted sample; B = 1:100 dilution).

### PCR Amplification

1. Prepare the PCR master mix as indicated above. Allow for 48 μl of mix per reaction, including the positive and negative control. *Note: It is best to make allowances for losses of volume by making enough master mix for one or two samples more than required.*
2. Place 48 μl of master mix into appropriately marked 0.6-ml microcentrifuge tubes. Distribute into the corresponding tubes 2 μl of each sample (undiluted, 1:100 dilution), making sure that you use a clean pipette tip for each sample.
3. Program the thermal cycler for a 1-s ramp and 1-min denaturation at 94°C, 1-s ramp and 1-min annealing at 60°C, and 1-s ramp and 1.5-min extension at 72°C for a total of 30 cycles.
4. Prepare 1.5% SeaKem agarose gel in TBE buffer-ethidium bromide. When cycling is completed, remove 10 μl from each tube and add to 2 μl of loading buffer. Load the entire amount onto the gel. Run the gel at a constant voltage of 100 V for approximately 30 to 45 min. The first dye front should have migrated approximately three-quarters the length of the gel.
5. Visualize the amplified product under UV light. The positive-control amplified product will be 1.6 kb, and the sample, if positive, will have an amplified product at 0.5 kb.

### Interpretation of Results

If the lane contains a band only at 1.6 kb for both dilutions and no band at 0.5 kb, the sample is assumed to be negative.

If the lane contains bands at 1.6 and 0.5 kb at either dilution, the sample is assumed to be positive.

If the lane contains a band at either dilution at 0.5 kb and not at 1.6 kb, the sample can be considered positive.

If the lane contains no bands at either 1.6 or 0.5 kb, the sample is unequivocal, and the procedure should be repeated at a lower dilution (1:10 and 1:1,000).

## Quality Control

### Positive Control for Amplification and Samples

Add 300 to 500 molecules of positive control per reaction.

### Negative Control for Amplification

2 μl of *Mycoplasma*-free water (i.e., water for injection or irrigation).

### Negative Sample Control

2 μl of fresh tissue culture medium.

## References

1. **Barile, M. F., and S. Razin.** 1979. *The Mycoplasmas*, vol. I. *Cell Biology*. Academic Press, Inc., New York.
2. **Bernet, C., M. Garret, B. de Barbeyrac, C. Bebear, and J. Bonnet.** 1989. Detection of *Mycoplasma pneumoniae* by using the polymerase chain reaction. *J. Clin. Microbiol.* **27:**2492–2496.
3. **Chanock, R. W., L. Hayflick, and M. F. Barile.** 1962. Growth on artificial medium of an agent associated with a typical pneumonia and its identification as a PPLO. *Proc. Natl. Acad. Sci. USA* **48:**41.
4. **Masover, G. K., and D. W. Buck.** 1983. Detecting mycoplasmas in cell cultures. *Med. Device & Diagn. Ind.* **5**(11)**:**43–48.
5. **Razin, S., and M. F. Barile.** 1985. *The Mycoplasmas*, vol. IV. *Mycoplasma Pathogenicity*. Academic Press, Inc., New York.
6. **Sambrook, J., E. F. Fritsch, and T. Maniatis.** 1989. *Molecular Cloning: A Laboratory Manual*, 2nd ed. Cold Spring Harbor Laboratory, Cold Spring Harbor, N.Y.

# PCR Detection of *Legionella pneumophila* and *L. dumoffii* in Water

# 1.14

JEFFERY S. LOUTIT AND LUCY S. TOMPKINS

## Background

At Stanford University Medical Center over the last 9 years, both *Legionella pneumophila* and *L. dumoffii* have been frequent causes of sternal wound infection, pleural effusion, pneumonia, and even endocarditis (9) in patients undergoing cardiothoracic surgery. In 1989 a probable link was made between contaminated water samples and infected patients (4). Before 1989, no epidemiologic link could be found between *L. pneumophila* culture-positive potable water and infected patients. In addition, *L. dumoffii* could not be cultured from any potable water source.

Cultivation of *Legionella* species from environmental water samples usually involves centrifugation and/or acid treatment of large water volumes and has been considered the standard against which other detection methods are compared. However, many investigators have been unable to confirm *Legionella* contamination of environmental and potable water samples by cultivation and have failed to pinpoint the reservoirs and mode of transmission of several nosocomial *Legionella* strains (10). Amplification of target DNA sequences before Southern hybridization probing with an internal DNA fragment theoretically offers a more sensitive and rapid alternative method of detecting a small number of viable, cultivatable bacteria or even "noncultivatable" or "nonviable" forms (1).

Our group has developed sensitive and species-specific detection systems for both *L. pneumophila* (8) and *L. dumoffii* (2) by using PCR and Southern hybridization (Table 1). Recent PCR and Southern hybridization analyses of water samples collected at Stanford University Medical Center over a 9-year period have been performed. Results indicate that this method of detection may not be any more sensitive than established culture techniques (3). However, *L. pneumophila*- and *L. dumoffii*-specific PCR tests of environmental water samples provide a rapid and sensitive alternative to current detection methods.

## Materials and Reagents

Aerosol-barrier pipette tips, 1 to 200 μl and 200 to 1,000 μl

"Digestion" buffer: 100 mM Tris-HCl, 1 mM EDTA (pH 8.5), 2% Laureth 12 (nonionic surfactant), 400 μg of proteinase K per ml (surfactant was obtained from PPG/Mazer Chemicals, Gurnee, Ill.; other chemicals were obtained from Sigma Chemical Co., St. Louis, Mo.)

**Table 1.** Oligonucleotides for PCR detection of *Legionella* spp.

| Species and Function | Name | Nucleotide sequence, 5′ to 3′ |
|---|---|---|
| *L. pneumophila* | | |
| Primer | LEG1 | GTC ATG AGG AAT CTC GCT G |
| Primer | LEG2 | CTG GCT TCT TCC AGC TTC A |
| Probe | LEG3 | GTC CGT TAT GGG GTA TTG ATC ACC A |
| *L. dumoffii* | | |
| Primer | LDBKS1 | ATA CAC GTG GTG GAG GTA C |
| Primer | LDBSK3 | GCG GGC AAT ATC TTG CAT C |
| Probe | LDBKS4 | GGA CAA TGC ACA GAA AAT CCT TGC C |

PCR master mix: 10 mM Tris (pH 8.3), 1.5 mM $MgCl_2$, 50 mM KCl, 200 μM each deoxyoligonucleoside triphosphate, 1.25 U of *Taq* polymerase (Perkin Elmer Cetus), 0.2 μM each primer LEG1 and LEG2 (*L. pneumophila*) or LDBKS1 and LDBSK3 (*L. dumoffii*) (per 50-μl reaction)

TAE buffer (50× stock, used as 1×): 242 g of Tris base (Sigma), 57.1 ml of glacial acetic acid (J. T. Baker Inc., Phillipsburg, N.J.), and 100 ml of 0.5 M EDTA (pH 8.0) per liter

Electrophoresis loading buffer (6× stock, used as 1×): 0.25% bromphenol blue, 0.25% xylene cyanol FF, and 30% glycerol in $H_2O$

Nylon membrane (Hybond-N; Amersham Corp., Arlington Heights, Ill.)

Denaturation solution: 0.25 N HCl, 0.6 M NaCl, 0.4 N NaOH

Renaturation solution: 0.5 M Tris-HCl (pH 7.5), 1.5 M NaCl

Prehybridization solution: 6× NaCl-EDTA-Tris (1× NaCl-EDTA-Tris is 0.15 M NaCl, 1 mM EDTA, and 0.03 M Tris-HCl [pH 8.0]), 0.5% Nonidet P-40, 100 μg of denatured salmon sperm DNA per ml

Hybridization solution: 6× NaCl-EDTA-Tris, 0.5% Nonidet P-40, 250 μg of yeast tRNA per ml, 50 ng of probe (LEG3 for *L. pneumophila* and LDBKS4 for *L. dumoffii*) per ml

Posthybridization solution: 6× SSC (1× SSC is 0.15 M NaCl plus 0.015 M sodium citrate)

## Methods

### Sample Collection

Collect 10 to 50 ml of environmental water sample in new sterile plastic containers, and store at 4°C until analysis.

### Sample Preparation

For preparation of template DNA prior to amplification, subject water samples to centrifugation and digestion (6, 11) by the following procedure.

1. Add 1 ml of water sample to a 1.5-ml microcentrifuge tube.
2. Centrifuge at maximum speed (ca. 12,000 rpm) in a microcentrifuge for 20 min.
3. Pour and pipette off remaining water, leaving ca. 30 to 50 μl in the bottom of the tube. *Note: Only rarely can a pellet be seen; therefore, take care not to remove any centrifuged bacteria. As an aid in locating the pellet, turn the neck*

*of the microcentrifuge tube outward to serve as a marker for the expected location of the pellet.*

4. Add 100 μl of digestion buffer (see above) for digestion of the pellet. Vortex briefly (3 to 5 s). Place tubes into a 55°C heating block for 2 to 3 h or overnight at 37°C.
5. Incubate digests for 10 min at 95°C. Centrifuge them briefly (5 to 10 s), and then place them in an ice bucket. *Note: The heating/snap-cooling step serves to inactivate the protease and denature the genomic target DNA.*
6. Spin at maximum speed in a microcentrifuge for 2 min; use a 10-μl aliquot of the supernatant per reaction.

## PCR Amplification

1. Prepare PCR master mix from the recipe above, allowing 40 μl per reaction. *Note: In preparing the master mix, it is useful to overestimate the volume by one to two reactions to allow for pipetting inaccuracies and other losses of volume.* Aliquot 40 μl of master mix into 0.5-ml microcentrifuge tubes. Add 10 μl of the digested water samples. *Note: Perform all pipetting operations, including preparation of the master mix, with aerosol-barrier pipette tips.*
2. Amplify in a programmable thermal cycler by using a three-step profile: 1 min of denaturation at 94°C followed by 1 min of annealing at 55°C and then 2 min of extension at 74°C, for a total of 35 cycles. After 35 cycles, perform a further 7-min extension step at 74°C, and then maintain the samples at 4°C until analysis.
3. Prepare 1% agarose gel in TAE buffer. Electrophoretically separate the reaction products (10 μl/ml) in TAE buffer; the bromphenol blue indicator should migrate approximately two-thirds the length of the gel.
4. Visualize the 0.8-kb (*L. pneumophila*) or 1.0-kb (*L. dumoffii*) amplification product after staining for 10 min with ethidium bromide in a UV light box.
5. Denature the gel by soaking twice in denaturation solutions (0.25 N HCl) for 15 min each and then for 30 min in NaCl-NaOH solution. Neutralize the gel by soaking in neutralization buffer for 30 min. Transfer reaction products onto nylon filters by Southern blotting as previously described (7).
6. Couple reaction products to the membrane by using a UV cross-linker set at a predetermined dosage of $1.2 \times 10^5$ μJ.

## Isotopic Detection

1. Prehybridize the nylon membrane with the prehybridization buffer for 3 h in a 65°C water bath with gentle agitation.
2. Prepare the internal probe (either LEG3 or LDBKS4) for hybridization by end labeling with [$\gamma$-$^{32}$P]dATP and polynucleotide kinase by previously described methods (5).
3. Hybridize in the hybridization solution overnight in a 42°C water bath with gentle agitation.
4. Following hybridization, wash the membrane in posthybridization solution for 5 min (repeat three times). Allow to dry on 3MM chromatography paper, and then expose plastic-wrapped membrane to Kodak X-Omat film overnight. *Note: The exposure time may be shorter or longer depending on the intensity of the signal.*

## Quality Control

### Positive Amplification Control

1 ng of *L. pneumophila* or *L. dumoffii* DNA (one reaction).

### Negative Amplification Controls

10 μl of distilled water added to PCR buffer (one reaction).

1 ml of sterile distilled water which underwent the same digestion processes as the tested water samples (one reaction). *Note: This is a negative control for the digestion process.*

### Sensitivity Control

1-ml water samples seeded with decreasing concentrations ($10^4$ to 1 CFU/ml) of *L. pneumophila* or *L. dumoffii* (four reactions).

## Comments

1. The presence of *L. pneumophila* or *L. dumoffii* is indicated by amplification of a 0.8- or 1-kb fragment, respectively, which hybridizes to their respective isotopically labeled internal probe.
2. Other nonspecific amplification products may be seen on an agarose gel, but these products will not hybridize to the internal probe.
3. The primers and internal probes were selected from random chromosomal fragments found to be specific for their respective organisms. Extensive testing of other *Legionella* species and water-borne organisms by using the described detection process did not reveal any cross-reactivity.
4. The sensitivity of this detection method for both *L. pneumophila* and *L. dumoffii* is very good; it is able to detect 1 to 10 CFU/ml of water, thereby eliminating the need for centrifugation, acid treatment, and culture of large volumes of water.

## References

1. **Hussong, D., R. R. Colwell, M. O'Brien, E. Weiss, A. D. Pearson, R. M. Weiner, and W. D. Burge.** 1987. Viable *Legionella pneumophila* not detectable by culture on agar media. *Bio/Technology* **5:**947–950.
2. **Loutit, J. S., and L. S. Tompkins.** 1990. Detection of *Legionella dumoffii* in water by DNA amplification and hybridization. *Program Abstr. 30th Intersci. Conf. Antimicrob. Agents Chemother.*, abstr. no. 1033.
3. **Loutit, J. S., and L. S. Tompkins.** 1992. Evaluation of a DNA amplification procedure for detection of *Legionella pneumophila* and *Legionella dumoffii* in water. *Abstr. 1992 Int. Symp. Legionella*, abstr. no. 1–32.
4. **Lowry, P. W., R. J. Blankenship, W. Gridley, N. J. Troup, and L. S. Tompkins.** 1991. A cluster of Legionella sternal wound infections due to postoperative topical exposure to contaminated tap water. *N. Engl. J. Med.* **324:**109–113.
5. **Sambrook, J., E. F. Fritsch, and T. Maniatis.** 1989. *Molecular Cloning: A Laboratory Manual*, 2nd ed. Cold Spring Harbor Laboratory, Cold Spring Harbor, N.Y.
6. **Shibata, D. K., N. Arnheim, and W. J. Martin.** 1988. Detection of human papilloma virus in paraffin-embedded tissue using the polymerase chain reaction. *J. Exp. Med.* **167:**225–230.

7. **Southern, E. M.** 1975. Detection of specific sequences among DNA fragments separated by gel electrophoresis. *J. Mol. Biol.* **98:**503–517.
8. **Starnbach, M. N., S. Falkow, and L. S. Tompkins.** 1989. Species-specific detection of *Legionella pneumophila* in water by DNA amplification and hybridization. *J. Clin. Microbiol.* **27:**1257–1261.
9. **Tompkins, L. S., B. J. Roessler, S. C. Redd, L. E. Markowitz, and M. L. Cohen.** 1988. Legionella prosthetic-valve endocarditis. *N. Engl. J. Med.* **318:**530–535.
10. **Tompkins, L. S., N. J. Troup, T. Woods, W. Bibb, and R. M. McKinney.** 1987. Molecular epidemiology of *Legionella* species by restriction endonuclease and alloenzyme analysis. *J. Clin. Microbiol.* **25:**1875–1880.
11. **Wright, D., and M. Manos.** 1990. Sample preparation from paraffin-embedded tissues, p. 153–158. *In* M. Innis, D. Gelfand, J. Sninsky, and T. White (ed.), *PCR Protocols: A Guide to Methods and Applications.* Academic Press, Inc., San Diego, Calif.

# PCR Detection of Toxin Genes in Strains of *Vibrio cholerae* O1

# 1.15

ØRJAN OLSVIK, TANJA POPOVIC, AND PATRICIA I. FIELDS

## Background

The causative organism of epidemic cholera is *Vibrio cholerae* O1, of the El Tor or Classical biotype and Inaba or Ogawa serotype. Only strains of *V. cholerae* O1 that produce cholera toxin have been associated with epidemics and pandemics in the past; therefore, production of cholera toxin has become an important marker for identifying isolates with epidemic potential (1, 5).

Cholera toxin belongs to a family of related enterotoxins that consist of two polypeptides (5). The A subunit is responsible for adenylate cyclase activation of small intestine epithelial cells, inducing the active secretion of water and salts which causes the tremendous fluid loss that is associated with cholera. The B subunit consists of five identical elements responsible for binding to the epithelial cell surface receptor, $GM_1$. The genes expressing A and B subunits are designated *ctxA* and *ctxB*, respectively. They are expressed as a single transcriptional unit, and the structural genes overlap by 1 base (6).

The production of cholera toxin should be verified in diagnosing cholera because nontoxigenic strains of *V. cholerae* are common, especially in the environment and because of the implications of reporting cholera for public health purposes (2). Cholera toxin production has traditionally been demonstrated by cell culture assays or immunological procedures (1). DNA hybridization assays applying polynucleotide or oligonucleotide probes labeled with radioisotopes, enzymes, or antigens have also been used (9). The PCR test for cholera toxin genes described here is a colony confirmation test designed to distinguish toxigenic from nontoxigenic *V. cholerae* strains taken from pure cultures (Fig. 1; Table 1) (4).

## Materials and Reagents

Dust/particle mask

Sterile water (make samples of 10 to 20 ml, open fresh for each assay)

10× PCR buffer: 100 mM Tris-HCl, 500 mM KCl, 15 mM $MgCl_2$, 0.1% (wt/vol) gelatin

12.5× deoxynucleoside triphosphates (dNTPs; 2.5 mM each of nucleotide in distilled water)

Primer CTX2 (100 μM in distilled water)

**Figure 1.** Positions of PCR primers for detection of the cholera toxin A gene.

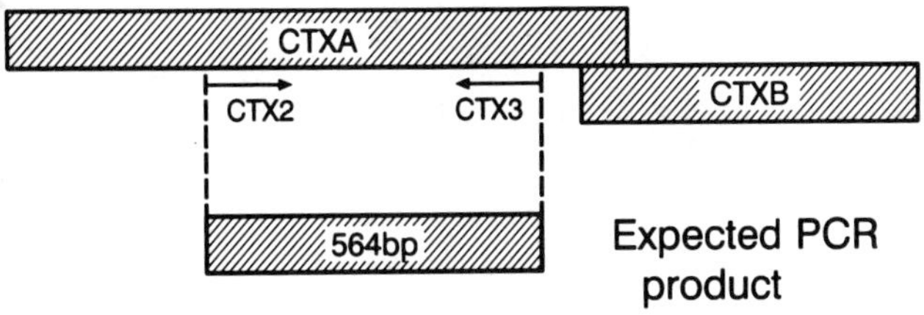


Primer CTX3 (100 μM in distilled water)

Ampli*Taq* (5 U/μl; The Perkin-Elmer Corp., Norwalk, Conn.)

SeaKem agarose (use 0.8% agarose gels in TBE)

10× gel-loading buffer: 0.5% sodium dodecyl sulfate (SDS), 10 mM EDTA, 50% glycerol, 0.1% bromphenol blue, 0.1% xylene cyanol

Quick-Spin columns (G-25 Sephadex) (Boehringer Mannheim)

Restriction endonucleases *Rsa*I and *Alu*I (BRL)

10× restriction buffer 1 (BRL)

30% polyacrylamide (29 parts acrylamide to 1 part bisacrylamide in water), stored in the dark at 4°C (use a face mask when preparing the stock solution as acrylamide is neurotoxic)

10% ammonium persulfate (in water), made fresh daily

TEMED (*N,N,N′,N′*-tetramethylethylenediamine; Sigma)

## Methods

### Sample Preparation

1. Suspend a toothpick of a colony from the control or the test strains in 0.5 ml of water to produce a concentration of $10^5$ to $10^6$ organisms per ml.
2. Boil the sample for 20 min to release DNA. *Note: Freezing and thawing twice can be used instead of boiling when the PCR procedure starts with a 5-min incubation at 95°C. Purified DNA can also be used as the template. A 1- to 10-ng portion of DNA is sufficient; too much DNA will inhibit the amplification reaction. It is important to use water rather than phosphate-buffered saline or any other phosphate-containing buffer as phosphate inhibits the PCR. Heme groups from blood agar plates also inhibit the PCR.*

### PCR Amplification

1. Use a 0.5-ml microtube for this PCR, which amplifies a 564-bp region of the *ctxA* gene. Prepare a master mix (2) of all reagents except the sample DNA,

**Table 1.** Primers for amplification of cholera toxin subunit A[a]

| Name | Nucleotide sequence, 5′ to 3′ | Location |
|---|---|---|
| CTX2 | CGG GCA GAT TCT AGA CCT CCT G | 73–94 |
| CTX3 | CGA TGA TCT TGG AGC ATT CCC AC | 614–636 |

[a] See references 4 and 6.

and place 49 μl in each tube before adding the sample template. This will reduce pipetting errors and produce more consistent concentrations of reagents.

2. Add 50 μl of reaction mix consisting of 38.75 μl of sterile water, 5 μl of 10× PCR buffer, 4 μl of 12.5× dNTP, 0.5 μl of each primer, 0.25 μl of *Taq* polymerase, and 1 μl of template DNA.
3. Overlay the reaction mix with 1 drop of sterile mineral oil before closing the tubes. Start the thermocycling with a preincubation step at 95°C for 5 min, and then use 30 cycles of 1 min at 95°C, 1 min at 60°C, and 1 min at 72°C; perform a final incubation at 72°C for 10 min. If the thermocycler has a 4°C cooling program, this can be used as an extension when the cycling ends outside normal working hours.
4. Prepare a 0.8% agarose gel with TBE buffer. If a microwave oven is used, make sure that the agarose is fully dissolved. Let the agarose cool to 50°C before pouring the gel, which should be completely solid before removing the comb and submerging it in TBE buffer.
5. Mix 10 μl of the PCR product found under the oil layer with 1 to 2 μl of 10× gel-loading buffer, and load in the wells by using a micropipette. Use appropriate positive and negative controls with each set of PCRs, including a molecular size standard on the gel.
6. Let the gel run until the bromphenol blue (purple color) has migrated about two-thirds of the way down the gel. Actual voltage and time will vary with the gel box used. The 564-bp *ctxA* fragment should migrate close to the bromphenol blue.
7. Stain the gel with ethidium bromide (1 drop of stock solution in 50 ml of water; this can be reused) for 20 min, and then destain in water for 10 to 20 min (use gloves, since ethidium bromide is carcinogenic).
8. Place the gel on the UV transilluminator, and take photographs for documentation.
9. To verify that the amplified DNA actually is from the *ctxA* gene, analyze the nucleic acid sequence of the amplicon by using restriction enzymes. Alternatively, hybridization with an internal probe after Southern blotting can be used, but this is too labor intensive for routine use.
10. Remove mineral oil from the sample by placing 1 drop of the PCR solution on an angled Parafilm strip. Mineral oil sticks to the film as the aqueous drop slides down the Parafilm.
11. The amplicons are easily purified by using a spin column (Boehringer Mannheim; or make your own by using CL6B Sephadex). Centrifuge the column at 600 × *g* for 5 min, add the sample (10 to 100 μl), repeat the centrifugation, and collect a pure sample of the same volume as was added.
12. Transfer 15 μl of the PCR solution to a tube, and add 5 μl of water, 2 μl of restriction buffer 1, and 1 μl of *Rsa*I or *Alu*I (or both). Incubate for 2 h at 37°C.
13. Separate the restricted fragments by electrophoresis in an acrylamide gel (this gives better resolution of the small DNA fragments). Make an 8% gel with 6.6 ml of 30% polyacrylamide, 15.6 ml of water, 2.5 ml of 10× TBE, 0.2 ml of 10% ammonium persulfate, and 20 μl of TEMED.
14. Pour the gel, and let it polymerize for 1 h before loading the samples (5 μl of

the PCR solution and 1 μl of 10× loading buffer). Let the bromphenol blue migrate only halfway down the gel before staining the DNA, as for the agarose gel.

## Quality Control

Control strains are *V. cholerae* O1 ATCC 14035 (Classical, Ogawa, toxin positive) and *V. cholerae* O1 ATCC 14033 (El Tor, Inaba, toxin negative). Restriction of the 564-bp *ctxA* amplicon with *Rsa*I gives three fragments of 480, 70, and 10 bp; the 10-bp fragment might not be visible. Restriction with *Alu*I gives two fragments of 500 and 65 bp. Combining the two enzymes generates four fragments of 415, 70, 65, and 10 bp.

## Comments

1. The cholera toxin gene is quite homogeneous in *V. cholerae* O1 strains, even though some heterogenicity exists (3, 5, 9). We recently sequenced the *ctxB* gene in 30 strains of both biotypes and serotypes isolated throughout the world over a period of 70 years (7). The primers used in the *ctxA* PCR assay are outside the variable regions, and the PCR detected *ctxA* in all toxin-positive strains tested (4). Several different bacterial species have been reported to possess toxins or proteins that are immunologically cross-reactive with cholera toxin (5). A nonspecific reaction (i.e., DNA amplicon of the incorrect size) was observed for some of the enterotoxigenic *Escherichia coli* strains at low annealing temperatures. However, the PCR product was not homologous to *ctxA* and was eliminated by increasing the annealing temperature to 60°C (4). Other bacterial species reported to produce proteins that cross-react with cholera toxin in immunologic tests were tested by PCR and found to be negative (4). In establishing the cholera toxin PCR, it was important to verify that the amplicon, in addition to being the correct size, contained the correct DNA sequence. For diagnostic purposes, cutting the amplicon with restriction enzymes such as *Rsa*I (three fragments) and *Alu*I (two fragments) is the easiest verification.
2. It is also important, in the evaluation of an assay, to confirm that a wide selection of organisms with the characteristic under examination give uniformly positive reactions. The *ctxA* PCR assay correctly identified all of the toxigenic *V. cholerae* strains isolated worldwide over the last 60 years (4). All nontoxigenic *V. cholerae* O1 strains tested were negative for *ctxA* by this PCR assay (4).
3. An important advantage of the *ctxA* PCR assay over the enzyme-linked immunosorbent assay (ELISA) is the rapid testing procedure. During the investigation of the Latin American epidemic, the presence of *ctxA* could be determined within 4 h, whereas the cholera toxin ELISA required at least 4 days. Important public health decisions were made on the basis of the PCR data; control strategies were, in some situations, implemented within 24 h (2).
4. The use of PCR is becoming increasingly important in the diagnosis of many infectious diseases. Shirai et al. (8) have also reported the development of a PCR assay for the cholera toxin gene. PCR assays have the potential to be not only more rapid but also more specific and possibly more sensitive than currently available tests for the detection and identification of many infectious

agents. New PCR assays must be thoroughly controlled and evaluated before their application to clinical specimens, and application directly on crude fecal, food, or environmental samples often requires special attention (10). We conclude that the *ctxA* PCR assay described above has proved to be a simple and fast alternative to immunologic assays that are currently used in microbiology laboratories for identifying toxigenic *V. cholerae* strains.

## References

1. **Almeida, R. J., F. W. Hickman-Brenner, E. G. Sowers, N. D. Puhr, J. J. Farmer III, and I. K. Wachsmuth.** 1990. Comparison of a latex agglutination assay and an enzyme-linked immunosorbent assay for detecting cholera toxin. *J. Clin. Microbiol.* **28:**128–130.
2. **Centers for Disease Control.** 1991. Cholera—Peru 1991. *Morbid. Mortal. Weekly Rep.* **40:**108–110.
3. **Fields, P. I., et al.** Unpublished data.
4. **Fields, P. I., T. Popovic, K. Wachsmuth, and Ø. Olsvik.** 1992. Use of polymerase chain reaction for detection of toxigenic *Vibrio cholerae* O1 strains from the Latin American cholera epidemic. *J. Clin. Microbiol.* **30:**2118–2121.
5. **Finkelstein, R. A.** 1988. Cholera, the cholera enterotoxins, and the cholera enterotoxin-related enterotoxin family, p. 85–102. *In* P. Owen and T. J. Foster (ed.), *Immunological and Molecular Genetic Analysis of Bacterial Pathogens.* Elsevier Science Publishers, Amsterdam.
6. **Mekalanos, J. J., D. J. Swartz, and G. D. N. Pearson.** 1983. Cholera toxin gene: nucleotide sequence, deletion analysis and vaccine development. *Nature* (London) **306:**551–557.
7. **Olsvik, Ø., J. Wahlberg, B. Petterson, M. Uhlén, T. Popovic, I. K. Wachsmuth, and P. I. Fields.** 1993. Use of automated sequencing of polymerase chain reaction-generated amplicons to identify three types of cholera toxin subunit B in *Vibrio cholerae* O1 strains. *J. Clin. Microbiol.* **31:**22–25.
8. **Shirai, H., M. Nishibuchi, T. Ramamurthy, S. K. Bhattacharya, S. C. Pal, and Y. Takada.** 1991. Polymerase chain reaction for detection of cholera enterotoxin operon of *Vibrio cholerae. J. Clin. Microbiol.* **29:**2517–2521.
9. **Wachsmuth, I. K., C. A. Bopp, and P. I. Fields.** 1991. Difference between toxigenic *Vibrio cholerae* O1 from South America and the US Gulf coast. *Lancet* **i:**1097–1098.
10. **Zywno, S. R., P. I. Fields, Ø. Olsvik, M. L. Motes, and A. DePaola.** Unpublished data.

# PCR Detection of Heat-Stable, Heat-Labile, and Shiga-Like Toxin Genes in *Escherichia coli*

## 1.16

Ørjan Olsvik and Nancy A. Strockbine

## Background

Protein exotoxins play an important role in the pathogenesis of diseases due to certain strains of *Escherichia coli*. Enterotoxigenic *E. coli* (ETEC) strains, important causes of diarrheal disease in humans and domestic animals, produce heat-labile enterotoxin (LT), heat-stable enterotoxin (ST), or both. Shiga-like toxin (SLT)-producing *E. coli* (SLTEC) strains cause a spectrum of human diseases ranging from nonbloody diarrhea to hemorrhagic colitis with the subsequent development of hemolytic uremic syndrome or thrombotic thrombocytopenic purpura in some individuals. SLTEC strains also cause disease in other animals: diarrhea in calves and edema disease in piglets.

Strains of *E. coli* that cause disease are difficult to distinguish from those that are part of the normal intestinal flora of humans and other animals. There are no selective or differential cultivation methods for diagnosing ETEC or, except for *E. coli* O157:H7, SLTEC infections. Serological characterization of O and H antigens will identify some strains belonging to these pathogenic groups; however, ETEC and SLTEC are more often, and probably more reliably, identified through the detection of their specific toxins or toxin genes. The toxins from these organisms have been detected by using animal, cell culture, immunologic, or DNA hybridization assays (1, 6, 11). In the past few years, PCR assays for the detection of LT, ST, and SLT genes have been described (2, 5, 9, 10). In this report, we describe PCR assays that detect the STI, LTI, SLT-I, and SLT-II genes and discriminate between variants of the STI (STIa and STIb) and LTI (LTIa and LTIb) genes. Because of their simplicity, rapidity, and specificity, PCR assays represent good alternatives to traditional methods for the diagnosis of ETEC and SLTEC infections.

## Materials and Reagents

Microtiter plates for the 96-well Stratagene cycler (Stratagene)

Sterile (autoclaved) water, stored at room temperature in 10- to 20-ml volumes, with a fresh aliquot used for each assay

10× PCR buffer: 100 mM Tris-HCl, 500 mM KCl, 15 mM $MgCl_2$, 0.1% (wt/vol) gelatin (Sigma), stored at −20°C in 100- to 300-μl aliquots

10× deoxynucleoside triphosphates (dNTPs): 2 mM each nucleotide in sterile distilled water, stored at −20°C in 100- to 300-μl aliquots

**Table 1.** Primer designation, sequence, location, and reference for PCRs used for detection of *E. coli* toxins

| Toxin | Primer | Nucleotide sequence, 5′ to 3′ | Location | Reference |
|---|---|---|---|---|
| STI | STI-1 | TTA ATA GCA CCC GGT ACA AGC AGG | 243–266 | 8, 10 |
| | STI-2 | CTT GAC TCT TCA AAA GAG AAA ATT AC | 127–144 | 8, 10 |
| STIa | STI-3ab | GAT TAC AAC AAA GTT CAC AGC AGT | 220–243 | 8, 10 |
| | STIIa-1 | ATT ACA TTA GAG ACT AAA AAG TGT GAT | 141–167 | |
| STIb | STI-3ab | GAT TAC AAC AAA GTT CAC AGC AGT | 220–243 | 8, 10 |
| | STIb-1 | ATC ACA CTA GAA TCA AAA AAA TGT AAC | 141–167 | 8, 10 |
| LTIa (p) | LTIa-1 | TCT CTA TAT GCA CAC GGA GC | 46–65 | 7, 9 |
| | LTI-2 | CCA TAC TGA TTG CCG CAA T | 349–367 | 7, 9 |
| LTIb (h) | LTIb-1 | TCT CTA TGT GCA TAC GGA GC | 46–65 | 7, 9 |
| | LTI-2 | CCA TAC TGA TTG CCG CAA T | 349–367 | 7, 9 |
| SLT-I | SLTI-1 | CAG TTA ATG TGG TGG CGA AG | 215–234 | 4 |
| | SLTI-3 | CTG CTA ATA GTT CTG CGC ATC | 1089–1109 | 4 |
| SLT-II | SLTII-1 | CTT CGG TAT CCT ATT CCC GG | 288–307 | 3 |
| | SLTII-2 | GGA TGC ATC TCT GGT CAT TG | 747–766 | 3 |

PCR master mix: 37.75 μl of sterile water, 5 μl of 10× PCR buffer, 5 μl of 10× dNTPs, 0.5 μl of each primer, 0.25 μl of *Taq* polymerase, and 1 μl of template DNA

Primers (100 μM in distilled water), stored separately at −20°C in aliquots of 10 to 50 μl (Table 1)

Ampli*Taq* (5 U/μl) (Perkin-Elmer Corp.)

10× Gel-loading buffer: 0.5% sodium dodecyl sulfate (SDS), 10 mM EDTA, 50% glycerol, 0.1% bromphenol blue, 0.1% xylene cyanol

Molecular size DNA markers: 1-kb ladder (0.1 μg/μl; GIBCO-BRL, Gaithersburg, Md.)

30% polyacrylamide (acrylamide/bisacrylamide, 29:1) in distilled water, stored in the dark at 4°C

10% ammonium persulfate (in distilled water), made fresh daily

TEMED (*N*,*N*,*N*′,*N*′-tetramethylethylenediamine; Sigma)

## Methods

### Sample Preparation

1. Suspend a toothpick of a colony in 0.5 ml of water to give a final concentration of $10^5$ to $10^6$ organisms per ml. This template DNA can be prepared from pure bacterial cultures cultivated on solid or in liquid medium. For bacteria grown in liquid medium, add 10 μl of an overnight broth culture to 0.5 ml of water.
2. Boil the bacteria for 20 min, or freeze-thaw them twice, to release template DNA. No further purification of the DNA is necessary. If template DNA is prepared from bacteria by freeze-thawing, be sure to incubate the samples for 5 min at 95°C prior to the amplification cycles to denature the DNA. Use only

**Figure 1.** PCR primers for detection of ST1 genes.

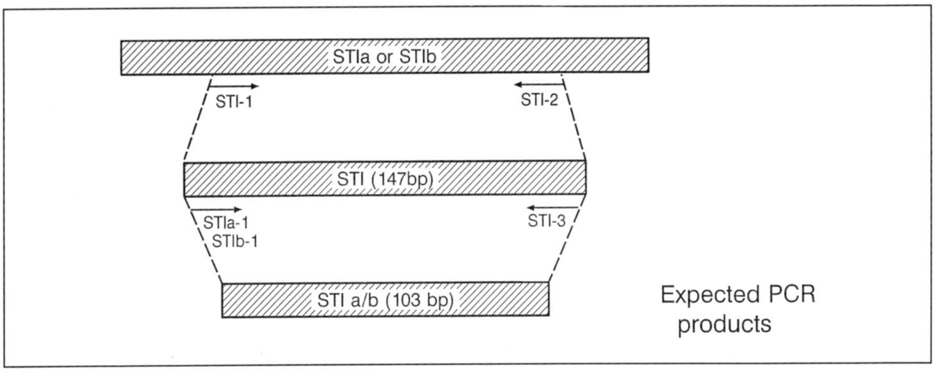


1 to 10 ng of chemically purified DNA; too much DNA can inhibit the reaction. Use water to dissolve sample DNA, and avoid phenol, phosphate-containing buffers, and heme-containing media, since these products also inhibit the reaction.

## PCR Amplification

1. Perform the PCR assays for STI, LTI, SLT-I, and SLT-II genes by using 0.5-ml microcentrifuge tubes or 96-well microtiter plates. *Note: Not all microtiter plates are suitable for PCR assays. Microtiter plates for use in thermal cyclers should be able to withstand high temperatures and should not release contaminants that could inhibit the reaction.* Mix a sufficient volume of PCR master mix for one or two more reactions than needed to allow for pipetting inaccuracies. Dispense 49 μl (or 50 μl if using a nonadjustable multipipette) to each tube or well before adding the sample template *Note: The master mix should not be frozen, because the DNA polymerase is inactivated by freezing-thawing and primers are more stable in a concentrated form.* If more than two primers are used at the same time, decrease the amount of water in the reaction by a volume equivalent to that occupied by the additional primers. Overlay the reactions with a drop of sterile mineral oil to prevent evaporation, and close the tubes or cover the microtiter plate with the lid supplied. Keep all reagents and tubes on ice before placing them in the thermocycler.
2. Amplify the desired sequences in a programmable thermal cycler according to the schemes shown in Fig. 1 to 3. The following cycling conditions are the same for amplification of the STIa, STIb, LTIa, LTIb, SLT-I, and SLT-II genes:

**Figure 2.** PCR primers for detection of the heat-labile enterotoxin gene.

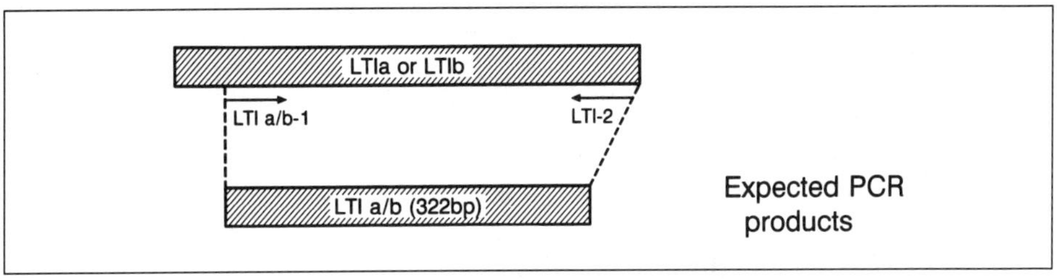

**Figure 3.** PCR primers for detection of SLT genes.

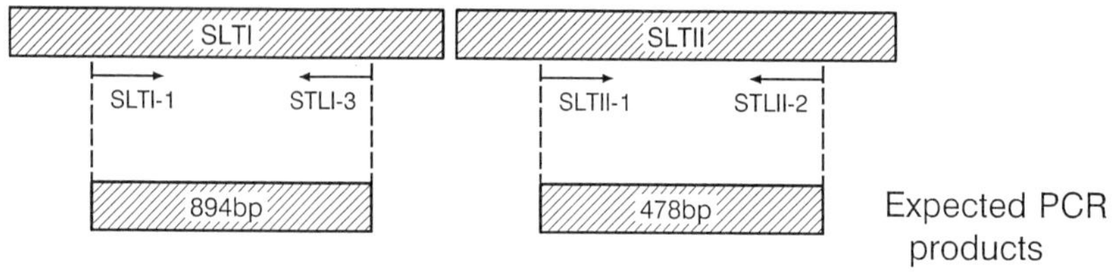


preincubation at 95°C for 5 min, then 30 cycles of 1 min at 95°C, 1 min at 60°C, and 1 min at 72°C, and a final incubation at 72°C for 10 min. If possible, hold the tubes at 4°C when the cycles end after normal working hours.

3. To distinguish between STIa and STIb genes, perform the nested PCR assay that is outlined in Fig. 1. Use primers STI-1 and STI-2 to produce a 147-bp amplicon from the STIa or STIb genes. Dilute the PCR product from the first reaction 1:100 in water, and use 1 μl of this as a template in subsequent reactions with the STIa-specific primer set (STIa-1 and STI-3) or the STIb-specific primer set (STIb-1 and STI-3). Although the PCR products generated in the second reaction will be the same size, the specificity of the reaction will result from the specific binding of the STIa-1 and STIb-1 primers to their respective toxin genes. Two rounds of amplification are needed to distinguish between these toxin genes. If discrimination between STIa and STIb is not necessary, we recommend performing only the first amplification reaction with primers STI-1 and STI-2, which amplify both toxin genes. We have observed multiple PCR fragments when the STIa- or STIb-specific primers are used in the first reaction, which makes interpretation of the results by gel electrophoresis difficult. New primers for the ST genes are now under development.
4. Amplify in two different reactions with primers LTIa-1 plus LTI-2 in one tube and LTIb-1 plus LTI-2 in another tube for detection and differentiation of the heat-labile enterotoxins. Amplification of LTIa and LTIb is carried out by a method analogous to that described above for discriminating between STIa and STIb. If it is not necessary to distinguish between LTIa and LTIb, mix all three primers together in one reaction.
5. Amplify the SLT-I and SLT-II genes with one reaction by mixing all four primers in the same tube. The two genes can be distinguished by differences in the size of their PCR products.
6. Examine the PCR products by agarose or polyacrylamide gel electrophoresis for the presence of DNA fragments of the appropriate size. The small fragments from the STI reaction are best visualized and sized after electrophoresis through an 8% polyacrylamide gel. The LTI and SLT fragments can be adequately separated and sized by electrophoresis through a 0.8% agarose gel.
7. Make an 8% polyacrylamide gel by mixing 6.6 ml of 30% acrylamide (*Caution: Acrylamide is neurotoxic, so use a face mask when preparing this solution*), 15.6 ml of water, 2.5 ml of 10× TBE, 0.2 ml of 10% ammonium persulfate, and 20 μl of TEMED. Pour the gel, and let it polymerize for 1 h before loading the samples (5 μl of sample plus 1 μl of loading buffer). Subject the PCR products to electrophoresis until the tracking dye has migrated approximately two-thirds

the length of the gel. Stain the gel for 10 to 20 min in an aqueous solution of ethidium bromide (5 to 10 μg of ethidium bromide per ml), and then destain it for 10 to 20 min in water alone. The stain solution can be reused for more than 1 month. Place the gel on a UV-transilluminator, and photograph for documentation. (*Caution: Wear gloves while handling ethidium bromide, as it is carcinogenic.*)

8. Prepare a 0.8% agarose gel in TBE. Electrophoretically separate the reaction products (10 μl per well) in TBE. Allow the bromphenol blue indicator to migrate approximately two-thirds the length of the gel. Fragments of approximately 500 bp will migrate with the bromphenol blue dye. On each gel, always use sufficient positive and negative controls, including the DNA ladder (0.2 μg per well).

## Quality Control

*E. coli* STIa (STp)-positive control strain ATCC 35401.
*E. coli* STIb (STh)-positive control strain ATCC 43896.
*E. coli* LTIa (LTh)-positive control strain ATCC 43886.
*E. coli* LTIb (LTp)-positive control strain HB101(pEWD 299).
*E. coli* SLT-I-positive control strain ATCC 43890.
*E. coli* SLT-II-positive control strain ATCC 43894.
*E. coli* SLT-I- and SLT-II-positive control strain ATCC 43895.
*E. coli* control strain negative for the above-mentioned toxins ATCC 25922.
*Vibrio cholerae* O1 classical, Ogawa, cholera toxin-positive control strain ATCC 14035.

In addition to the control strains, we highly recommend including a PCR test without any added DNA template to detect contamination of reagents by previous PCR products.

## Comments

1. The PCR assays described in this chapter are for use on pure bacterial cultures and have not been fully evaluated for use on food or fecal samples.
2. The SLT-I PCR assay will also amplify an 894-bp fragment from *Shigella dysenteriae* type 1 since the nucleotide sequences of the genes for Shiga toxin from *S. dysenteriae* type 1 and SLT-I from *E. coli* are virtually identical (12). The SLT-I amplicon can be digested with *Hin*dIII, giving two fragments of 516 and 378 bp. The SLT-II amplicon of 478 bp can be digested with *Eco*RV into two fragments of 88 and 390 bp.
3. Because of substantial sequence homology between the heat-labile (LTI) genes of *E. coli* and the cholera toxin genes of *V. cholerae* O1, the LTI PCR assay will amplify the cholera toxin gene if the annealing temperature falls below 58°C. For this reason, we recommend including a control strain for cholera toxin in the PCR assays. The LTIa amplicons can be digested with *Hha*I, giving two fragments of almost equal size. This *Hha*I (GCGC) site is changed in LTIb strains to GCGA, preventing this cleavage of the amplicon (7, 8).
4. We have successfully mixed primers for SLT-I and SLT-II in the same PCR. We have not yet succeeded in amplifying the STI and LTI genes in the same reaction.

5. The presence of STI, LTI, SLT-I, and SLT-II genes is indicated by amplification of the appropriate-sized fragments (Fig. 1 to 3). To verify the identity of a fragment, one could perform a subsequent PCR by using the first PCR product as template with a set of nested internal primers to test for amplification of a second fragment of a predicted size. Alternatively, hybridization with specific DNA probes or restriction endonuclease analysis of the primary PCR products can be used to verify the correctness of the test.

## References

1. **Greenberg, R. B., and R. L. Guerrant.** 1986. *E. coli* heat-stable enterotoxin, p. 115–151. *In* F. Dorner and J. Drews (ed.), *Pharmacology of Bacterial Toxins.* Pergamon Press, New York.
2. **Hornes, E., Y. Wasteson, and Ø. Olsvik.** 1991. Detection of *Escherichia coli* heat-stable enterotoxin genes in pig stool specimens by an immobilized, colorimetric nested polymerase chain reaction. *J. Clin. Microbiol.* **29:**2375–2379.
3. **Jackson, M. P., R. J. Neill, A. D. O'Brien, R. K. Holmes, and J. W. Newland.** 1987. Nucleotide sequence analysis and comparison of the structural genes for Shiga-like toxin I and Shiga-like toxin II encoded by bacteriophages from *Escherichia coli* 933. *FEMS Microbiol. Lett.* **44:**109–114.
4. **Jackson, M. P., J. W. Newland, R. K. Holmes, and A. D. O'Brien.** 1987. Nucleotide sequence analysis of the structural genes for Shiga-like toxin I encoded by bacteriophage 933J from *Escherichia coli. Microb. Pathog.* **2:**147–153.
5. **Karch, H., and T. Meyer.** 1989. Single primer pair amplifying segments of distinct Shiga-like-toxin genes by polymerase chain reaction. *J. Clin. Microbiol.* **27:**2751–2757.
6. **Karmali, M. A.** 1989. Infections by verocytotoxin-producing *Escherichia coli. Clin. Microbiol. Rev.* **2:**15–38.
7. **Leong, J., A. C. Vineland, and W. S. Dallas.** 1985. Nucleotide sequence comparison between heat-labile toxin B subunit cistrons from *Escherichia coli* of human and porcine origin. *Infect. Immun.* **48:**73–77.
8. **Moseley, S. L., J. W. Hardy, M. I. Huq, P. Escheverria, and S. Falkow.** 1983. Isolation and nucleotide sequence determination of a gene encoding a heat-stable enterotoxin of *Escherichia coli. Infect. Immun.* **39:**1167–1174.
9. **Olive, D. M.** 1989. Detection of enterotoxigenic *Escherichia coli* after polymerase chain reaction amplification with a thermostable DNA polymerase. *J. Clin. Microbiol.* **27:**261–265.
10. **Olsvik, Ø., E. Rimstad, E. Hornes, N. Strockbine, Y. Wasteson, A. Lund, and K. Wachsmuth.** 1991. A nested PCR followed by magnetic separation of amplified fragments for detection of *Escherichia coli* Shiga-like toxin genes. *Mol. Cell. Probes* **5:**429–435.
11. **Robertson, D. C., J. L. McDonel, and F. Dorner.** 1986. *E. coli* heat-labile enterotoxin, p. 77–113. *In* F. Dorner and J. Drews (ed.), *Pharmacology of Bacterial Toxins.* Pergamon Press, New York.
12. **Strockbine, N. A., M. P. Jackson, L. M. Sung, R. K. Holms, and A. D. O'Brien.** 1988. Cloning and sequencing of the genes for *Shigella dysenteriae* type 1. *J. Bacteriol.* **170:**1116–1122.

# PCR Detection of *Shigella* Species and Enteroinvasive *Escherichia coli*

1.17

Gary K. Schoolnik

## Background

*Shigella* species and the enteroinvasive *Escherichia coli* (EIEC) are closely related gram-negative bacterial pathogens of the human intestine. Both typically cause dysentery, a diarrheal illness characterized by blood- and mucus-containing stools, tenesmus, abdominal cramping pain, and fever.

The pathological hallmark of dysentery due to the shigellae and EIEC is the ability of these organisms to enter and replicate within colonic epithelial cells (4). This is followed by inflammation of the lamina propria, leading to the shedding of epithelial cells, neutrophils, and blood into the lumen of the intestine. This process can be simulated by the production of keratoconjunctivitis after inoculation of these organisms into the eye of a guinea pig (the Sereny test). Both organisms also penetrate cultured epithelial cell monolayers in vitro (3). Together, these histopathological and in vitro experimental observations have led to the classification of shigellae and EIEC as agents of invasive diarrhea.

The virulence of these organisms is determined by multiple chromosomal and genetic loci. At least four such loci are carried on a large (120- to 140-MDa) plasmid; a homolog of this plasmid is present in all virulent strains of the four *Shigella* species and each of the EIEC serotypes examined thus far (8). One of these plasmid-encoded loci, termed *inv* (invasion), has provided the basis for the PCR assay described below (6, 12). For the sequences of the primers and probe used in this assay, see Table 1.

## Materials and Reagents

Sterile phosphate-buffered saline (PBS; pH 7.4)

Lysis buffer 1: 50 mM Tris-HCl (pH 8.0), 20% (wt/vol) sucrose, 50 mM EDTA, and 100 μg of lysozyme (Sigma Chemical Co., St. Louis, Mo.) per 75 μl of buffer

Lysis buffer 2: 50 mM NaCl, 1% (wt/vol) sodium dodecyl sulfate (SDS), and 800 μg of proteinase K (Boehringer, Indianapolis, Ind.) per 300 μl of buffer

Tris-EDTA buffer: 10 mM Tris HCl (pH 8.0), 1 mM EDTA

PCR master mix, as specificed in the GeneAmp kit (Perkin Elmer Cetus, Norwalk, Conn.): 0.4 mol of each deoxynucleoside triphosphate, 2.5 U of *Taq* polymerase, and 0.2 μg each of primers *Sh*-1 and *Sh*-2 (per 100-μl reaction)

**Table 1.** Oligonucleotides for detection of *Shigella* species and EIEC

| Function | Name | Nucleotide sequence, 5′ to 3′ |
|---|---|---|
| Primer | *Sh*-1 | CTG GAT GGT ATG GTG AGG |
| Primer | *Sh*-2 | GGA GGC CAA CAA TTA TTT CC |
| Probe | *Sh*-AP | CCA TCT ATT AGA ATA CCT GTG |

0.3 M NaOH

Nylon membrane (Hybond NT; Amersham Corp., Arlington Heights, Ill.)

5× SSC (20× SSC is 0.3 M sodium citrate [pH 7.0] and 3 M NaCl) containing 5× Denhardt's, 0.5% sodium dodecyl sulfate (SDS), and 0.1 mg of yeast tRNA per ml

*Sh* alkaline phosphatase-labeled (*Sh*-AP) DNA probe, 4 ng per membrane

2× SSC–0.5% SDS

1× SSC–0.1% SDS

1× SSC–0.5% Triton X-100

Nitro Blue Tetrazolium

5-Bromo-4-chloro-3-indolyl phosphate (BCIP)

## Methods

### Sample Collection

Collect a stool sample by rectal swab or from a stool specimen. The sample may be stored frozen at −20°C for 1 month prior to analysis.

### Sample Preparation

1. Resuspend the stool swab in 4 ml of PBS.
2. Allow heavy particulate to settle to the bottom of the tube (15 to 30 min).
3. Remove 1 ml of stool suspension for DNA extraction, and dispense into a 1.5-ml microcentrifuge tube. Centrifuge at maximum speed (ca. 16,000 × *g*) in a microcentrifuge for 1 min.
4. Resuspend the pellet in 75 μl of lysis buffer 1, and incubate the reactants at 37°C for 30 min.
5. Add 300 μl of lysis buffer 2, and incubate the reactants for an additional 60 min at 37°C.
6. Centrifuge at 14,000 × *g* for 5 min in a microcentrifuge to remove undigested particulate.
7. Transfer ~300 μl of the supernatant to a new 1.5-ml microcentrifuge tube, and precipitate the DNA by the addition of absolute ethanol. Collect precipitate by centrifugation (16,000 × *g* for 1 min).
8. Discard the supernatant, and dissolve the DNA pellet in 500 μl of Tris-EDTA buffer. *Note: In some cases the pellet may have to be incubated at 4°C overnight in the Tris-EDTA buffer before it is fully dissolved.* Dilute the resuspended and dissolved pellet 1:50 in the Tris-EDTA buffer. Take 50 μl of this for PCR.

### PCR Amplification

1. Prepare the PCR master mix as specified by the manufacturer, and dispense it into new 500-μl microcentrifuge tubes (50 μl per tube). Add 50 μl of the extracted DNA with positive-displacement pipettors.
2. Amplify in a programmable thermal cycler by using 26 cycles of a three-step profile: 94°C for 1 min, 43°C for 2 min, and 72°C for 3 min.

### Detection of the PCR-Amplified Product

1. Incubate 10 μl of the PCR-amplified DNA in 0.3 M NaOH for 5 min at 70°C. Apply to a nylon membrane.
2. Incubate the membrane for 15 min at 42°C in 5× SSC containing 5× Denhardt's solution, 0.5% SDS, and 0.1 mg of yeast tRNA per ml.
3. Add 4 ng of the *Sh*-AP probe, and continue to incubate for 15 min at 42°C.
4. Wash the membrane four times at room temperature in 2× SSC–0.5% SDS for 10 min, 1× SSC–0.1% SDS for 10 min, 1×–SSC–0.5% Triton X-100 for 5 min, and 1× SSC for 1 min.
5. Remove the membrane, and then incubate with Nitro Blue Tetrazolium and BCIP for 4 h at room temperature.

## Quality Control

### Positive Amplification Control

1 ng and 1 pg of DNA from recombinant *E. coli* HB101 (*ial*) which harbors pBR325 containing a 2.5-kbp *Hin*dIII fragment from an EIEC 140-MDa invasion plasmid (12); the cloned *Hin*dIII fragment contains an invasion-associated locus (*ial*) common to all EIEC strains and *Shigella* species and from which the primers (*Sh*-1 and *Sh*-2) and the *Sh*-AP probe used here were derived (6) (one reaction each).

### Negative Amplification Control

10 μl of distilled water added to PCR buffer (four reactions).

### Positive Fecal Specimen Control

100 CFU of the *E. coli* HB101 (*ial*) positive control strain homogenized with 1 ml of a frozen, noninfected stool specimen (one reaction).

### Negative Fecal Specimen Control

The same noninfected stool specimen homogenate without the added positive control strain (two reactions).

### Specificity Control

An avirulent *Shigella sonnei* strain lacking the ca. 140-MDa invasion plasmid and thus lacking the *inv* locus corresponding to the *Sh*-1 and *Sh*-2 PCR primers and the *Sh*-AP probe used in the assay described above (one reaction).

## Comments

1. On PCR amplification, the primer sequences chosen from within the *ial* locus yield a 320-bp fragment from all four *Shigella* species and from each of the 14 tested EIEC strains. This fragment can be readily detected by using a 1.5% agarose gel stained with ethidium bromide. The presence of a fragment of this exact size as a product of PCR amplification with the primers described here thus constitutes an alternative detection system. The sequence of the amplified 320-bp fragment was determined and compared with all other DNA sequences in several gene banks; this led to the identification of three unique sequences, which on further analysis resulted in the selection of the two primer sequences and the *Sh*-AP DNA probe sequence used here.
2. The *Sh*-AP DNA probe was prepared by the method of Jablonski et al. (9) and coupled to alkaline phosphatase (AP) by using a single modified base containing a 12-carbon linker with a terminal primary amine. Alternative labeling procedures, including end labeling with chemiluminescent reporter molecules, can be used.
3. With a strain of *Shigella flexneri* added in various numbers to samples of normal stools from three persons, the sensitivity of the PCR assay described here was found to be approximately 10 CFU per 1-ml specimen after 26 cycles of amplification.
4. The ca. 140-MDa invasion plasmid in *Shigella* species can be rapidly lost on in vitro culture, even at frequencies as high as 50% per generation per cell for some *S. sonnei* strains (6). Thus, the assay described here is not applicable for *Shigella* strains that have been stored or passaged in vitro for long periods, since loss of the invasion plasmid will yield an avirulent strain that gives a false-negative result in this PCR assay. Because all virulent *Shigella* strains present in a freshly passed fecal specimen harbor the invasion plasmid, this PCR assay is applicable for most clinical purposes. In contrast, of 14 EIEC strains passaged in vitro and analyzed by plasmid gel electrophoresis, all still harbored the invasion plasmid and all were positive by this PCR assay (6).
5. It has been possible to conduct the above-described PCR assay under field conditions in a developing-country setting by using three water baths set at 94, 43, and 72°C instead of a thermocycler (5), an instrument which might be too expensive for some Third World laboratories.
6. Addition of primers corresponding to the heat-stable (ST) and heat-labile (LT) enterotoxin genes of enterotoxigenic *E. coli* (ETEC) to the *Shigella ial* primers described here has permitted the simultaneous amplification of three virulence genes (the *ial* locus and the ST and LT genes) from a single stool specimen, using the same reaction tube (5). Then 10-μl aliquots of the amplified product can be spotted onto separate nylon filters, which are probed with the AP-labeled DNA probe corresponding to each of these genes.
7. This PCR assay will not differentiate between a dysentery syndrome due to shigellae and one due to EIEC, nor will it provide *Shigella* species or serotype information. The usefulness of the assay clinically comes from its capacity to rapidly detect shigellae and EIEC directly in a stool specimen, thus serving to eliminate as etiologic considerations *Campylobacter* species, salmonellae, and *Entamoeba histolytica*, each of which can cause a similar clinical syndrome (except of course for the unusual occurrence of a simultaneous infection with

a *Shigella* sp. or EIEC plus one or more of the other species noted above). For epidemiological investigations and as a means of predicting antibiotic susceptibility patterns, it may be desirable to isolate and characterize (by species and serotype) the *Shigella* species from a stool specimen that has been shown to be positive for a *Shigella* species or EIEC by the PCR assay described here.

## References


1. **Centers for Disease Control.** 1988. *Shigella, Annual Tabulation 1988.* Centers for Disease Control, Atlanta.
2. **Cheasty, T., and B. Rowe.** 1983. Antigenic relationships between the enteroinvasive *Escherichia coli* O antigens and *Shigella* O antigens. *J. Clin. Microbiol.* **17:**681–684.
3. **Clerc, P. L., and P. J. Sansonetti.** 1987. Entry of *Shigella flexneri* into HeLa cells: evidence for directed phagocytosis involving actin polymerization and myosin accumulation. *Infect. Immun.* **55:**2681–2688.
4. **Formal, S. B., T. L. Hale, and P. J. Sansonetti.** 1983. Invasive enteric pathogens. *Rev. Infect. Dis.* **5:**S702–S707.
5. **Frankel, G., J. A. Giron, J. Valmassoi, and G. K. Schoolnik.** 1989. Multi-gene amplification: simultaneous detection of three virulence genes in diarrhoeal stool. *Mol. Microbiol.* **3:**1729–1734.
6. **Frankel, G., L. Riley, J. A. Giron, J. Valmassoi, A. Friedmann, N. Strockbine, S. Falkow, and G. K. Schoolnik.** 1990. Detection of *Shigella* in feces using DNA amplification. *J. Infect. Dis.* **161:**1252–1256.
7. **Gomes, T. A. T., M. R. F. Toledo, L. R. Trabulsi, P. K. Wood, and J. G. Morris.** 1987. DNA probes for identification of enteroinvasive *Escherichia coli. J. Clin. Microbiol.* **25:**2025–2027.
8. **Hale, L., and S. B. Formal.** 1987. Pathogenesis of *Shigella* infections. *Pathol. Immunopathol. Res.* **6:**117–127.
9. **Jablonski, E., E. W. Moomaw, R. H. Tullis, and J. L. Ruth.** 1986. Preparation of oligodeoxynucleotide-alkaline phosphatase conjugates and their use as hybridization probes. *Nucleic Acids Res.* **14:**6115–6128.
10. **Morris, G. K., J. A. Kochler, and E. J. Gangarosa.** 1970. Comparison of media for direct isolation and transport of *Shigella* from fecal specimens. *Appl. Microbiol.* **19:**434–437.
11. **Silva, R. M., M. R. F. Toledo, and L. R. Trabulsi.** 1980. Biochemical and cultural characterization of invasive *Escherichia coli. J. Clin. Microbiol.* **11:**441–444.
12. **Small, P. L., and S. Falkow.** 1986. Development of a DNA probe for the virulence plasmid of *Shigella* spp. and enteroinvasive *Escherichia coli*, p. 121–124. *In* L. Lieve, P. F. Bonventre, J. A. Morello, S. D. Silver, and W. C. Wu (ed.), *Microbiology—1986.* American Society for Microbiology, Washington, D.C.

# PCR Detection of *Helicobacter pylori*

## 1.18

JOAN L. VALENTINE

## Background

*Helicobacter pylori* (formerly *Campylobacter pylori*) is a microaerophilic spiral gram-negative rod implicated in active chronic gastritis and peptic ulcer disease. Assays for the diagnosis of *H. pylori* include microbiologic culture, histologic examination, rapid urease tests, urea breath tests, serology, and nucleic acid probe technology (2). More recently, nucleic acid amplification technology has been applied to the detection and differentiation of this organism; these methods include PCR assays based on the gene sequences encoding urease structural genes (3, 5), a species-specific antigen (4), and 16S rRNA (1). This chapter describes a PCR assay developed from a randomly selected sequence of *H. pylori* chromosomal DNA used to detect *H. pylori* in clinical specimens (8).

The PCR assay is based on an 8-kb recombinant plasmid, pCPY202, containing a 1.9-kb fragment of *H. pylori* chromosomal DNA randomly selected from an *H. pylori* library (6). Preliminary studies demonstrated the cloned fragment to have high sensitivity and specificity for the identification of *H. pylori*. A portion of the *H. pylori* fragment was sequenced, and two 20-bp oligomers, CAM-2 and CAM-4, were selected for use in a PCR assay (Fig. 1; Table 1). The expected product of amplification of the target sequence with these primers is 203 nucleotide pairs in length. A third 20-base oligonucleotide designated CAM-3, complementary to a region of DNA between the two primers, was synthesized for use as a probe to identify the amplified *H. pylori* DNA in the reaction products (Fig. 1; Table 1). The PCR products may be analyzed by agarose gel electrophoresis, Southern hybridization, or slot blot hybridization. The assay has been successfully applied to gastric biopsy and gastric aspirate samples (8).

## Materials and Reagents

Denaturation solution: 0.5 M NaOH, 1.5 M NaCl

Disposable pellet pestle with 1.5-ml microcentrifuge tube (Kontes, Vineland, N.J.)

Electrophoresis loading dye: 0.25% bromophenol blue, 0.25% xylene cyanol FF, and 25% Ficoll (type 400) in water

Neutralization solution: 1 M Tris-Cl (pH 8.0), 1.5 M NaCl

Nylon membrane (Nytran; Schleicher & Schuell, Keene, N.H.)

**Figure 1.** PCR primers and probe for detection of *H. pylori.*

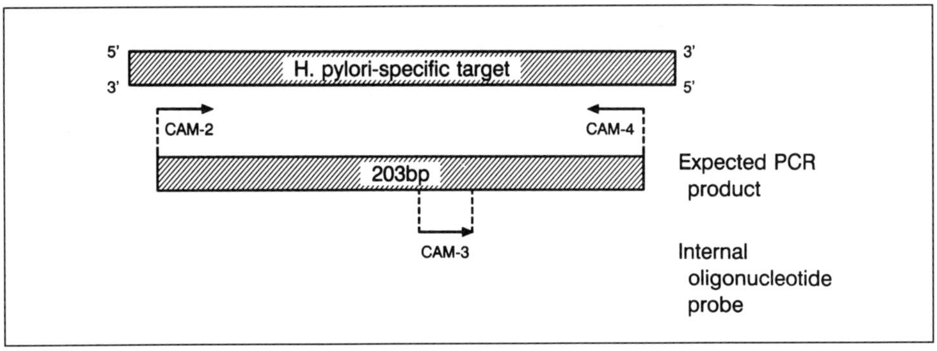


Oligonucleotide primers CAM-2 and CAM-4

Oligonucleotide probe CAM-3

PCR master mix: 0.5 μM (each) oligonucleotide primers CAM-2 and CAM-4 (per 50-μl reaction); 200 μM (each) dATP, dCTP, dTTP, and dGTP; 1× reaction buffer (50 mM KCl, 10 mM Tris-Cl [pH 8.3], 1.5 mM MgCl, 0.01% [wt/vol] gelatin); 1.25 U of *Taq* polymerase (Perkin-Elmer Corp., Norwalk, Conn.)

Phenol, buffered with an equal volume of 0.1 M Tris-hydrochloride (pH 7.5)

Prehybridization solution: 3× SSPE, 5× Denhardt's (0.1% bovine serum albumin, 0.1% polyvinylpyrrolidone, 0.1% Ficoll), 0.5% sodium dodecyl sulfate (SDS), 0.25 mg of tRNA per ml

Sephadex G50 (Sigma, St. Louis, Mo.), 20 g in 500 ml of 10 mM Tris-hydrochloride–1 mM EDTA (pH 8.0)

TAE buffer: 0.04 M Tris-acetate, 0.001 M EDTA

Tissue extraction buffer: 50 mM Tris-hydrochloride (pH 7.2)–1 mM EDTA containing 1% Laureth-12 (PPG/Mazer Chemicals, Gurnee, Ill.) and 100 μg of proteinase K per ml

## Methods

### Sample Requirements

Gastric tissue biopsy specimen (approximately 20 to 30 $mm^2$ in size) transported in 0.5 ml of sterile 0.85% normal saline. May be frozen at −70°C for several months.

Gastric aspirate specimen (minimum sample size, 500 μl) transported in a sterile container. May be frozen at −70°C for several months.

**Table 1.** Oligonucleotides for PCR detection of *H. pylori*

| Function | Name | Nucleotide sequence, 5′ to 3′ |
|---|---|---|
| Primer | CAM-2 | TAA CAA ACC GAT AAT GGC GC |
| Primer | CAM-4 | CAT CTT GTT AGA GGG ATT GG |
| Probe | CAM-3 | CGC TCT TTA GTT TTG GAG CG |

## Sample Preparation

*Note: Include tubes of water as blanks during the tissue extraction process to use as contamination controls during the PCR run.*

***Gastric Biopsy Specimens***

1. Mince gastric tissue (by using sterile Kontes pellet pestle) in a 1.5-ml microcentrifuge tube.
2. Add 100 μl of tissue extraction buffer, vortex, and incubate at 55°C for 2 h.
3. Extract by mixing with an equal volume of phenol-chloroform-isoamyl alcohol (25:24:1), centrifuge for 5 min, and remove the aqueous layer to a new microcentrifuge tube.
4. Extract a second time by mixing with an equal volume of chloroform-isoamyl alcohol (24:1), centrifuge for 5 min, and remove the aqueous layer to a new centrifuge tube.
5. Precipitate the DNA by adding 1/10 volume of 3 M sodium acetate followed by 2.5 volumes of cold ethanol. Freeze the sample at −70°C for 15 to 30 min.
6. Microcentrifuge for 10 min at 16,000 × *g*.
7. Carefully remove the supernatant with a sterile disposable pipette. Avoid disturbing the DNA pellet.
8. Wash the remaining DNA pellet with 70% ethanol, microcentrifuge, and carefully remove the supernatant.
9. Reconstitute the resulting DNA pellet in 50 μl of sterile TE buffer.
10. Use 10 μl of the DNA solution for PCR amplification.

***Gastric Aspirate Specimens***

*Note: To avoid exposure of H. pylori DNA to the acidic environment of the specimen when the bacterial cells are lysed, the gastric aspirate specimens must be deacidified with sodium hydroxide prior to processing for PCR.*

1. To eliminate the acidity of the gastric aspirate, add 20 μl of 1 N NaOH to 250 μl of gastric aspirate in a 1.5-ml microcentrifuge tube.
2. Add 250 μl of tissue extraction buffer, vortex, and incubate at 55°C for 2 h.
3. Follow steps 3 to 10 above as for biopsy specimens.

## PCR Amplification

1. Perform amplification in 50-μl volumes. Prepare the PCR master mix in a single tube according to the recipe, and place 40 μl into each 0.5-μl microcentrifuge tube.
2. Overlay each tube with 2 or 3 drops of sterile light mineral oil.
3. Pipette 10 μl of sample DNA into the reaction mix in each tube. *Note: When adding sample DNA, immerse the pipette tip below the mineral oil layer.*
4. Subject tubes to 40 cycles of amplification in a DNA thermal cycler. Each cycle consists of a 1-min denaturation step at 95°C, a 30-s annealing step at 42°C, and a 1-min extension step at 72°C.
5. Store tubes at 4°C prior to analysis of PCR products.

*Note: Several negative controls consisting of distilled water or human placental DNA should be interspersed within the PCR run. A recommendation is to place a distilled-water blank and a human placental DNA blank after every five clinical samples within the run. These controls should be carefully monitored for evidence of carryover or contamination in the final hybridization results.*

## PCR Product Analysis

### *Southern Hybridization*

1. Prepare a 3% Nusieve–1% agarose gel in TAE with 0.5 μg of ethidium bromide per ml.
2. Load each well with 10 μl of the amplified product mixed with 3 μl of loading dye.
3. The molecular size standard lane contains 2 μl of 123-bp ladder mixed with 8 μl of distilled $H_2O$ and 3 μl of loading dye.
4. Electrophoretically separate the reaction products at 100 V in TAE buffer containing 0.5 μg of ethidium bromide per ml until the dye has migrated approximately two-thirds of the length of the gel.
5. Examine the agarose gel under UV light for the presence of the 203-nucleotide-pair product. *Note: A lower band may be seen on agarose gel electrophoresis or Southern hybridization; this represents specific single-stranded PCR product.*
6. Perform Southern transfer using standard techniques (7) by denaturing the agarose gel in 500 ml of denaturation solution for 15 min. Change to a second 500-ml aliquot of denaturation solution for an additional 15 min. Neutralize the gel for 30 min in neutralization solution.
7. Transfer overnight to a nylon membrane filter as described previously (7).
8. Expose the filter to UV light for 2 min on a transilluminator, and air dry.

### *Slot Blot Analysis*

1. Denature 10 μl of PCR product in 0.5 N NaOH–25 mM EDTA, and apply to a nylon membrane filter (prewetted in 20× SSPE) by using a slot blot apparatus.
2. Rinse each slot twice with 200 μl of 20× SSPE.
3. Expose the filter to UV light for 2 min, and air dry.

### *Hybridization of Filters*

1. End label CAM-3 oligonucleotide with [$^{32}$P]ATP and T4 polynucleotide kinase to a specific activity of $10^8$ to $10^9$ cpm/μg.
2. Purify the labeled probe on a 10-ml G50 Sephadex column with TE buffer as the column buffer.
3. Prehybridize the filter in a roller tube for 1 h at 42°C in 10 ml of prehybridization solution.
4. Add $^{32}$P-labeled CAM-3 to the prehybridization solution (final concentration, 5 ng/ml).
5. Hybridize for 1 h at 42°C in a roller tube.
6. Wash the filter at room temperature in 3× SSPE for 5 min.
7. Wash the filter three times at room temperature in 1× SSPE–0.1% SDS for 5 min each.
8. Wash the filter at 50°C for 10 min in 5× SSPE–0.1% SDS.
9. Expose the filter to X-Omat autoradiographic film with intensifying screens for 1 to 4 h. *Note: Prolonged exposure of slot blot films may lead to false-positive*

*results. Slot blot results should be recorded after an interval no longer than 2 h of autoradiographic exposure.*

## Quality Control

### Positive Amplification Controls

10 pg of pCPY202 (one reaction each).
10 pg of *H. pylori* ATCC 43504 DNA (one reaction each).

### Negative Amplification Controls

Distilled water.
10 pg of human placental DNA.

### Specificity Control

10 pg of *Wolinella succinogenes* ATCC 29543 DNA (one reaction each).

## Comments

1. The sensitivity of the assay permits the detection of as few as 100 bacterial cells when 10-fold dilutions of an *H. pylori* ATCC 43579 bacterial suspension were extracted and amplified as described for the biopsy tissues (8).
2. The detection of the 203-nucleotide-pair fragment in the PCR products is specific for *H. pylori*. The assay is nonreactive with other closely related strains such as *Helicobacter mustelae*, *Wolinella succinogenes*, *Wolinella recta*, *Succinomonas amylolytica*, *Campylobacter jejuni*, and *Campylobacter coli* (8).
3. For detection of *H. pylori* in fresh gastric biopsy material, the assay has a sensitivity of 93% (Southern and slot blot hybridization) and a specificity of 100% when compared with microbiologic culture and histologic examination (8).
4. The sensitivity of the assay for detection of *H. pylori* in gastric aspirates was 85% (Southern and slot blot hybridization) when compared with standard culture and histologic examination of the corresponding biopsy tissue (8). Deacidification of gastric aspirates prior to lysing the cells is critical to obtaining this sensitivity.
5. The diagnosis time is 12 h if the products are analyzed by slot blot hybridization with the oligoprobe. However, Southern hybridization requires overnight transfer.
6. Hammar et al. (4) recently described a PCR assay that amplified sequences of a species-specific antigen of *H. pylori*. The assay involved a two-step thermal profile which combines the annealing and extension steps, resulting in shorter amplification times. This method may also be applicable to the presently described PCR assay to decrease amplification times.
7. This PCR assay had only limited success in detection of *H. pylori* in paraffin-embedded tissues when compared with the results obtained with corresponding fresh tissues. Further studies are required to optimize this assay for use with preserved specimens.

## References


1. **Clayton, C., K. Kleanthous, and S. Tabaqchali.** 1991. Detection and identification of *Helicobacter pylori* by the polymerase chain reaction. *J. Clin. Pathol.* **44:**515–516.
2. **Dick, J. D.** 1990. *Helicobacter* (*Campylobacter*) *pylori*: a new twist to an old disease. *Annu. Rev. Microbiol.* **44:**249–269.
3. **Foxall, P. A., L. T. Hu, and H. L. T. Mobley.** 1992. Use of polymerase chain reaction-amplified *Helicobacter pylori* urease structural genes for differentiation of isolates. *J. Clin. Microbiol.* **30:**54–58.
4. **Hammar, M., T. Tyszkiewicz, T. Wadstrom, and P. W. O'Toole.** 1992. Rapid detection of *Helicobacter pylori* in gastric biopsy material by polymerase chain reaction. *J. Clin. Microbiol.* **30:**54–58.
5. **Ho, S., J. A. Hoyle, F. A. Lewis, A. D. Secker, D. Cross, N. P. Mapstone, M. F. Dixon, J. L. Wyatt, D. S. Tompkins, G. R. Taylor, and P. Quirke.** 1991. Direct polymerase chain reaction test for detection of *Helicobacter pylori* in humans and animals. *J. Clin. Microbiol.* **29:**2543–2549.
6. **Mobley, H. L. T., L. E. Rosenthal, A. F. Trofa, and B. D. Jones.** 1989. Optimization of detection of *Campylobacter pylori* by urease and DNA hybridization, p. 127–131. *In* F. Megraud and H. Lamouliatte (ed.), *Gastroduodenal Pathology and Campylobacter pylori: Proceedings of the First Meeting of the European Campylobacter pylori Study Group.* Elsevier Science Publishing Co., New York.
7. **Sambrook, J., E. F. Fritsch, and T. Maniatis.** 1989. *Molecular Cloning: A Laboratory Manual*, 2nd ed. Cold Spring Harbor Laboratory, Cold Spring Harbor, N.Y.
8. **Valentine, J. L., R. R. Arthur, H. L. T. Mobley, and J. D. Dick.** 1991. Detection of *Helicobacter pylori* by using the polymerase chain reaction. *J. Clin. Microbiol.* **29:**689–695.

# PCR Detection of Toxigenic *Clostridium difficile*

# 1.19

Paul H. Gumerlock, Yajarayma J. Tang, and
Joseph Silva, Jr.

## Background

Toxigenic *Clostridium difficile* is the bacterial pathogen responsible for antibiotic-associated colitis (8). The disease can develop rapidly in patients who are taking antibiotics which suppress the normal anaerobic intestinal flora, allowing *C. difficile* to increase in numbers and to produce its toxins. Two toxins are known to be produced by *C. difficile*: toxin A, an enterotoxin, and toxin B, a potent cytotoxin. These toxins are known to damage the colonic mucosa and cause intraluminal fluid accumulation (6).

*C. difficile* colitis is currently diagnosed in the clinical laboratory by demonstrating the presence of its toxins, since culturing of the organism is difficult and imprecise in many clinical laboratories and frequently takes 3 to 5 days. Toxin B causes cytotoxicity in fibroblastic tissue culture assays (1). False-negative results are a problem with this assay and are thought to be due to degradation of the toxin protein by proteases present in human stools. A number of enzyme-linked immunosorbent assays are available for toxin A detection (1, 7, 11). Although these tests are rapid and sensitive, the major disadvantage is the occurrence of indeterminate readings and unconfirmed positive results. We recently reported a PCR-based assay for the detection of the 16S rRNA gene of *C. difficile* (3). In addition, by incorporating toxin B gene amplification, toxigenic and nontoxigenic strains can be discriminated (4). We have used this system to monitor the course of infection in patients during and after therapy (5). Our results indicate that this PCR-based test is more sensitive than either the cytotoxicity toxin assay or direct anaerobic culturing of *C. difficile.*

## Materials and Reagents

Microcentrifuge tubes, 2 ml (Micro Test Tube "EZ"; Bio-Rad Laboratories)

Microcentrifuge tubes, 0.6 ml

Sterile-filtered pipette tips

Anaerobic jars (BBL)

Anaerobic generators ($H_2$ + $CO_2$)

Benchkote (Fisher Scientific)

Brain heart infusion (BHI) medium (Difco)

Lysozyme (stock of 100 mg/ml)

Proteinase K (stock of 20 mg/ml)

TEN buffer (50 mM Tris-HCl [pH 8], 10 mM EDTA, 150 mM NaCl; Sigma)

5% sodium dodecyl sulfate (SDS) in TEN

5× SSPE–0.5% SDS (1× SSPE is 150 mM NaCl, 10 mM $NaH_2PO_4 \cdot H_2O$, and 50 mM EDTA [pH 7.4])

TE buffer: 100 mM Tris (pH 7.4), 1 mM EDTA

n-*Taq* polymerase (Perkin-Elmer Corp.)

10× PCR buffer (1× PCR buffer is 50 mM KCl, 20 mM Tris-HCl, 2.5 mM $MgCl_2$, 100 μg of bovine serum albumin per ml [pH 8.4])

123-bp DNA ladder (GIBCO-BRL, Grand Island, N.Y.)

10 mM each 2′-deoxynucleoside 5′-triphosphates (dNTPs) (Pharmacia, Alameda, Calif.)

10× loading buffer (0.167 g of bromophenol blue, 10 ml of $H_2O$, 10 ml of glycerol)

0.4 N NaOH

Biodyne B Membrane (Pall BioSuport, East Hills, N.Y.)

Biotin deoxyuridine triphosphate (Biotin dUTP; Boehringer Mannheim)

50 mM Tris-HCl (pH 7.4)

Terminal deoxytransferase (Tdt; USB, Cleveland, Ohio)

Streptavidin-horseradish peroxidase (5 μg/ml; Boehringer Mannheim)

ECL kit reagents 1 and 2 (Amersham)

Buffer B: 100 mM NaCl, 1 M urea, 5% Triton X-100, 1% dextran sulfate

10× buffer C: 1 M sodium citrate (pH 5)

## Methods

### Sample Collection

Collect stool specimens in sterile 5-fluid-ounce Sage Specimen containers. They may be stored at −20°C for up to 3 months until ready for testing.

### Sample Preparation

1. Grow *C. difficile* anaerobically at 37°C in 5 ml of prereduced BHI broth for 36 to 48 h.
2. Harvest cells by centrifugation at 6,000 × *g* for 20 min at 4°C.
3. Wash the cell pellet in 10 ml of 50 mM Tris-HCl (pH 7.4). Centrifuge for 10 min at 6,000 × *g* at 4°C.
4. Suspend the cell pellet in 1.5 ml of TBE, and transfer it to a 2-ml microcentrifuge tube.
5. Add lysozyme to a final concentration of 0.5 mg/ml. Mix by inverting three times, and incubate the mixture for 30 min at 37°C.
6. Add 20% SDS to a final concentration of 0.5%.
7. Centrifuge at 200 × *g* for 5 to 10 min to remove broken cells and debris.
8. Transfer the supernatant to a new microcentrifuge tube, and extract it with an equal volume of phenol-$CHCl_3$-isoamyl alcohol (invert vigorously for 30 s to 1 min). Centrifuge for 2 min at 8,900 × *g* at 4°C.

9. Transfer the aqueous layer to a new microcentrifuge tube, and extract as above until a clean interface is obtained. *Note: Three to five extractions are usually needed.*
10. Precipitate the DNA by adding 2 volumes of cold absolute ethanol.
11. Freeze at −70°C for >1 h (or −20°C for >3 h or overnight).
12. Centrifuge for 30 min at 11,650 × *g* at 4°C to pellet the precipitated DNA.
13. Remove the supernatant, and dry the residual pellet in a Speed Vac.
14. Dissolve the DNA pellet in 150 to 200 μl of TE buffer.
15. Determine the yield by reading the $A_{260}$, setting 1 optical density unit as equal to 50 μg/ml.

### DNA Extraction from Stool Specimens

1. Resuspend 100 mg of formed stools or 200 μl of watery stools in 0.75 ml of TEN buffer in a 2-ml microcentrifuge tube.
2. Add 5 μl of 100-mg/ml lysozyme. Invert three times to mix, and incubate for 1 h at 37°C.
3. Add 50 μl of 5% SDS in TEN and 2.5 μl of 20-mg/ml proteinase K. Invert to mix, and incubate for 30 min at 55°C.
4. Boil for 10 min (you can use a heating block at 95 to 100°C).
5. Centrifuge for 5 min at 180 × *g* to remove large particles.
6. Transfer the supernatant to a new 2-ml microcentrifuge tube.
7. Extract with an equal volume of phenol-$CHCl_3$-isoamyl alcohol (invert 15 times).
8. Centrifuge for 2 min at 8,900 × *g*, and transfer the aqueous layer to a new microcentrifuge tube.
9. Extract as above until a clean interface is obtained. *Note: Four or five extractions are usually needed.*
10. Add 2 volumes of cold absolute ethanol to precipitate DNA.
11. Freeze at −70°C for >1 h (or −20°C for >3 h or overnight).
12. Pellet the DNA by centrifugation at 11,650 × *g* for 30 min at 4°C. Wash with cold 70% ethanol. Centrifuge as above for 10 min, and remove the supernatant. *Note: Pellets are often not visible from the stool specimens.*
13. Dry the pellet in a Speed Vac.
14. Dissolve the pellet in 50 μl of TE buffer.
15. Quantitate the DNA as above.

### PCR Amplification

1. Prepare the 10× PCR buffer.
2. Prepare each 100-μl reaction to contain 10 to 20 pmol of the selected primers (Table 1), 0.1 mM each dNTP, 10 μl of 10× PCR buffer, 2 U of thermostable recombinant *Taq* polymerase, and 500 ng of DNA. Bring to 100 μl with sterile distilled $H_2O$. *Note: To reduce possibilities of cross-contamination, all pipetting operations of pre-PCR reagents should be done in a "clean bench" or sterile hood with dedicated pipettors and sterile-filtered pipette tips.*
3. Cover reaction mixtures with 150 μl of light mineral oil.